PHYSICAL DATA ACQUISITION FOR DIGITAL PROCESSING

Components, Parameters, and Specifications

GAYLE F. MINER
DAVID J. COMER

both of Brigham Young University

PRENTICE HALL, Englewood Cliffs, New Jersey 07632

Library of Congress Cataloging-in-Publication Data

Miner, Gayle F.
  Physical data acquisition for digital processing : components, parameters, and specifications / Gayle F. Miner & David J. Comer.
     p. cm.
  Includes bibliographical references and index.
  ISBN 0-13-209958-6
  1. Computer input-output equipment. 2. Automatic data collection systems. I. Comer, David J. II. Title.
TK7887.5.M55 1992
621.39'85—dc20   91-32795
                                                              CIP

Acqusitions editor: Peter Janzow
Production editor: Merrill Peterson
Supervising editor and interior design: Joan L. Stone
Copy editor: Nikki Herbst
Cover designer: Joe DiDomenico
Prepress buyer: Linda Behrens
Manufacturing buyer: Dave Dickey

© 1992 by Prentice-Hall, Inc.
A Simon & Schuster Company
Englewood Cliffs, New Jersey 07632

All rights reserved. No part of this book may be reproduced, in any form or by any means, without permission in writing from the publisher.

Printed in the United States of America
10 9 8 7 6 5 4 3 2 1

ISBN 0-13-209958-6

Prentice-Hall International (UK) Limited, *London*
Prentice-Hall of Australia Pty. Limited, *Sydney*
Prentice-Hall Canada Inc., *Toronto*
Prentice-Hall Hispanoamericana, S.A., *Mexico*
Prentice-Hall of India Private Limited, *New Delhi*
Prentice-Hall of Japan, Inc., *Tokyo*
Simon & Schuster Asia Pte. Ltd., *Singapore*
Editora Prentice-Hall do Brasil, Ltda., *Rio de Janeiro*

Contents

PREFACE   ix

0 INTRODUCTION TO DATA ACQUISITION SYSTEMS, THE DATA ACQUISITION PROBLEM, AND DATA CODING   1

0.1   Introduction   1
0.2   The Unifying Impetus of Physical Data Signal Processing   1
0.3   The Important Questions Designers Should Ask   2
0.4   Basic Components of a Signal Processing System   3
0.5   The Data Acquisition Problem   6
0.6   Digital Signals: Of Bits, Nibbles, Bytes, and Words   7
0.7   Binary Coding Schemes   9
      0.7.1 The Natural Binary Fractional Code: Unipolar Signals, 9
      0.7.2 Binary-Coded Decimal (BCD) or 8-4-2-1 BCD: Unipolar Signals, 10
      0.7.3 2-4-2-1 Binary-Coded Decimal: Unipolar Signals, 12
      0.7.4 Gray Code: Unipolar and Bipolar Signals, 14
      0.7.5 Complementary Coding Schemes: Unipolar Signals, 16
      0.7.6 Sign-Magnitude Coding: Bipolar Signals, 16
      0.7.7 Offset Binary Coding: Bipolar Signals, 17
      0.7.8 Two's Complement Coding: Bipolar Signals, 20
      0.7.9 One's Complement Coding: Bipolar Signals, 20
      0.7.10 Conversion between Types of Coding, 21
      0.7.11 Additional Types of Coding, 22
      0.7.12 Summary of Coding Schemes, 23

General Exercises   25
References   27

1 TRANSDUCERS, TRANSDUCER PARAMETERS (SPECIFICATIONS), AND THEIR EFFECTS ON DIGITAL PROCESSING   28

1.1   Introduction   28
1.2   The General Concept of a Transducer and Its Basic Elements   28
1.3   The Concept of Error and Its Relationship to Digital Signal Processing   32
1.4   Transducers and Sensors for Data Acquisition   34
    1.4.1 Capacitive Sensors, 34
    1.4.2 Inductive Sensors, 37
    1.4.3 Resistive Sensors, 39
    1.4.4 Sensors That Make Direct Physical Variable to Electric Variable Conversion, 42
1.5   Transducer Specifications, an Introduction   45
    1.5.1 An LVDT, 46
    1.5.2 A Pressure Transducer, 46
    1.5.3 A Strain Gauge Accelerometer, 47
    1.5.4 A Piezoelectric Accelerometer, 47
    1.5.5 A Capacitance Angle Transducer, 47
1.6   Transducer Specification: Linearity   48
1.7   Transducer Specification: Hysteresis (Friction, Backlash)   58
1.8   Transducer Specification: Time Response   59
1.9   Transducer Specification: Frequency Response   66
    1.9.1 Ripple Factor, 71
    1.9.2 3-dB Frequencies, 75
    1.9.3 Concluding Comments on Frequency Response, 78
1.10 Transducer Specification: Resolution (Threshold)   84
1.11 Transducer Specification: Precision (Repeatability)   87
1.12 Transducer Specification: Accuracy   91
1.13 Transducer Specification: Sensitivity   94
1.14 Transducer Specification: Selectivity   99
1.15 Transducer Specification: Drift   104
1.16 Transducer Specification: Noise and Noise Figure   105
1.17 General Discussion and Conclusions   114
    General Exercises   114
    References   117

2 AMPLIFIERS AND THEIR INTEGRATION INTO SIGNAL PROCESSING SYSTEMS 119

- 2.1 Introduction 119
  - 2.1.1 Amplifier Gain, 120
  - 2.1.2 Amplifier Impedances, 121
- 2.2 Amplifiers and the Transmission Line Problem 123
  - 2.2.1 Distortion Caused by Transmission Line Mismatch, 124
  - 2.2.2 Distortion Caused by Transmission Line Dispersion, 132
- 2.3 Amplifier Types and Their General Characteristics 135
- 2.4 Characteristics and Applications of Ideal Operational Amplifiers 137
  - 2.4.1 Inverting Amplifier Configuration, 139
  - 2.4.2 Noninverting Amplifier Configuration, 144
- 2.5 Operational Amplifier Specifications 147
- 2.6 Operational Amplifier Open Loop Gain Value 151
- 2.7 Operational Amplifier Frequency Response 157
  - 2.7.1 Effect of Op Amp Frequency Response on Amplifier Designs, 157
  - 2.7.2 Effect of Op Amp Frequency Response on the Bit Size N of the Data Words, 162
- 2.8 Operational Amplifier Slew Rate 167
- 2.9 Operational Amplifier Input Signals: Differential and Common Mode Voltages 177
- 2.10 Operational Amplifier Common Mode Rejection 184
- 2.11 Operational Amplifier Current Offset and Voltage Offset 195
- 2.12 Operational Amplifier Temperature Characteristics 202
- 2.13 Operational Amplifier Input Impedances 205
- 2.14 Other Amplifier Parameters 207
- 2.15 Instrumentation Amplifiers 209
- 2.16 Applications of Instrumentation Amplifiers 217
- 2.17 Isolation Amplifiers 224
- 2.18 Instrumentation Amplifier Offset Drift 226
- 2.19 Selection of Instrumentation Amplifiers or Operational Amplifiers 228
- 2.20 Error Analysis and Error Budgets for Amplifiers 228

General Exercises 232
References 235
Appendix 2.1 Operational Amplifiers 237

3 FILTERS — 245

- 3.1 Aliasing 246
- 3.2 Introduction to Filters 252
  - 3.2.1 *The Need for Filters, 252*
  - 3.2.2 *Filter Specifications, 254*
  - 3.2.3 *Filter Functions, 255*
- 3.3 Synthesis Fundamentals 255
  - 3.3.1 *Normalization/Denormalization, 255*
  - 3.3.2 *Finding Poles from the Magnitude Response, 257*
- 3.4 Transfer Functions for Filters 261
  - 3.4.1 *Butterworth (Maximally Flat) Response, 261*
  - 3.4.2 *Chebyshev (Equiripple) Response, 264*
  - 3.4.3 *Linear Phase (Bessel) Filters, 269*
- 3.5 Presampling Filters 276
- General Exercises 280
- References 282

4 SYNTHESIS OF FILTERS — 283

- 4.1 Passive Synthesis 283
  - 4.1.1 *Properties of LC Networks, 284*
  - 4.1.2 *Synthesis of LC Driving Point Functions, 285*
- 4.2 Synthesis of Transfer Functions 290
  - 4.2.1 *Two-Port Parameters, 290*
  - 4.2.2 *Transfer Function Synthesis, 293*
  - 4.2.3 *Zero Shifting to Obtain Finite Transmission Zeros, 300*
- 4.3 Low-Pass to High-Pass Transformation 304
- 4.4 Active Synthesis 305
  - 4.4.1 *Bach Filters, 307*
  - 4.4.2 *Band-Pass Filters, 312*
  - 4.4.3 *Switched-Capacitor Filters, 316*
- General Exercises 316
- References 318

5 BASIC ERROR CONSIDERATIONS IN DATA ACQUISITION SYSTEMS — 319

- 5.1 Simple Quantization 319
  - 5.1.1 *Quantum Interval and Error, 319*
  - 5.1.2 *Span of an Analog Signal, 322*
- 5.2 Effect of Errors on Word Size 323
  - 5.2.1 *Quantization Error, 323*
  - 5.2.2 *Noise Considerations, 326*
  - 5.2.3 *Signal-to-Noise Ratio, 330*
- 5.3 Selection of Converter Size 331

Contents   vii

    5.4   Overview of Data Acquisition Systems   332
    5.5   Waveform Reconstruction   333
           General Exercises   344
           References   344

6 THE DIGITAL-TO-ANALOG CONVERTER   345

    6.1   Performance Parameters of the DAC   346
           6.1.1   *Number of Bits, 346*
           6.1.2   *Settling Time, 346*
           6.1.3   *Conversion Factor, 347*
           6.1.4   *Accuracy or Linearity, 347*
           6.1.5   *Resolution, 347*
    6.2   Design of the DAC   348
           6.2.1   *Small Word Size, Binary-Weighted DACs, 350*
           6.2.2   *The Ladder Configuration, 354*
    6.3   A Practical DAC   356
    6.4   Other Types of DAC   358
           6.4.1   *BCD DACs, 358*
           6.4.2   *Bipolar DACs, 359*
    6.5   The Glitch Problem in DACs   361
           General Exercises   362
           References   363

7 THE ANALOG-TO-DIGITAL CONVERTER   365

    7.1   Performance Parameters of the ADC   365
           7.1.1   *Number of Bits, 366*
           7.1.2   *Conversion Time, 366*
           7.1.3   *Quantum Interval or LSB Value, 366*
           7.1.4   *Accuracy, 366*
    7.2   Types of ADC   367
           7.2.1   *Parallel Converter, 367*
           7.2.2   *Successive Approximation ADC, 370*
           7.2.3   *Iterative Converter, 374*
           7.2.4   *The Ramp, Staircase, and Tracking Converters, 378*
           7.2.5   *The Dual-Slope Digital Voltmeter, 380*
           General Exercises   383
           References   384

8 SAMPLE-AND-HOLD CIRCUITS   385

    8.1   Uses of the Sample-and-Hold Circuit   385
    8.2   When to Use the S/H Module   389
    8.3   Operating Characteristics of Sample-Hold Devices   390
    8.4   Fundamental Configurations and Circuit Operation   397

8.5 Control Circuitry for S/H Module: An Example 404
8.6 Applications of Sample-Hold Modules and Other Systems 406
General Exercises 408
References 409

APPENDIX: LABORATORY EXERCISES AND DEMONSTRATIONS   ***410***

INDEX   ***413***

Preface

The impact of computers on control systems and physical data monitoring fields has produced engineers whose dominant interests are in the digital processing area. One effect of this specialization in processing has been to produce digital designs and algorithms that have been divorced from any considerations of the capabilities of the analog data gathering and converting modules. Invariably such designs and algorithms lead to unreasonable demands on the analog portions of a system, with the resulting "patching together" of an analog system from which "proper" input digital data can be obtained.

On the other hand, there are engineers whose interests are predominantly analog. They frequently do not appreciate the requirements of the digital system. They occasionally select components for analog processing which "look good," with the result that digitally derived information may be obtained that was not in the original physical signals.

The purpose of this text is to provide a correct understanding of the interfacing requirements of importance when connecting the analog and digital domains. With this understanding engineers on either end of the system can properly communicate their needs and capabilities in a design. This will provide some common understanding of the specifications and limitations that can reasonably be imposed by either party in an overall system design. Another goal is to help engineers realize that an initial system definition should involve skills from both analog and digital areas whenever physical variables are to be processed digitally.

Digital signal processing is outside the intent and scope of this text. Because of this, no processing algorithms or techniques are given. Our main concern is how to correctly generate the digital words from physical variables that are to be used in subsequent digital processing.

Since complete data acquisition systems (DAS) are now available—and even on a single chip—one may wonder why a detailed study of the various components is even necessary. The material of this text will assist the engineer in making a proper assessment of the specifications of a DAS module. Also, the engineer will be able to determine the most important parameters to be considered for a given application, and even those parameters that are necessary but not provided by the manufacturers.

One of the more subtle purposes of the text is to develop in engineers an appreciation for properly defined, stated, and interpreted specifications and parameters. This is, of course, important in all realms of engineering activity. But, when one begins to compare the products of several manufacturers vying for a contract, the nuances of the definitions become painfully apparent. The authors hope the ideas and problems presented will help both those engineers who develop specifications and those who use them to be aware of the necessity of giving careful and precise thought to those definitions.

A perusal of the table of contents reveals a wide range of subject matter, so that mention of the breadth and depth of the text is appropriate. Since a DAS uses many components and modules in a system configuration (either as discrete blocks or on a chip), we have tried to give some mention of the more important components. Each chapter of the text is properly the subject of an entire book (and in fact many are). There is a conscientious attempt to give enough detailed discussion of each element, component, and configuration so that the parameters and specifications of each have physical bases for definitions. Any need for details beyond those given can be satisfied by consulting the references at the end of each chapter. Finally, and of most significance, we develop specific relationships between each parameter (specification) and the number of bits in the data word used to represent the physical variable. This is one of the more important features of the text.

There are exercises available so that one can apply the concepts presented. Most of these have been developed as we have read journal articles and advertisements, examined manufacturers literature, evaluated the designs of others, and been involved in system definition and design. The exercises are of two types. The first are those that appear at the end of each chapter section. These are intended to reinforce the ideas in that specific section. The second type are those at the end of each chapter, denoted as general exercises. These latter exercises give one the opportunity to determine which concepts to apply in a given situation, a very important part of the engineering process in practice. It should be pointed out that—to the consternation of teacher, student, and grader—a few exercises and problems have either more information or insufficient information given than needed to solve them uniquely. Such problems may occasionally have more than one viable solution, depending upon which definition of a specification or parameter is used. The teacher is at liberty to "tie down" the problem to remove such variability. Many of the exercises contain design features.

Appreciation is extended to the Electrical and Computer Engineering Department at Brigham Young University for the opportunity to work in an environment that encourages change when needed and provides the occasion to pursue an idea via a special topics course.

Finally, to our students we give thanks for having been patient while being subjected to developmental notes in a true laboratory test of this material. The text was found adequate for a three credit hour, one semester course. We have frequently assigned laboratory work on various concepts of the course, usually in lieu of a class period so that the ideas could be tested, both on original designs and commercial units. We usually had three such periods. Unfortunately, an already full curriculum usually precluded additional lab work. Demonstrations were occasionally used as a compromise experience, with students often volunteering to present them. Some of these laboratory exercises and demonstrations are given in the Appendix.

There were also many lively discussions of the concepts and exercises. These discussions further strengthened our suspicions that, first, specifications frequently are not carefully defined or understood, and, second, that intuitive notions of specifications and parameters do not always suffice. The questions and suggestions of those students have contributed materially to whatever merits the text may have.

As for background for this text, we assume that the reader has had introductory electronics and circuits courses that included discussion of basic discrete component circuits and fundamental applications of operational amplifiers. A rudimentary knowledge of binary number systems is assumed, although a modest introduction to the various types of digital coding is included in the preliminary chapter.

Gayle F. Miner
David J. Comer
Provo, Utah

0
Introduction to Data Acquisition Systems, the Data Acquisition Problem, and Data Coding

0.1 INTRODUCTION

The advents of computers and microprocessors along with sophisticated digital signal processing techniques have produced applications in every facet of instrumentation, process control, and physical variable analysis. This digital emphasis is well founded and proper for many reasons, among which are high noise immunity, ease of data storage, and relative ease of data transmission. Additionally, there are many digital data processing algorithms available, such as the fast Fourier transform (FFT), which provide efficient data analysis.

The overall purpose of this chapter is to present the general concept of the acquisition of physical data in a form that can be used by a digital computer. So, when we speak of data acquisition we refer to physical data, not data already digitally coded. First, the analog-digital dichotomy is discussed, and the discussion is concluded with two fundamental questions to which analog and digital designers should respond. Next, the basic components of a data acquisition system are described on a block diagram level along with discussions of the errors that contaminate the data. Then the data acquisition problem is defined. The final sections describe some of the popular digital coding formats.

0.2 THE UNIFYING IMPETUS OF PHYSICAL DATA SIGNAL PROCESSING

One result of the "intrusion" of digital techniques into the signal analysis field has been the development of a field of digital signal processing that has tended to divorce the processing schemes from the analog or physical variable. An examination of essentially any digital signal processing textbook reveals that the data to be processed are assumed to be available in digital format. Mention is seldom made of the source of those digital data values, how well they really represent the intended original data source, or what information one can legitimately hope or even attempt to extract from those values. Thus an artificial gulf has developed between the analog system designers and the digital system designers. Both fields began to mature almost independently, when in fact the desire and need to digitally process data should have produced a closer alliance between them. However, advances in integrated circuit theory and manufacturing soon became the focal point for both analog and digital designers. Once chips were completed, independent applications of those chips followed—the desired interaction between the two fields did not continue to develop.

When one is involved with the processing of signals derived from physical variables, the techniques of both analog and digital processing are forced to be compatible, since it is the signal provided by the analog system that ultimately provides the digital data. We see that the division of signal processing efforts is really illusory, since an engineer must be acquainted with the requirements of both types of processing if a viable system is to be developed.

The primary purpose of this text is to help engineers develop confidence in the selection and design of modules of any complexity that are used to generate digital data from physical variables such as temperature, pressure, and rotation. The relationships between the number of bits used to represent physical data values and the specifications and parameters of analog modules provide the unifying link needed. Consequently, an engineer who considers digital design the forte can be aware of the fact that analog processing at some points may offer advantages over digital techniques and can understand the limitations imposed and what demands can be made on analog techniques. Similarly, for the analog designer an understanding of the module and component specifications results, so that the effects of those specifications on the digital processing can be predicted. Additionally, when one is assigned to develop a data acquisition system to be placed on a chip, all of the important parameters and performance characteristics such as temperature effects can be properly specified.

0.3 THE IMPORTANT QUESTIONS DESIGNERS SHOULD ASK

In view of the foregoing discussion, one may wonder what is of fundamental importance in the design of a signal processing system for physical variables. We propose here the two questions analog and digital designers should keep foremost in their minds as they develop a processing system. These are:

For the analog designer:

What specifications must I have for this particular electronic module so that it will be compatible with both the properties of the physical variable, in terms of such parameters as frequency content or amplitude range, and the number of bits the digital designer desires to use to represent data values in the processing?

For the digital designer:

What are the bit rates and how many bits for data representation am I justified in requesting for digital signal processing in order to set physically realizable specifications on analog modules?

From the nature of these questions one should also see the importance of involving both analog and digital engineers in the initial formulation of a processing system. This will ensure that modules selected will not be either under- or over-specified in terms of their performance parameters, including both analog and digital modules.

Other benefits accrue from properly answering the preceding questions. One is that it is possible to analyze a given error source in a processing system or module and determine whether or not the error will have any significant effect on the desired results. A second is that designers develop an appreciation for the definitions of device parameters and operating specifications. In fact, it is hoped that this appreciation will lead to more precise definitions of parameters and specifications whenever one is placed in a position of having to characterize a newly designed module.

0.4 BASIC COMPONENTS OF A SIGNAL PROCESSING SYSTEM

The signal processing system, to be explored in this text, monitors a physical variable and then determines one or more properties of the variable in order that some action may be taken. This concept is shown in Fig. 0-1. Among the variable properties are amplitude, frequency content, transient components, and overall time variation. Some typical input and output quantities are listed in the figure. As indicated in the figure, our main interest here is in those signal processors that accept physical variables, process them—including digital processing—and generate at the output either another physical variable or some other analog signal.

The development of the digital computer and the concomitant numerical processing techniques followed by the subsequent development of minicomputers and microcomputers (here including microprocessors) have led to the rapid expansion of digital processing of signals. The desirable features of high signal-to-noise ratio and ease of computational algorithm development justify this approach. Many times, however, the effects of devices that operate on the real-world variable to produce the digital signals are overlooked. There are few digital variables in nature. Digital codes are a human invention to allow precision and speed in data manipulation and efficient data storage in computer systems. Most natural variables are continuous analog quantities that cannot be directly used by a

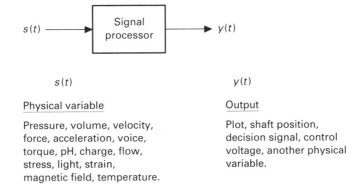

Figure 0-1 The signal processing concept.

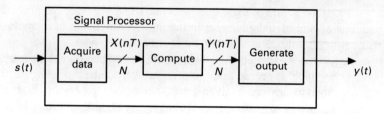

Figure 0-2 Separation of the signal processor into analog and digital sections.

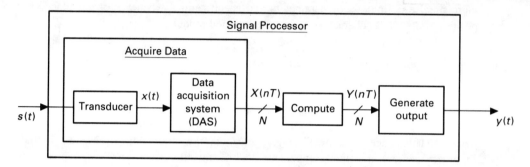

Figure 0-3 Generic block diagram of a signal processor.

digital system. In order to take advantage of computer capabilities, many analog variables are measured and then converted to digital code for processing by a digital system. The system may then use the processed digital signal directly or convert this signal to analog form in order to drive some analog actuator. Thus the first step is to take the general block diagram of Fig. 0-1 and divide it to show the basic units required to accomplish the processing task. This is shown in Fig. 0-2. An explanation of each item in the figure follows.

The function of the unit that acquires the data is to convert the physical variable to digitally coded sample values denoted by $X(nT)$. These values occur only at discrete values of time given by the argument list nT, where T is the time between samples and n the integer sample number. Each value of $X(nT)$ is represented by a binary number of N bits. Various methods for coding (formatting) the binary numbers are given later. The $X(nT)$ are data words. A string of data words constitutes the digitized representation of the physical variable $s(t)$.

Following digital processing by the computer, an output data string of words $Y(nT)$ is produced. In actuality, both the number of bits and time between samples of Y may be different than those of $X(nT)$, but that will not be explored here. The values of $Y(nT)$ are then used to generate the desired output $y(t)$. In some cases $Y(nT)$ may be the desired output.

It is common practice to subdivide the box labeled "Acquire data" into two sections: a transducer and a data acquisition system (DAS). This refinement is shown in Fig. 0-3. The transducer converts the physical variable into an analog quantity that represents the physical variable. The form must be compatible with the input requirements of the data acquisition system and is usually a voltage, current, or charge. We call the physical

Chap. 0   Introduction to Data Acquisition Systems   5

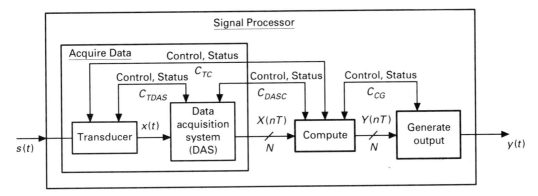

Figure 0-4 Signal processor control line options.

variable measured by the transducer the measurand, and this variable is denoted $s(t)$. The converted value applied to the data acquisition system is denoted $x(t)$.

In order to perform digital computations, the analog signal $x(t)$ must be converted to the digital format of the computer. This is the function of the data acquisition system. Such data acquisition systems are available in several forms, as follows:

1. A single chip with a small number (say four) of peripheral driving or signal conditioning elements such as amplifiers and filters. As the solid-state art has advanced, even the amplifiers and filters (frequency tunable) appear on the chip.
2. A system designed by the engineer that consists of several chips and other active and passive elements.
3. A complete printed circuit board with modules already compatibly designed to produce from the analog signal the required digital inputs for the computer.

In this text we use the third form of system, since operational details of that form will be important whether one is designing an IC package or selecting a complete off-the-shelf unit. This approach will also help greatly in understanding the specifications and performance limitations of data acquisition. The basic configuration shown in Fig. 0-3 is somewhat simplistic, since it implies that there are no interactions between the blocks other than data flow. In reality there will be considerable interaction between the blocks. In Fig. 0-4 is shown a more realistic signal processor system which provides such interaction control signals to allow for proper synchronization and timing among the elements. Such control signals allow for the usual handshaking schemes as well as for status and condition signaling.

As a final note we also emphasize that the configurations shown in Figs. 0-3 and 0-4 are frequently not as clear-cut as the elements may lead one to believe. In some cases a device that one manufacturer may place on a data acquisition chip or board may be physically integrated with the transducer by another manufacturer or built into the computer. Regardless of the physical configuration of a system, however, one can usually identify the signals $x(t)$ and $X(nT)$ at some points, although they may not always be accessible for monitoring.

Exercises

E0.4a Select a physical variable $s(t)$ and list some important properties of that variable. Then tell why each property is important. Be precise in your answers.

E0.4b List three real-world variables other than pressure that could be measured with a pressure transducer. Show the physical configuration of each system.

E0.4c For the physical variables listed in Fig. 0-1, construct a table that lists the variables in one column, an example of where that variable might occur in practice in the second column, and an example of an output variable $y(t)$ and its use with respect to the physical variable in the third column.

0.5 THE DATA ACQUISITION PROBLEM

This text is primarily concerned with the first, second, and last blocks of Fig. 0-4. Considerable emphasis is on the sources of error that contribute to the contamination of the signals at each point in the processor. The data acquisition problem can be more clearly illustrated by dividing the DAS block into two sections, as shown in Fig. 0-5. The signal output of the amplification and filtering block, $x_{BL}(t)$, is a band-limited transducer output signal. This band-limiting is required by the Nyquist sampling theorem, which states that a signal must be sampled at a rate that is at least twice the highest frequency in that signal. If $x_{BL}(t)$ has f_h as its highest frequency component, then $x_{BL}(t)$ must be sampled greater than $2f_h$, otherwise a phenomenon called aliasing occurs. If this condition is not satisfied, frequency components higher than $f_{sample}/2$ are folded down (aliased) into the lower spectral portions of the input signal. This is discussed in more detail in the chapter on filters. Additional signal degradation due to the insertion of noise at all stages of the processor also occurs.

The signals shown in Fig. 0-5 are described as follows:

$s(t)$: The desired measurand, which can be corrupted by noise and extraneous signals (a specific type of noise).

$x(t)$: Transducer output consisting of signal plus transducer errors.

$x_{BL}(t)$: Band-limited signal, which is the amplified transducer signal plus amplified transducer errors plus conditioning circuit errors.

$X(nT)$: Output digital code representing amplified transducer signal plus amplified transducer errors plus conditioning errors plus conversion errors.

$Y(nT)$: Computer output digital code. Computations may have been corrupted by acquisition errors and computer processing round-off errors.

$y(t)$: Processor output consisting of desired signal plus conditioning errors plus conversion errors.

With the preceding background, the data acquisition problem can be stated as follows:

For a signal processor (that may be given or that is to be designed), how are the various errors identified and quantified so that for a given value of N (number of bits) the effects of those errors may be minimized?

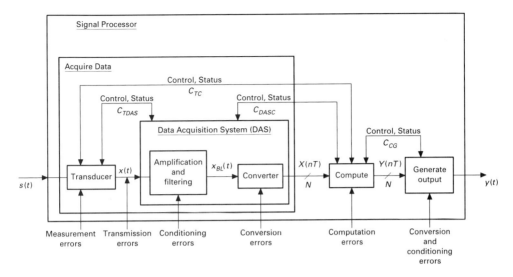

Figure 0-5 Sources of errors in a signal processor.

Exercises

E0.5a For the basic signal processor shown in Fig. 0-5, show a possible timing signal sequence for the various signals C that will convert a physical variable $s(t)$ into a real-world $X(nT)$. Do this for just one sample of $s(t)$, that is, for one value of n.

E0.5b For the two blocks shown within the data acquisition system block of Fig. 0-5, suggest what elements you would use in the blocks and some general specifications for them.

E0.5c Suppose that the data words (of 8 bits each) are coming into a computer every $T = 0.1$ ms, a sample rate on the real-world variable of 10 kHz. What is the greatest slope the time signal of the real-world variable may have if the maximum signal (unipolar) is V_S volts and we want to track the signal amplitude as close as possible? Repeat for 16-bit words.

0.6 DIGITAL SIGNALS: OF BITS, NIBBLES, BYTES, AND WORDS

Many types, sizes, and definitions of binary digit strings may be found coursing their way through a digital computer. The smallest unit, the bit (binary digit), has a value of "0" or "1" and is the basic unit. We will not go into a discussion of the binary number system, for this background will be assumed.

Now if we concatenate 4 bits, for example as 1011, we have what is often referred to as a nibble. Concatenating 2 nibbles into an 8-bit string we obtain a byte. Examples of bytes would be 10101100 and 11111111.

In a digital processing unit the bit string with which the unit works is called a word, although at present there is a tendency to use the term word to refer to a 16-bit (2-byte) configuration. A 16-bit word is shown in Fig. 0-6 with its component parts.

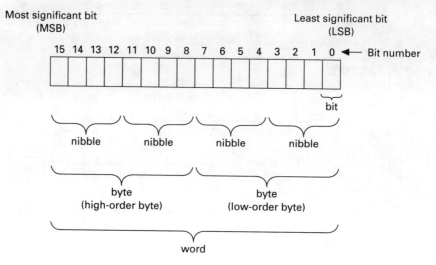

Figure 0-6 Bit configuration within a 16-bit word, including the most significant bit (MSB) and the least significant bit (LSB).

Note that according to the basic definition of word given, a machine that uses an 8-bit string as its basic unit would have a word consisting of 1 byte. In general we speak of an N-bit word, indicated on a schematic diagram by a slash across the data line. See Fig. 0-5.

Suppose that one took a pair of diagonal cutters and went into a digital system or processor at an arbitrary point and cut a wire and recorded the word bits coming off that line. If a string such as 10011010 came out, one would ask: What does this word mean? Without going into a lot of detail, we broadly classify a bit string into one of two categories:

1. Instruction word
2. Data word

(There are systems that have an option of carrying the data within the instruction, but for simplicity here we treat them as distinct.)

The instruction word is used to direct the computer's attention to a particular area within memory and to indicate the type of action that is to be taken—add, store, and so forth.

The data word is the digital representation of a signal originally derived, in our study here, from the real-world variable or some number to be combined in some way with another data word that came from the real-world variable.

Although these distinctions are extremely simplistic, they serve to indicate that it is important to know how the devices that convert from the real-world variable to the digital word affect the bits that appear in the data word. Put in other terms, one could ask: How many bits am I justified in using to represent data to use in a digital computer for the devices that appear between the real-world variable and the computer? An equally

Chap. 0 Introduction to Data Acquisition Systems

important question is the converse, which would be: If I want to use N bits to represent digital data words (which are values of some real-world variable), what are the electrical and mechanical specifications I must demand of devices between the variable and the computer? Answers to these questions are thoroughly explored in this text.

There are two commonly used methods for transmitting the digital words in a system. These are

1. **Parallel mode:** All bits in the word appear simultaneously on a set of N lines, 1 bit per line, plus a ground line.
2. **Serial mode:** Bits are sent sequentially in time, requiring only one line, plus a ground line.

A modification of the serial mode would be the sequential transmission of nibbles or bytes of a word. In this serial mode if the voltage level returns to ground (zero) between bits it is called a return-to-zero (RZ) mode, whereas if the level changes only when a bit changes it is called a non-return-to-zero (NRZ) mode.

0.7 BINARY CODING SCHEMES

The function of the converter block in Fig. 0-5 is to convert the band-limited time signal (most frequently a voltage) into a sequence of binary words. Each word corresponds to the value of $x_{BL}(t)$ at a particular time nT, where T is the interval between samples and n is the sample number from some starting time $t = t_o$. Thus the words would correspond with times $t_o, t_o + T, t_o + 2T, t_o + 3T, \ldots$, the values being denoted by $X(0), X(T), X(2T)$, and so on. Each X is a binary word of N bits.

This section describes several schemes presently used to arrange the bits in a word. The process of constructing the word is called coding, and the way the bits are arranged within the word is the coding scheme.

There are two cases that need to be considered. The first is the situation where the signal $x_{BL}(t)$ is always positive or always negative, which is called the unipolar case. The second case is where $x_{BL}(t)$ can have both positive and negative values, and is called the bipolar case. In either case the coding scheme is set up so that the binary word represents a fraction of the full-scale (largest) value that $x_{BL}(t)$ may have. Typical ranges for $x_{BL}(t)$ are: 0 to +5 V, 0 to +10 V, –5 V to +5 V, and –10 V to +10 V.

0.7.1 The Natural Binary Fractional Code: Unipolar Signals

If we let b_i represent a bit (0 or 1), we can construct an N-bit fractional word as

$$0.b_1 b_2 b_3 \ldots b_N \tag{0-1}$$

where b_1 is the most significant bit (MSB) and b_N is the least significant bit (LSB). The corresponding base two number would be

$$b_1 \times \frac{1}{2^1} + b_2 \times \frac{1}{2^2} + b_3 \times \frac{1}{2^3} + \ldots + b_N \times \frac{1}{2^N} \quad (0\text{-}2)$$

For example,

$$0.101101_2$$

is

$$\frac{1}{2^1} + \frac{1}{2^3} + \frac{1}{2^4} + \frac{1}{2^6} = 0.703125_{10}$$

This means that the binary-coded word represents a little over 7/10 of the maximum value. For a unipolar system operating over the range 0 to 10 V, this would mean that the word

$$X(nT) = 0.101101$$

corresponds to

$$x_{BL}(t_n) = 7.03125 \text{ V}$$

Since the binary words always represent fractions, it is common practice to omit the binary point and use the shorthand form

$$b_1 b_2 b_3 \ldots b_N \quad (0\text{-}3)$$

Table 0-1 is an example of 4-bit word coding along with the corresponding voltages for a unipolar 0 to 10 V system. From the table there are four important observations:

1. It is not possible to code the full-scale value (10 V in the case shown) of the system. The maximum value coded is full scale −1 LSB.
2. The code word with only the MSB as 1 corresponds to the midrange system value, or $\frac{1}{2}$ of the full-scale (FS) value.
3. Any analog processing must preserve the signal value precision to about 0.6 V to give the true value, and this is only for $N = 4$!
4. For an N-bit system the LSB value is $2^{-N} \times FS$. Table 0-2 shows the LSB values for a 10-V full-scale system.

Table 0-2 also indicates how much noise can be tolerated on an input signal, at least to a first approximation. For a 10-V system using 12-bit data words, noise must be below 2.4 mV.

0.7.2 Binary-Coded Decimal (BCD) or 8-4-2-1 BCD: Unipolar Signals

The binary-coded decimal (BCD) system is often used in digital meters and thumbwheel switches. In this code each decimal digit is represented by a pure binary equivalent. Since there are ten decimal digits (0 to 9), it takes 4 bits for each decimal position. A few

Chap. 0   Introduction to Data Acquisition Systems

TABLE 0-1 BINARY FRACTIONAL CODES FOR 4-BIT WORDS AND THE CORRESPONDING VOLTAGE VALUES FOR A UNIPOLAR, 0 TO 10 V SYSTEM ($N = 4$)

Decimal fraction	Binary code ($N = 4$)				Binary fraction	Value of x_{BL}, volts (0 to 10 V system)
	b_1(MSB) (1/2)	b_2 (1/4)	b_3 (1/8)	b_4 (1/16)		
0	0	0	0	0	0.0000	0.000
$\frac{1}{16} = 2^{-4}$ (LSB)	0	0	0	1	0.0001	0.625
$\frac{2}{16} = \frac{1}{8}$	0	0	1	0	0.0010	1.250
$\frac{3}{16} = \frac{1}{8} + \frac{1}{16}$	0	0	1	1	0.0011	1.875
$\frac{4}{16} = \frac{1}{4}$	0	1	0	0	0.0100	2.500
$\frac{5}{16} = \frac{1}{4} + \frac{1}{16}$	0	1	0	1	0.0101	3.125
$\frac{6}{16} = \frac{1}{4} + \frac{1}{8}$	0	1	1	0	0.0110	3.750
$\frac{7}{16} = \frac{1}{4} + \frac{1}{8} + \frac{1}{16}$	0	1	1	1	0.0111	4.375
$\frac{8}{16} = \frac{1}{2}$ (MSB)	1	0	0	0	0.1000	5.000
$\frac{9}{16} = \frac{1}{2} + \frac{1}{16}$	1	0	0	1	0.1001	5.625
$\frac{10}{16} = \frac{1}{2} + \frac{1}{8}$	1	0	1	0	0.1010	6.250
$\frac{11}{16} = \frac{1}{2} + \frac{1}{8} + \frac{1}{16}$	1	0	1	1	0.1011	6.875
$\frac{12}{16} = \frac{1}{2} + \frac{1}{4}$	1	1	0	0	0.1100	7.500
$\frac{13}{16} = \frac{1}{2} + \frac{1}{4} + \frac{1}{16}$	1	1	0	1	0.1101	8.125
$\frac{14}{16} = \frac{1}{2} + \frac{1}{4} + \frac{1}{8}$	1	1	1	0	0.1110	8.750
$\frac{15}{16} = \frac{1}{2} + \frac{1}{4} + \frac{1}{8} + \frac{1}{16}$	1	1	1	1	0.1111	9.375

bits are wasted, since the binary count only needs to cover 0000 to 1001 for each position.

Since 4 bits are required, it is convenient to refer to each group of 4 bits as a quad and to refer to the first quad as the most significant quad (MSQ), down to the smallest decimal value, called the least significant quad (LSQ). This scheme for coding 3-digit decimal numbers is shown in Table 0-3.

Each decimal value is generated from the quad as follows. Let m be the quad number, 1 being the MSQ.

$$m\text{th } quad \text{ } decimal \text{ } value = \frac{b_1 \times 8 + b_2 \times 4 + b_3 \times 2 + b_4 \times 1}{10^m} \tag{0-4}$$

TABLE 0-2 BINARY FRACTIONAL CODING AND VOLTAGE VALUES FOR LSB FOR N-BIT WORDS

N	Decimal fraction $= \dfrac{1}{2^N}$	Binary fraction code	LSB value for 10 V FS (volts)
1	1/2	1	5
2	1/4	01	2.5
3	1/8	001	1.25
4	1/16	0001	0.625
5	1/32	00001	0.3125
6	1/64	000001	0.15625
7	1/128	0000001	0.078125
8	1/256	00000001	0.0390625
9	1/512	000000001	0.01953125
10	1/1024	0000000001	0.009765625
11	1/2048	00000000001	0.0048828125
12	1/4096	000000000001	0.00244140625
13	1/8192	0000000000001	0.001220703125
14	1/16384	00000000000001	0.0006103515625
15	1/32768	000000000000001	0.00030517578125
16	1/65536	0000000000000001	0.000152587890625

For example, if the coded word is 0001100000110101, we first write the quads and then convert each quad to its decimal value, as follows:

MSQ			LSQ
0001	1000	0011	0101
1	8	3	5

Finally, divide each quad by the decimal place power of 10 to obtain

$$0.1 + 0.08 + 0.003 + 0.0005 = 0.1835$$

For a 10-V unipolar system, this means that

$$x_{BL} = (10)(0.1835) = 1.835 \text{ volts}$$

0.7.3 2-4-2-1 Binary-Coded Decimal: Unipolar Signals

The only difference between 2-4-2-1 BCD and 8-4-2-1 BCD is that the first (most significant) bit, b_1, in each quad has weight 2 rather than 8 as in Eq. (0-4). Thus the corresponding quad value of quad number m would be

$$m\text{th quad decimal value} = \frac{b_1 \times 2 + b_2 \times 4 + b_3 \times 2 + b_4 \times 1}{10^m} \qquad (0\text{-}5)$$

For example, for a three-quad 2-4-2-1 BCD word

$$001101100100$$

Chap. 0   Introduction to Data Acquisition Systems

TABLE 0-3   A 3-DIGIT BINARY-CODED DECIMAL SYSTEM

	Decimal fraction of full-scale value	BCD (MSQ) Quad 1 (X.1)	Quad 2 (X.01)	(LSQ) Quad 3 (X.001)	Value of x_{BL}, volts (0 to 10 V system)
0.000	0.000 + 0.000 + 0.000	0000	0000	0000	0
0.001	0.000 + 0.000 + 0.001	0000	0000	0001	0.01
0.002	0.000 + 0.000 + 0.002	0000	0000	0110	0.02
.	.	.	.	.	.
0.010	0.000 + 0.010 + 0.000	0000	0001	0000	0.1
0.011	0.000 + 0.010 + 0.001	0000	0001	0001	0.11
0.012	0.000 + 0.010 + 0.002	0000	0001	0010	0.12
.	.	.	.	.	.
0.020	0.000 + 0.020 + 0.000	0000	0010	0000	0.20
0.021	0.000 + 0.020 + 0.001	0000	0010	0001	0.21
0.022	0.000 + 0.020 + 0.002	0000	0010	0010	0.22
.	.	.	.	.	.
0.030	0.000 + 0.030 + 0.000	0000	0011	0000	0.3
.	.	.	.	.	.
0.035	0.000 + 0.030 + 0.005	0000	0011	0101	0.35
.	.	.	.	.	.
0.100	0.100 + 0.000 + 0.000	0001	0000	0000	1
0.101	0.100 + 0.000 + 0.001	0001	0000	0001	1.01
0.102	0.100 + 0.000 + 0.002	0001	0000	0010	1.02
.	.	.	.	.	.
0.120	0.100 + 0.020 + 0.000	0001	0010	0000	1.20
.	.	.	.	.	.
0.200	0.200 + 0.000 + 0.000	0010	0000	0000	2.0
.	.	.	.	.	.
0.235	0.200 + 0.030 + 0.005	0010	0011	0101	2.35
.	.	.	.	.	.
0.355	0.300 + 0.050 + 0.005	0011	0101	0101	3.55
.	.	.	.	.	.
0.998	0.900 + 0.090 + 0.008	1001	1001	1000	9.98
0.999	0.900 + 0.090 + 0.009	1001	1001	1001	9.99

TABLE 0-4 A 2-DIGIT 2-4-2-1 BINARY-CODED DECIMAL SYSTEM

Decimal fraction of full-scale (FS) value	2-4-2-1 BCD	Value of x_{BL} (volts) for 0–10 V system
0.00	00000000	0
0.001 = 0.0 + 0.01	00000001	0.1
0.02 = 0.0 + 0.02	00000010	0.2
0.3 = 0.0 + 0.03 = (0.0) + (0.02 + 0.01)	00000011	0.3
0.04 = 0.0 + 0.04	00000100	0.4
0.05 = 0.0 + 0.05 = (0.0) + (0.04 + 0.01)	00000101	0.5
0.06 = 0.0 + 0.06 = (0.0) + (0.04 + 0.02)	00000110	0.6
0.07 = 0.0 + 0.07 = (0.0) + (0.04 + 0.02 + .01)	00000111	0.7
0.08 = 0.0 + 0.08 = (0.0) + (0.02 + 0.04 + 0.02)	00001110	0.8
0.09 = 0.0 + 0.09 = (0.0) + (0.02 + 0.04 + 0.02 + 0.01)	00001111	0.9
.	.	.
.	.	.
.	.	.
0.35 = 0.3 + 0.05 = (.2 + .1) + (.04 + .01)	00110101	3.5
.	.	.
.	.	.
.	.	.
0.87 = 0.8 + 0.07 = (.2 + .4 + .2) + (.00 + .04 + .02 + .01)	11100111	8.7
.	.	.
.	.	.
.	.	.
0.99 = 0.9 + 0.09 = (.2 + .4 + .2 + .1) + (.2 + .4 + .2 + .1)	11111111	9.9

the decimal equivalent is

$$\frac{0 \times 2 + 0 \times 4 + 1 \times 2 + 1 \times 1}{10^1} + \frac{0 \times 2 + 1 \times 4 + 1 \times 2 + 0 \times 1}{10^2} + \frac{0 \times 2 + 1 \times 4 + 0 \times 2 + 0 \times 1}{10^3}$$

or,

$$0.3 + 0.06 + 0.004 = 0.364$$

An example of a two-quad 2-4-2-1 BCD coding system is given in Table 0-4.

This system, while seemingly unorthodox from a number system point of view, offers a couple of interesting features. First, the maximum value (FS − 1 LSB) is now represented by all ones. The second reason is that in the design of D/A converters (to be studied in a later chapter) the range (value spread) of components is reduced if 2-4-2-1 BCD is used.

0.7.4 Gray Code: Unipolar and Bipolar Signals

In pure binary counting, as the count proceeds from one decimal amount to the next, more than one bit position may change value. For example, the change

3 → 4

Chap. 0   Introduction to Data Acquisition Systems   15

TABLE 0-5 COMPARISON OF 4-BIT BINARY AND GRAY CODES

Decimal fraction	Gray code	Binary code
0	0 0 0 0	0 0 0 0
1/16	0 0 0 $\underline{1}$[a]	0 0 0 1
2/16	0 0 $\underline{1}$ 1	0 0 1 0
3/16	0 0 1 $\underline{0}$	0 0 1 1
4/16	0 $\underline{1}$ 1 0	0 1 0 0
5/16	0 1 1 $\underline{1}$	0 1 0 1
6/16	0 1 $\underline{0}$ 1	0 1 1 0
7/16	0 1 0 $\underline{0}$	0 1 1 1
8/16	$\underline{1}$ 1 0 0	1 0 0 0
9/16	1 1 0 $\underline{1}$	1 0 0 1
10/16	1 1 $\underline{1}$ 1	1 0 1 0
11/16	1 1 1 $\underline{0}$	1 0 1 1
12/16	1 $\underline{0}$ 1 0	1 1 0 0
13/16	1 0 1 $\underline{1}$	1 1 0 1
14/16	1 0 $\underline{0}$ 1	1 1 1 0
15/16	1 0 0 $\underline{0}$[b]	1 1 1 1

[a] Underlined bits are those that change as number increases from preceding value.
[b] Note 15/16 goes to 0.

would be

$$011 \rightarrow 100$$

Due to finite switching times during the transition, one bit may be delayed in transition, leading to false code values. In the example above we might have

$$3 \rightarrow 4$$

going as

$$011 \rightarrow 101 \rightarrow 100 \text{ (rightmost bit delayed in transition)}$$

so the transitions give the impression of the sequence

$$3 \rightarrow 5 \rightarrow 4$$

The Gray code avoids this by having only one bit at a time change. Table 0-5 shows this code; the underlined bit is the one that changed from the preceding number. This type of code is particularly good for counters or sequential systems. A device that encodes shaft position directly (called a shaft position encoder) is a very good example of this type of system.

The Gray code is also called a reflected or reflecting code because there is a line that divides the complete code sequence into mirror images for all bits except the MSB. A 3-bit Gray code would reflect as follows:

0	0	0		
0	0	1		
0	1	1		
0	1	0	↑	Reflection
—	—	—	—	—
1	1	0	↓	line
1	1	1		
1	0	1		
1	0	0		

Because of this reflection property, the Gray coding scheme can also be used for bipolar signals if one is willing to sacrifice the MSB as a sign indicator. Thus all values below the reflection line could be used to represent negative values, but in the example above only 2 bits of resolution would result. This also is a reasonable result, since bipolar effectively doubles the total voltage span, −10 V to +10 V versus 0 to +10 V, for example, so the voltage increment per bit (the LSB) increases. One could, of course, add another bit to preserve the resolution. This will be discussed later.

0.7.5 Complementary Coding Schemes: Unipolar Signals

Complementary coding methods simply complement the data word. This is accomplished by changing all ones to zeros and all zeros to ones. The complementary code can be used with any of the codes that have been discussed as well as the bipolar schemes to be presented shortly. This coding is useful when complementary logic is being used.

In these types of codes, the decimal number "0" is represented by all ones; for example, in a 4-bit word, we would have

$$0_{10} = 1111$$

and the midscale code would be 0111 rather than 1000. Two popular complementary code schemes are shown in Table 0-6, for 3-bit natural binary and one-quad BCD.

0.7.6 Sign-Magnitude Coding: Bipolar Signals

The output signals of many transducers are inherently bipolar in nature. As the physical variable increases and decreases about some mean value, the transducer responds with positive and negative outputs.

Perhaps the most obvious method for handling bipolar signals is to add a bit to be used as a polarity or sign indicator. As noted previously, if we just used the MSB of the original code we cut the resolution in half for −10 V to +10 V as compared with 0 to +10 V. Negative values have a one for the sign bit, positive values have a zero for the sign bit. This scheme is shown in column three in Table 0-7. (All entries in this table assume 3-bit data values plus sign, for −10 V to +10 V operation.)

If N is the total number of bits in the word, then the LSB fraction of FS is

$$\frac{1}{2^{N-1}} \tag{0-6}$$

TABLE 0-6 COMPLEMENTARY CODING FOR NATURAL BINARY FRACTIONAL CODING AND ONE-QUAD BINARY-CODED DECIMAL SYSTEMS, WITH CORRESPONDING VALUES OF THE INPUT VOLTAGES (0 TO 10 V)

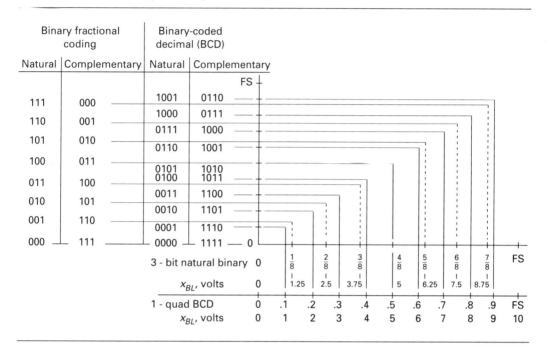

This code is particularly attractive because only 1 bit changes in going from a positive value to a negative value. The shortcomings of this coding scheme include the following:

1. There are two codes for zero.
2. The converter circuitry is somewhat more complex.
3. Digital computations cannot be performed directly on the numbers.

0.7.7 Offset Binary Coding: Bipolar Signals

For offset binary coding we make the most negative value of the system have the all-zeros code word, and then increase in natural binary count so that the all-ones code word corresponds to the most positive system value minus 1 LSB. This is shown in column four of Table 0-7. For an N-bit word the LSB value is

$$\frac{total\ span\ of\ voltage}{2^N}\ \text{volts} \qquad (0\text{-}7)$$

or, as a fraction,

$$\frac{1}{2^N} \qquad (0\text{-}8)$$

TABLE 0-7 BIPOLAR CODING SCHEMES

Decimal fraction	x_{BL}, volts	Sign – magnitude	Offset binary	Two's complement	One's complement
8/8	10	no code	no code	no code	no code
7/8	8.75	0111	1111	0111	0111
6/8	7.50	0110	1110	0110	0110
5/8	6.25	0101	1101	0101	0101
4/8	5.00	0100	1100	0100	0100
3/8	3.75	0011	1011	0011	0011
2/8	2.50	0010	1010	0010	0010
1/8	1.25	0001	1001	0001	0001
0^+	0	0000	1000	0000	0000
0^-	0	1000	1000	0000	1111
–1/8	–1.25	1001	0111	1111	1110
–2/8	–2.50	1010	0110	1110	1101
–3/8	–3.75	1011	0101	1101	1100
–4/8	–5.9	1100	0100	1100	1011
–5/8	–6.25	1101	0011	1011	1010
–6/8	–7.50	1110	0010	1010	1001
–7/8	–8.75	1111	0001	1001	1000
–8/8	–10	no code	0000	1000	no code

For the usual case where the positive and negative extremes are the same, the total span of voltage is simply twice the unipolar range, so the LSB value can be reduced to

$$\frac{total\ span\ of\ voltage}{2^{N-1} \times 2} = \frac{unipolar\ span}{2^{N-1}} \qquad (0\text{-}9)$$

The preceding equation shows that for bipolar coding we effectively lose 1 bit of resolution. This is to be expected, since we are representing twice the voltage span with the same number of bits.

From column four of the table one can see that offset binary is much like a sign + magnitude coding. This accounts for the effective loss of 1 bit in the code word also.

This type of coding has the advantage of being very easy to implement, even with converters that are unipolar. Suppose we are converting from x_{BL} to the digital code, that is, from analog to digital (A/D). If the unipolar converter has a range of one-half the bipolar range (same maximum value), then offset coding can be obtained by attenuating x_{BL} by a factor of two and adding a DC bias of one-half the unipolar range. For example, if a bipolar system operates over the range –10 V to +10 V, and if a particular value of x_{BL} is, say, 7 volts, the input to a unipolar converter operating from 0 to +10 V would be the value

$$\frac{x_{BL}}{2} + \frac{unipolar\ range}{2} = \frac{7}{2} + \frac{10}{2} = 8.5\ V \qquad (0\text{-}10)$$

Note this can also be considered as offsetting the signal by the unipolar range and dividing by two.

Chap. 0   Introduction to Data Acquisition Systems   19

For the inverse case where we have an offset binary-coded word and want to use a unipolar converter to produce an analog output (digital to analog, D/A), as would be necessary to go from $Y(nT)$ to $y(t)$ in Fig. 0-5, we proceed as follows. Suppose, as before, that the unipolar converter operates over half the bipolar range. First it is necessary to double the value (weight) of each bit (because we have only $N-1$ magnitude bits) and then offset the result by the negative of the converter operating range. From Table 0-7, for example, if we have a 4-bit converter operating from 0 to +10 V, we know that only 3 bits are for magnitude, so we would need to double the bit weight and then subtract 10 volts. As a specific case (see Table 0-7), suppose an input offset binary code is

$$1011_2 \rightarrow +3.75 \text{ V}$$

The required converter LSB value is

$$2 \times \frac{10 \text{ V}}{2^4} = 1.25 \text{ V}$$

Thus for the input code 1011_2 sent to the unipolar converter,

$$voltage\ output = 2^3 \times 1.25 + 0 \times 1.25 + 2^1 \times 1.25 + 2^0 \times 1.25 - 10 = 3.75 \text{ V}$$

One obvious problem with this is that if the −10 V offset is inserted after the D/A converter, the converter must actually be capable of generating a 20-V (less 1 LSB) signal. This means the unipolar converter must have a range equal to the total bipolar voltage span, or 20 volts in this case. There are other ways of avoiding this problem, but they will not be explored here.

In addition to the ease of implementation described above, some other advantages of offset binary coding are as follows:

1. The code is computer compatible since it is a natural binary sequence.
2. One can easily obtain the two's complement by simply complementing the MSB. This is a computationally useful scheme.
3. There is an unambiguous code for zero.

Some disadvantages are as follows:

1. A major number of bit transitions can occur when going to zero, since all bits change, i.e., 0111 to 1000 coming from negative values can yield many glitches dynamically (D/A).
2. There are linearity problems statically at major transitions (such as 1000 → 0111) because the transition is the difference between two large numbers.
3. Zero errors may be greater than with sign − magnitude, because the zero analog level is usually obtained by taking a difference between the MSB (1/2 full range) and a bias (1/2 full range), which again is two large numbers for D/A converters.

0.7.8 Two's Complement Coding: Bipolar Signals

This coding scheme is implemented by using the natural binary coding for positive values and the two's complement of the positive natural binary code for the negative values. The two's complement is obtained by complementing all bits in the natural binary code (*zeros* → *ones*, *ones* → *zeros*) and adding 1 LSB.

Example

Suppose we have a 4-bit system which uses sign + magnitude. The natural binary fraction code for 5/8 would be

$$\frac{5}{8} = \frac{4}{8} + \frac{1}{8} \rightarrow 0.101 \rightarrow 101$$

Appending the sign bit yields $+\frac{5}{8} \rightarrow 0101$. To obtain $-\frac{5}{8}$, we change the code as follows:

$$-\frac{5}{8} \rightarrow \overline{0101} + 0001 = 1010 + 0001 = 1011$$

The 4-bit coding for two's complement is shown in column three of Table 0-7. Note that the only difference between this and offset binary coding is that the MSB is complemented.

The usefulness of this type of coding is that for computations, the process of subtraction becomes addition.

Example

Perform $\frac{7}{8} - \frac{6}{8}$ using two's complement coding using Table 0-7. From the table we obtain:

$$\frac{7}{8} - \frac{6}{8} = \frac{7}{8} + \left(-\frac{6}{8}\right) \rightarrow 0111 + 1010 = 10001.$$

We neglect any MSB carry above the 4 bits:

$$\therefore \frac{7}{8} - \frac{6}{8} \rightarrow 0001 \rightarrow \frac{1}{8} \text{ (of full scale)}$$

The disadvantages of two's complement coding are the same as those of offset binary coding.

0.7.9 One's Complement Coding: Bipolar Signals

One's complement coding is implemented by using the sign – magnitude form for positive values and by complementing the magnitude bits of the sign – magnitude form for negative values.

Chap. 0   Introduction to Data Acquisition Systems   21

Example

For the case where the value of x_{BL} is $+\frac{3}{8}$ of full scale, we would have

$$+\frac{3}{8} = +\left(\frac{1}{4}+\frac{1}{8}\right) \rightarrow +(010+001) = +(011) \rightarrow 0011$$

For the case where the value of x_{BL} is $-\frac{3}{8}$ of full scale, we would have

$$-\frac{3}{8} = -\left(\frac{1}{4}+\frac{1}{8}\right) \rightarrow -(010+001) = 1011 \rightarrow 1100$$

A complete listing of this coding technique is given in column six of Table 0-7 for a 4-bit word size.

As with the two's complement coding, numbers can be subtracted by adding the one's complement for the number being subtracted. However, for one's complement work any carry occurring at the sign bit position is brought around and added at the LSB position (called end-around-carry).

For example, suppose we have the subtraction of

$$\frac{7}{8} - \frac{3}{8}$$

Using Table 0-7 for one's complement coding, we get the following:

$$\frac{7}{8} - \frac{3}{8} = \frac{7}{8} + \left(-\frac{3}{8}\right) \rightarrow 0111 + 1100 = 1)0011$$

The extra left carry bit is brought around and added, to give

$$0011 + 0001 = 0100 \rightarrow +\frac{1}{2}$$

There are two disadvantages of this code type:

1. Zero has two codes, 0000 and 1111.
2. This code is a little harder to put into hardware than is two's complement code.

0.7.10 Conversion between Types of Coding

Conversions between the various types of coding discussed in the preceding subsections can be accomplished using the processes given in Table 0-8.

TABLE 0-8 CONVERSION BETWEEN CODING SCHEMES

To convert to → from ↓	Sign magnitude	Offset binary	Two's complement	One's complement
Sign magnitude	—	Complement MSB. If new MSB = 0, complement other bits and add 000....0001	If MSB = 1, complement other bits and add 000...0001	If MSB = 1, complement other bits
Offset binary	Complement MSB. If new MSB = 1, complement other bits and add 000...0001	—	Complement MSB	Complement MSB. If new MSB = 1, add 111...1111
Two's complement	If MSB = 1, complement other bits and add 000...0001	Complement MSB	—	If MSB = 1, add 111...1111
One's complement	If MSB = 1, complement other bits	Complement MSB. If new MSB = 0, add 000...0001	If MSB = 1, add 000...0001	—

0.7.11 Additional Types of Coding

There are about as many different types of coding as there are people who have thought about the problem of coding. The preceding sections have presented only the more standard ones. Each of the codes has its particular advantages under a restricted set of operating conditions. Additional popular categories of coding are:

1. Modified Coding: In this type of coding the sign bit (MSB) only is complemented.
2. Complementary Coding: In these coding schemes all the bits in any one of the codes discussed earlier are complemented. These are usually used to take advantage of state-of-the-art techniques in integrated circuit technology and digital signal processing.

Chap. 0   Introduction to Data Acquisition Systems

0.7.12 Summary of Coding Schemes

A graphical summary of the more popular coding schemes is given in Fig. 0-7. The plots show the relationships between the LSB, MSB levels and the corresponding fractions and values of $x_{BL}(t)$. The band-limited subscript indicator BL has been omitted for clarity.

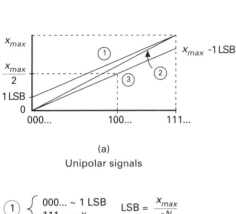

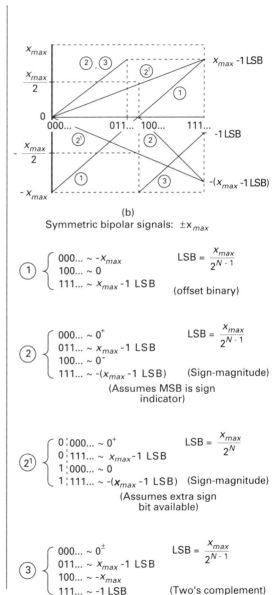

Figure 0-7 Data coding schemes related to input signals for unipolar and bipolar signals.

Exercises

E0.7a For a unipolar 0–5 V system, what is the largest voltage that can be coded for an 8-bit binary fraction coding?

E0.7b Construct a table similar to Table 0-1 for $N = 5$ and a unipolar 0 to 10 V system. What does the number 10000 correspond to in voltage?

E0.7c For an amplifier with the input at ground (0 V) potential, the output is ideally zero. However, the output voltage drifts from the zero value as temperature changes. A typical operational amplifier may drift 10 µV/°C. For a 0–5 V unipolar system, what is the allowable temperature change for 12-, 14-, and 16-bit words ($N = 12, 14, 16$) so the output just changes 1 bit voltage value?

E0.7d Write the following decimal numbers in BCD:

$$0.37624 \quad 0.10100$$
$$0.20102 \quad 0.76543$$

E0.7e The following numbers are in BCD. Obtain the decimal equivalents and determine the voltage values of each for a 0–5 V system.

$$01001001011000101001$$
$$10010110100110000110$$

E0.7f Work exercise 0.7d for 2-4-2-1 BCD.

E0.7g Work exercise 0.7e for 2-4-2-1 BCD.

E0.7h For the 2-4-2-1 BCD system, give a rationale for using 0011 rather than 1001 to represent 0.3 in the MSQ.

E0.7i Using the Gray code, construct the count sequence for a 5-bit word and show the reflection line. Begin with 0000 0000, use every other number in each quad as count increases.

E0.7j Develop a table similar to Table 0-6 for two-quad BCD. Assume unipolar 0 to 10 V operation. Begin with 0000 0000, use every other number in each quad as count increases.

E0.7k Develop a table similar to Table 0-6 for 4-bit natural binary fractional coding.

E0.7l Show that the digital system in Fig. E0.7l converts a 4-bit Gray code to 3-bit natural binary code.

E0.7m Suppose a system has a signal range on x_{BL} of –10 V to +10 V, and an engineer wants to code the data with a converter that operates from 0 to +5 V. How should the x_{BL} signals be scaled for offset binary coding? Develop a general equation corresponding to Eq. (0-10) that gives the scaling value for an x_{BL} signal ranging from $-V_B$ to $+V_B$ to a unipolar A/D converter operating from 0 to $+V_U$. Does your result reduce to Eq. (0-10) for $V_U = V_B$?

E0.7n Develop a table similar to Table 0-7 that shows the Modified Coding and Complementary Coding words for a 4-bit ($N = 4$) system, for all bipolar code schemes.

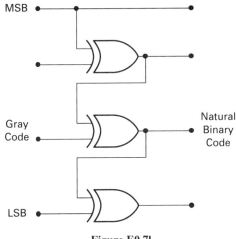

Figure E0.71

GENERAL EXERCISES

0.1 In the section on coding it was pointed out that all codes were in fractional format; that is, the output was always some fraction of the maximum or reference voltage. Suppose one has a converter which has an N-bit word in and a voltage out, as shown in Fig. PR0.1, a digital to analog converter (D/A). Thus the output will be some fraction of E_{REF}. Effectively the output is E_{REF} multiplied by the digital input fraction. If E_{REF} were also variable, we could use such a D/A converter to obtain a precise amount (fraction) of any voltage applied at E_{REF}. For a 3-bit natural binary coding, plot E_o versus E_{REF} (0 to +10 V) with the value of the input word as a parameter (000, 001, etc.). This will give a family of lines on the plot.

0.2 Repeat problem 0.1 for the bipolar case (–10 V to +10 V) using 4-bit sign – magnitude coding words as the parameter (0000, 0001, etc.).

0.3 Suppose one has a converter that has a voltage in and an N-bit word out as shown in Fig. PR0.3, an analog to digital converter (A/D). Since the output code is a fraction, the code word is actually the ratio of E_{in} to E_{REF}. Thus the A/D may be used as a divider. If E_{REF} and E_{in} are both allowed to vary, we have a device that can compute the ratio of two voltages. For 4-bit sign – magnitude coding, plot the output code fraction value versus E_{in} using E_{REF} as a parameter (2, 4, 8, 10 V) where $E_{REF}^{MAX} = \pm 10V$. Is there any limit on the value of E_{in}? Assume converter has all 1's as $E_{REF} - 1$ LSB.

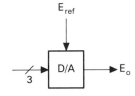

Figure PR0.1

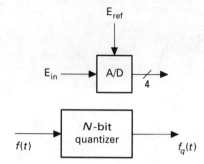

Figure PR0.3

Figure PR0.11

0.4 Obtain a manufacturer's data sheet for a data acquisition system on a chip, and draw a block diagram that shows the main system components contained thereon.

0.5 A 10-V peak sine wave is to be digitized. The sine wave has high-frequency noise of 25-mV peak added to it. Explain the merits of 8-bit and 12-bit data words for coding sign + magnitude.

0.6 Suppose one has a digital computer that uses 8 bits to represent data. If a unipolar input signal of maximum value 10 V is to be converted, what is the smallest input voltage change that can be detected, and what is the largest voltage value that the data words may represent?

0.7 A flashmeter is a device used by a photographer to measure the light received by a subject. The light is generated by one or more Xenon electronic strobes. The flashmeter is placed by the subject, and the strobes are positioned for optimum lighting effect. The meter system is to ultimately provide the photographer with a readout of the required aperture setting for proper film exposure. Draw a simple block diagram system from input transducer to readout, showing also any other required inputs, that will measure the light energy received by the subject during the strobing interval and yield the required aperture setting readout.

0.8 Describe a digital coding system that would use all 1s to represent x_{max} and 1000...(MSB) to represent $x_{max}/2$. (Hint: What will 0000...represent?)

0.9 For ±5 volt bipolar offset binary coding, prepare a table that shows the value of each data word for $N = 4$.

0.10 Name three or more advantages that two's complement coding has over offset binary coding in processing bipolar signals. Be sure to consider the arithmetic scheme likely to be used by the digital processor. How difficult is it to convert from offset binary to two's complement?

0.11 The quantizer box shown in Fig. PR0.11 has as its input a continuous function of time (such as a voltage) and has as its output a discrete level representation of the input, the number of levels being determined by a number N of bits in a binary coding scheme, as indicated in Fig. 0-7. Suppose that $N = 3$ and that $f(t)$ is a sine wave that has a peak-to-peak amplitude of 10. Draw the quantizer output for the coding cases 1, 2, and 3 given in Fig. 0-7 (indicate code bit string at each level) for quantizers that operate in the following two ways:

1. Output increases to next bit level whenever $f(t)$ value is exactly at that level.
2. Output increases to next bit level whenever $f(t)$ value is within 1/2 LSB of that level.

REFERENCES

1. VanDoren, A. H., *Data Acquisition Systems*, Reston Publishing Company (Prentice Hall), Reston, VA, 1982.
2. Garrett, P. H., *Analog Systems for Microprocessors and Minicomputers*, Reston Publishing Company (Prentice Hall), Reston, VA, 1978.
3. Arbel, A. F., *Analog Signal Processing and Instrumentation*, Cambridge University Press, Cambridge, England, 1980.
4. Wobschall, D., *Circuit Design for Electronic Instrumentation*, McGraw-Hill, New York, 1979.
5. Morrison, R., *Instrumentation Fundamentals and Applications*, John Wiley and Sons, New York, 1984.
6. Carr, J. J., *Microcomputer Interfacing Handbook*, TAB Books, Blue Ridge Summit, PA, 1980.
7. Barney, G. C., *Intelligent Instrumentation: Microprocessor Applications in Measurement and Control*, Prentice Hall International, Englewood Cliffs, NJ, 1985.
8. Oppenheim, A. V., and R. W. Schafer, *Digital Signal Processing*, Prentice Hall, Englewood Cliffs, NJ, 1975.
9. Gordon, B., "Noise Effects on Analog to Digital Conversion Accuracy," *Computer Design*, Part I: March 1974, Part II, April 1974.
10. Bibbero, R. J., *Microprocessors in Instruments and Control*, John Wiley and Sons, New York, 1977.
11. Sheingold, D. H. (ed.), *Analog-Digital Conversion Notes, Part I*, Analog Devices, Norwood, MA, 1977.
12. Clayton, G. B., *Data Converters*, Halsted Press (John Wiley and Sons), New York, 1982.

1

Transducers, Transducer Parameters (Specifications), and Their Effects on Digital Processing

1.1 INTRODUCTION

The first important step in the processing sequence is the conversion of the physical variable into an electrical variable that can be used by the data acquisition system (see Fig. 0-5). A transducer is the unit that provides this conversion. The primary concern in this chapter is the effects of the transducer (and its associated parameters) on the physical variable, $s(t)$, as those effects relate to the subsequent digital processing. It is not the intent of the text to present an exhaustive study of the specific types of transducers and their applications to specific measurement problems, because that depth is covered in separate books on that subject [1]–[6]. However, we devote a section to some examples of transducers so that the theory which follows may be discussed with respect to a particular device and/or application.

Following a general discussion of transducers and transducer types are sections that discuss each transducer parameter in detail. Each parameter is carefully defined and then related to the number of bits N in the data words $X(nT)$ shown in Fig. 0-5. Some suggestions are given for correcting and properly accounting for the parameter effects. Terms relating to the static, dynamic, and noise characteristics of transducers are treated.

1.2 THE GENERAL CONCEPT OF A TRANSDUCER AND ITS BASIC ELEMENTS

Perhaps the most comprehensive definition of a transducer is any unit that appears between any two or more signals. On this basis, the largest block in Fig. 0-5, labeled Signal processor, would be a transducer, since it appears between $s(t)$ and $y(t)$. Such a

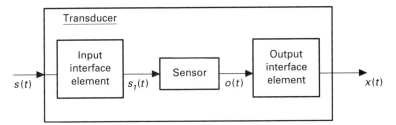

Figure 1-1 General components of a transducer.

transducer would probably have more variables to describe it than could be mentally dealt with at one time, so in the spirit of engineering we break our system into smaller "transducers." The smaller blocks become manageable and more easily described. Nevertheless, the general definition of a transducer is important, because it allows the results of this chapter to be applied to many different kinds of input-output devices. Thus even if we limit our definition to a more restricted class of elements, the parameters discussed can be thought of in more general kinds of applications.

For purposes of this chapter we define a transducer as a unit that has a physical variable as an input and some analog output such as voltage, charge, or current. While this still includes general devices, we intend to use the term for single-function devices.

A transducer may be conveniently represented by three elements as shown in Fig. 1-1. The heart of the transducer is the *sensor*. The terms sensor and transducer are often used interchangeably, but we will use them consistently as indicated. This is the element that responds to a physical variable by producing an output $O(s)$ which is most frequently proportional to $s(t)$. The output may be an electrical quantity or some quantity that can be conveniently monitored electrically. As one example, consider a sonar hydrophone constructed from a piece of piezoelectric material such as $BaTiO_3$ (Barium Titanate). When a sound pressure wave, $s(t)$, impinges on the material, the resulting deformation produces a voltage, $O(t)$, between two faces of the material. The voltage amplitude depends upon the amplitude of the sound pressure.

A second example of a sensor is a resistance temperature element. When the resistive material is subject to a temperature change, $s(t)$, the value of the resistance, $O(t)$, changes. In this case, the output is not a current or voltage directly, but the resistance change can be measured electrically.

The transducer block in Fig. 1-1 contains an input interface element that has three primary functions. One is to provide improved coupling between the measurand $s(t)$ and the sensor. This function is frequently referred to as a "matching" function. The second purpose of the interface element is to protect the sensor from undesirable environmental effects. The hydrophone provides a good example of the first function. For a sound wave in sea water the coupling of the acoustic pressure wave to the piezoelectric material can be improved by placing a material, called rho-c rubber, in front of the sensor. This, of course, also provides protection of the sensor against moisture. Using the resistance temperature device example, there may be only a small coating on the outside to protect the element from the environment. Finally, a third function of the interface element is to convert $s(t)$ to another physical variable, $s_1(t)$, required by a sensor that cannot respond to $s(t)$ directly. This condition frequently occurs when a sensor is used to monitor a variable

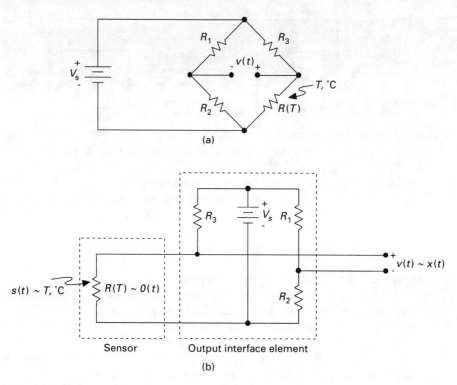

Figure 1-2 Wheatstone bridge used as an output interface element. (a) Conventional representation. (b) Redrawn to identify transducer components.

other than that for which it was originally designed. For example, suppose one has a sensor whose output varies with a linear motion of a shaft. If this sensor is to be used to measure angular displacement, some interface device is required that converts angular motion to linear motion.

The output block in the transducer is called the output interface element. This element also has three important functions. The first is to provide improved or more efficient coupling between the sensor and the unit that follows (the DAS in Fig. 0-5), the so-called matching function. The element is also used to provide protection from the environment as the input element did. The third function of the element is to convert the sensor output $O(t)$ into an electrical or optical variable. A good example of this last function is provided by the resistance temperature device. Some technique for converting the resistance change to a voltage change is required. One popular method is to use a Wheatstone bridge, as shown in Fig. 1-2. The bridge components constitute the output interface element. In the figure the bridge is redrawn to emphasize the various elements of the transducer.

In some data systems there may be significant distances between any of the units shown in Fig. 0-5, even many kilometers. Thus any of the lines connecting the blocks may consist of a simple printed circuit board trace, a coaxial or twisted pair transmission line, a fiber optic cable, or even a radio link. With this extended point of view, $x(t)$ may be an optical or a radio signal, not merely a voltage, current, or charge. The implication of this is that the output interface unit may consist of an optical or radio transmitter. Appropriate receivers must be provided at the block inputs also. In this text we will treat $x(t)$ as a voltage in most cases.

There are at least three classes of systems that use transducers. Perhaps the simplest is strictly a measurement system. In this instance, the analog measurand is sensed with the output electrical quantity driving a display. A digital thermometer is an example of such a measuring system. The input to the transducer is temperature while the output may be an analog voltage. This analog voltage is converted to a digital equivalent signal that drives the display.

A second class of system is used for control. In this case the measurand produces an output that controls some quantity. The quantity being controlled often influences the measurand. For example, a temperature-controlled oven senses oven temperature which the transducer converts to output voltage. This voltage is used to determine the time or duty cycle of the heating element voltage which, in turn, determines the oven temperature.

A third class of system is a signal processing system. The measurand produces an output quantity that is processed or transformed in some useful way and then used appropriately. A telephone line carrying the electrical analog of the sound pressure variations of a speaker's voice may also carry unwanted noise. A signal processing system may digitally filter this line signal to eliminate or minimize the noise and then use this filtered signal to reproduce the original speech.

As an additional note, one should be aware that there are transducers that convert directly from the physical variable to the digital data word. For example, a shaft position encoder may have a plate connected to a shaft, the plate having holes or slots which are sensed by light beams to generate zeros and ones directly.

An effort to produce a standard for transducer nomenclature was undertaken in the late 1960s. Such a standard was proposed in 1969 [1] and later adopted as American National Standard, ANSI MC 6.1-1975. Since the adherence to standards in the United States is voluntary, not all manufacturers use the proposed nomenclature. We will attempt to do so as we further discuss the characterization of transducers.

The description of a transducer is generally based on several of the following considerations [2].

1. What does the transducer measure (that is, what is the measurand)?
2. What is the operating or transduction principle of the electrical or output portion of the transducer?
3. What sensing device in the transducer responds to the measurand?
4. What special features must be considered to use the transducer?
5. What is the range of measurand values over which the sensor is designed to operate?
6. What departures from ideal behavior does the transducer exhibit?

A typical description for a transducer given by a manufacturer is as follows:

The 8694 is a piezoelectric triaxial accelerometer that measures up to 500g with a sensitivity of 4 mV/g. The linearity (endpoint) is ±1%.

This transducer measures acceleration using a piezoelectric crystal. It has a range of 0 to 500 g with an output of 4 mV for each g of acceleration. The specification of linearity characterizes the nonideal behavior of the device.

Exercises

E1.2a For the Wheatstone bridge shown in Fig. 1.2(a), develop the relationship between the resistors that makes $V(T)$ equal to zero at a reference temperature T_0 (that is, $V(T_0) = 0$). For $V_S = 10$ V, what ΔR is required (in terms of resistor values) in order to have $\Delta V(T) = 1$ V at any temperature in the range $\pm 25°$ C from T_0?

E1.2b Design an alternate method for converting a resistance change into a current, voltage, or charge, for a resistance temperature device.

E1.2c List three physical variables other than pressure which could be measured using a pressure transducer. For each variable, suggest a possible input interface element.

1.3 THE CONCEPT OF ERROR AND ITS RELATIONSHIP TO DIGITAL SIGNAL PROCESSING

From the preceding section we see that right from the start, distortion and contamination of the representation of $s(t)$ by $x(t)$ are introduced. See Fig. 0-5. Noise and other external quantities also appear on the analog representation $x(t)$. In this text we examine each of the system components in turn and evaluate the effects of component nonideality on the subsequent signal processing.

In order to relate errors to digital signal processing, we need to develop some criterion that quantifies the errors in terms of the number of bits N in the data word. This is explained next. Suppose that a given variable V has a range of values from V_{min} to V_{max}. The span of the variable is the magnitude of the total variation of V, which we will denote by V^{span}. For example, if $V_{min} = 1$ and $V_{max} = 8$,

$$V^{span} = |8 - 1| = 7$$

For $V_{min} = -3$ and $V_{max} = 10$,

$$V^{span} = |10 - (-3)| = 13$$

For $V_{min} = -6$ and $V_{max} = -2$,

$$V^{span} = |-2 - (-6)| = 4$$

In general,

$$V^{span} = |V_{max} - V_{min}| \tag{1-1}$$

If V_{max} and V_{min} are of the same sign, the variable V is said to be unipolar; otherwise it is bipolar. If $V_{min} = -V_{max}$ the system is called symmetric bipolar. For most of our work we will assume that unipolar signals are positive. Nevertheless, the magnitude bars are important.

Now if we want to represent V using N-bit data words, each bit will represent a value (called the least significant bit or LSB value) given by

$$\text{LSB } value = \frac{V^{span}}{2^N} \qquad (1\text{-}2)$$

This is frequently called the quantum interval, denoted by q. In order to keep the physical concept evident, we will use LSB directly in the first part of the text and q in later chapters. Nonlinear coding, where the LSB value is not constant, is considered in Chapter 5. By definition, we say that a digital word will not increment by one until the variable V has changed by more than 1/2 LSB value. This means that an error in the value of V is insignificant as long as

$$|\,error\,| \leq 1/2 \text{ LSB} = \frac{V^{span}}{2^{N+1}} \qquad (1\text{-}3)$$

where $|\,error\,| = |\,actual\ value\ of\ variable - ideal\ value\ of\ variable\,|$.

This equation is used in practice as follows. The error produced by a device or device parameter is first calculated. That value is then compared with the 1/2 LSB value. If the error satisfies Eq. (1-3), that error can be ignored. If not, then correction measures must be implemented, either in the design of the acquisition system or in the computer processing.

A natural extension of the concept of error is that of the accumulation of error. The total error is obtained from what is called the error budget, which is a list of all identifiable errors. Since a transducer has some dozen possible parameters that contribute to the error budget, one might ask how to combine the effects of all the errors. If each parameter error contributes 1/2 LSB, it would seem that before long the error exceeds the signal value. Additionally, each block in the signal processor introduces its errors. This problem is handled by first minimizing the individual errors—using zero set, matching, or compensation—and then combining the remaining or residual errors according to

$$\text{root sum square (RSS) error} = \left(\sum_{i=1}^{m} |\,error_i\,|^2 \right)^{1/2} \qquad (1\text{-}4)$$

The ideal situation is to have the RSS error less than 1/2 LSB. The justification for this rather crude total error expression is that some errors will be positive and some will be negative. Thus a simple summing of the error magnitudes would not be as accurate as the RSS value given by the preceding equation, although the sum would provide a maximum error limit.

1.4 TRANSDUCERS AND SENSORS FOR DATA ACQUISITION

In this section we present some of the more popular sensors used in data acquisition. Since a given sensor can be used with many measurands, we feel it appropriate to describe the sensor and then suggest ways in which the sensor might be coupled to a measurand with an appropriate interface element. An alternate approach is to take a given physical variable and suggest sensors that can be used [1,5,6]. In either case it is clearly impossible to be exhaustive, because the creativity of engineers will forever add to the lists of transducers, sensors, and interface elements. By listing the sensors it is easier to refer to them in subsequent sections of this chapter.

The list of physical variables that can be sensed by various transducers is fortunately quite lengthy. By sensing we mean that the device produces some output quantity that depends on an input variable in a predictable way. Normally, the output variable is either a voltage or current or a quantity that can be easily converted to one of these two electrical quantities. Transducers having an electrical output variable are the most important in use, and we will restrict all further discussion to these types of devices. The input variable is a physical variable such as temperature, pressure, force, torque, mass, linear or angular displacement, linear or angular velocity, linear or angular acceleration, stress, strain, fluid flow, viscosity, moisture, volume, heat flow, light intensity, color, nuclear radiation, magnetic field, electric field, or some other natural quantity.

1.4.1 Capacitive Sensors

A capacitive sensor allows the measurand to produce a capacitance change. For a parallel plate capacitor, the capacitance is given by (neglecting fringing)

$$C_s = \varepsilon \frac{A}{d} \tag{1-5}$$

where ε is the dielectric constant, A is the area of a plate, and d is the plate separation. The most popular capacitive transducer allows one plate to move relative to a second, fixed plate. This movement modifies d as the measurand varies. A second method of transduction varies the common (overlap) area between the plate by sliding one plate relative to the other. In this method, d remains constant. The third possible capacitive transduction method causes the dielectric constant to vary while keeping A and d constant. This can be accomplished by allowing the dielectric to move parallel to the plates. The dielectric may be a solid or a liquid, depending on the application.

Capacitive transducers are frequently used in linear displacement or angular displacement measurements. Displacement transducers generally allow a plate or set of plates to move parallel to a second plate or set of plates. The common area between plates is then a function of the displacement.

Capacitive level sensing of liquids can also be accomplished. The dielectric constant between two long electrodes partially immersed in a liquid changes as the liquid level changes, resulting in a capacitance variation with liquid level.

Capacitive pressure transducers allow one plate to move toward or away from another plate or set of plates. This results in a change in plate separation with applied pressure. One version of pressure transducer consists of two fixed plates on either side of

Chap. 1   Transducers, Transducer Parameters (Specifications)

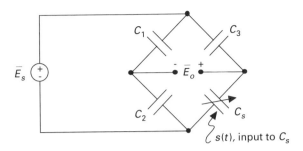

Figure 1-3 Wheatstone bridge configuration for converting capacitive sensor change (C_s) into a voltage using a sinusoidal source.

the movable plate forming two capacitors. As pressure moves the center plate toward a fixed plate, capacitance between these plates increases. Capacitance between the center plate and the other fixed plate simultaneously decreases. These two pressure-variable capacitors can serve as two elements in a bridge circuit that uses an AC excitation signal.

An important pressure transducer is the sound-pressure sensor or capacitor microphone. These devices are capable of measuring very low pressures, and they also have high-frequency responses. The capacitor microphone can achieve good response over a range of frequencies from 0.01 Hz to 20 kHz.

There are several methods for converting the capacitance change into a voltage. One of the popular configurations is the Wheatstone bridge arrangement shown in Fig. 1-3. The source $\overline{E}_s$ is usually a sinusoidal source, and the output $\overline{E}_o$ depends upon the degree of unbalance of the bridge. The output is given by

$$\overline{E}_o = \left[\frac{C_3}{C_3 + C_s} - \frac{C_1}{C_1 + C_2} \right] \overline{E} \qquad (1\text{-}6)$$

There are many other bridge arrangements that will perform the same interface function [7]. These include resonance effects when used with inductors in bridges.

Another interface element could be an oscillator whose frequency of oscillation is determined by C_s. The frequency would then be related to the measurand.

As an example of the application of the capacitive sensor we consider the fluid level detector shown in Fig. 1-4. When the liquid level rises, the effective dielectric constant between the two plates will increase. As a first approximation we use the equivalent circuit of the (b) part of the figure. The total capacitance is the parallel combination of C_o (air dielectric, ε_o) and C_1 (liquid dielectric, ε_1), given by

$$C_s = C_o + C_1 = \frac{\varepsilon_o A_o}{d} + \frac{\varepsilon_1 A_1}{d} \qquad (1\text{-}7)$$

Since the total plate area is constant, $A = A_o + A_1$,

$$C_s = \frac{\varepsilon_o (A - A_1)}{d} + \frac{\varepsilon_1 A_1}{d} = \left[\varepsilon_o A + (\varepsilon_1 - \varepsilon_o) A_1 \right] \qquad (1\text{-}8)$$

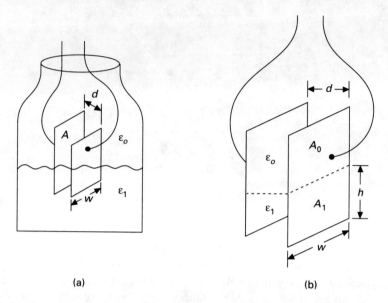

Figure 1-4 (a) Application of a capacitive sensor as a liquid level detector. (b) The equivalent system.

From the figure, $A_1 = hw$ so that

$$C_s = \frac{1}{d}\left[\varepsilon_o A + (\varepsilon_1 - \varepsilon_o)hW\right]. \qquad (1\text{-}9)$$

The capacitor value is thus directly related to the height of the liquid on the plates. There are, however, some practical problems with such bridges. Any temperature change affects the capacitance, and induced AC noise is not filtered out. See problem 1.17.

Exercises

E1.4.1a Derive Eq. (1-6) for the capacitive sensor. Determine the slope $d\bar{E}_o/dC_s$ and sketch what it would look like qualitatively (slope versus C_s). Also sketch $|\bar{E}_o|$ versus C_s.

E1.4.1b Suppose a capacitive sensor is used to detect pressure, P, by varying d (d decreasing with increasing P) in Eq. (1-5). Sketch a graph of C_s versus P. If exercise E1.4.1a has been worked, sketch what $|\bar{E}_o|$ versus P would look like.

E1.4.1c A capacitive level sensor is part of a series LC network driven by a fixed frequency sinusoidal source, as shown in Fig. E1.4.1c. Make a sketch of $|\bar{E}_o|$ as a function of h. The operating frequency of the generator is

$$\omega_g = \left(\frac{d}{\varepsilon_o AL}\right)^{1/2} \text{ rad/s}$$

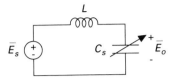

Figure E1.4.1c Circuit for capacitive level detector.

1.4.2 Inductive Sensors

Inductive sensors are conveniently divided into two basic types: the single-winding configuration and the transformer configuration. The single-winding sensor is the common two-terminal inductor from circuit theory, for which the inductance is given by

$$L_s = \mu F(geometry) N^2 \tag{1-10}$$

where μ is the permeability of the material around and within the windings, N is the number of turns, and $F(geometry)$ is a factor whose value depends upon the physical configuration of the inductor (toroid, solenoid), length, and diameter. For these sensors it is most often convenient to change μ by using a movable core material, usually made of iron or ferrite. As the core is inserted into the windings, the effective value of μ increases. Two possibilities are shown in Fig. 1-5. Another possibility is to construct the inductor out of a spring material and use the length change to vary the value of L_s. This class of sensor is most frequently used in a bridge circuit for an interface unit.

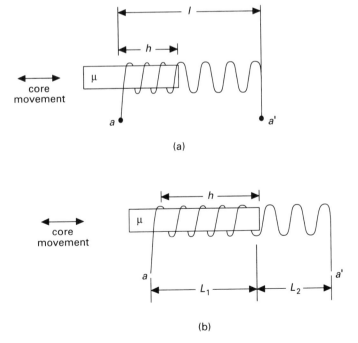

Figure 1-5 (a) Single-winding inductive sensor using core movement. (b) The electrical equivalent.

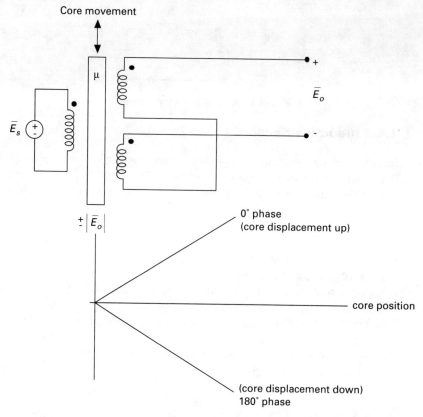

Figure 1-6 Linear variable differential transformer (LVDT) and its characteristic.

The second type of sensor, the transformer configuration, is also called a reluctance sensor. Fig. 1-6 shows the most popular configuration, called the linear variable differential transformer (LVDT). When the core is centered, equal voltages appear across each secondary. With the secondaries connected as shown, e_{out} consists of the difference voltage of the two secondaries. As the core moves upward, the voltage of the upper winding increases while that of the lower winding decreases. The differential output increases as the core is moved either direction from center, but the phase indicates which direction the core has moved. The addition of rectifiers to each secondary winding converts the output signals to a voltage that goes positive for upward core movement and negative for downward movement.

Various types of LVDT are available to cover a wide range of core movement. One LVDT may produce usable outputs for a movement of a few hundredths of an inch, while another may produce signals that vary linearly with core movement up to 10 inches. In addition a rotary variable differential transformer (RVDT) is also manufactured to relate rotary movement to output voltage. The output voltage can be linear, with movement of the shaft for angles of ±60°.

A DC-LVDT that produces an AC output and is powered by a DC source can be used in situations wherein AC power is unavailable. This transducer uses integrated electronics to convert the input DC voltage to an AC voltage by means of an oscillator. This AC voltage then drives the primary winding. A precision amplifier-rectifier is used to convert the output AC signal to a DC value. Inductive devices suffer from spurious movements of the system which move the core.

A sensor related to the two types discussed above is the electromagnetic induction sensor. In this device a magnetic field generated by a magnetized core or current loop is caused to move past a wire or turns of a wire, which motion generates an induced emf. The induced emf is related to the physical variable itself (if the variable is a magnetic field) or by an interface unit that converts the physical variable to a moving magnetic field. Since the induction devices usually depend upon motion, any unwanted vibrations generate noise in the output.

Exercises

E1.4.2a For Fig. 1-5(a), develop an equation for the inductance seen at the terminals a–a' as a function of h. The total number of turns is N. Hint: Use inductors in series and Eq. (1-10). Consider carefully all changing parameters as the core moves. You will not obtain an expression for inductance as an explicit function of h because $F(geometry)$ is not known.

E1.4.2b For Fig. 1-6, show how rectifiers could be added to the system to produce a DC output voltage that would yield the characteristic shown in the figure as a DC response.

1.4.3 Resistive Sensors

These sensors convert the measurand into a resistance change. The resistance variation may result from mechanical motion (as the wiper arm on a rheostat or potentiometer), stress (piezoresistive), humidity, or temperature.

A potentiometer is frequently used to measure position by interfacing the position of a shaft to the wiper arm of the potentiometer. As the shaft rotates the resistance setting changes. This sensor is the basis for position monitoring in automatic control systems. A voltage or current source can be used to indicate the position of the wiper arm (see exercise E1.4.3a).

One of the important resistive stress transducers is the strain gauge. These are usually bonded to a moving or stressed surface so that the fine wires from which the gauges are made undergo resistance changes. The formula for a uniform cross section material resistance is

$$R_S = \rho \frac{l}{A} \tag{1-11}$$

where ρ is the material resistivity, l is the length of the sensor, and A is the cross sectional area. A typical system is a strain gauge and is shown in Fig. 1-7. As stress is applied to the material, the wire is also stressed, so that l increases and A decreases. The dominant effect is due to the change in l. This effect is frequently referred to as the piezoresistive effect.

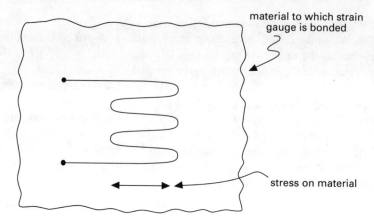

Figure 1-7 Strain gauge bonded to material to measure material displacement due to stress.

Humidity may have several effects. If the moisture is absorbed into the surface of the sensor, the resistivity may change, or the length or area may change due to a swelling or shrinking of the sensor. For both this case and the strain gauge the Wheatstone bridge provides a convenient and sensitive interface to generate a voltage.

A resistance constructed from a material having a temperature-dependent resistivity value is an important sensor. Equation (1-11) shows that the resistance varies directly with ρ, although ρ may not be a linear function of temperature. This general class of sensors has the name resistance temperature detector (RTD). Any resistor exhibits this temperature dependence, but semiconductor materials are also used to construct what are called thermistors, and these are very sensitive sensors. Metallic resistors have temperature coefficients (TC) of about 0.4% change per °C, while thermistors may have TCs of 5% per °C. Thermistors do not have linear resistance versus temperature characteristics, but they are nearly so for $\delta T < 50°$ C.

A semiconductor PN junction is another temperature-dependent resistance. As Fig. 1-8 shows, an increase in the temperature of a junction diode results in a significant change in the terminal volt-ampere characteristic. Since DC resistance is the ratio of v to i, this is a highly nonlinear effect. The figure shows the effects of temperature on the diode resistance for applied voltage sources and for applied current sources.

Another frequently used type of resistive sensor is the photoresistor. These are also known by the names optoelectric or photoelectric sensors. The materials from which these devices are made have resistivities that vary with the intensity and wavelength of the light being monitored. Cadmium Sulfide was the first type of material that had wide application in photoresistors, but in recent years doped germanium and silicon have become commonplace. Most of the wavelength-sensitive sensors are constructed of compound semiconductors and are often configured as diode heterojunctions. Semiconductor diodes and transistors, usually of silicon, are the major devices in use for most light measurements. In practice these devices are often used in a voltage divider configuration to convert the resistance change into a voltage, as shown in Fig. 1-9. The resistor R_{limit} is used to limit the total current through the sensor. This resistor serves two functions: to limit the power dissipation in the sensor to rated value and to keep the sensor resistance from changing due to temperature variation of resistance (called self-heating error).

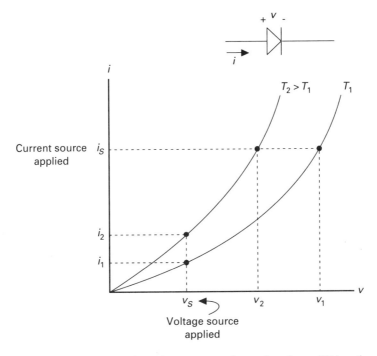

Figure 1-8 Temperature-dependent response of a semiconductor PN junction.

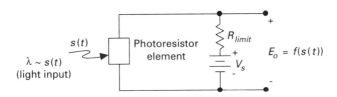

Figure 1-9 Voltage divider configuration for converting photoresistor element resistance variation to a voltage variation.

Exercises

E1.4.3a Discuss the merits or demerits of the four methods shown in Fig. E1.4.3a for converting the potentiometer position to a voltage output. Consider the effects of the current and voltage sources on the $s(t)$ to e_o relationship, units to which the output e_o may be attached.

E1.4.3b Devise a method for using a strain gauge to measure acceleration, so that input is $\bar{a}$ and output is voltage. Assume linear acceleration in one dimension.

E1.4.3c Obtain the volt-ampere characteristic of a diode at two temperatures T_1 and T_2. For an applied constant voltage source of 0.5 V, estimate the diode resistances at T_1 and T_2. Estimate the percentage of resistance change per °C, and compare with resistance wire and thermistor values given in the text.

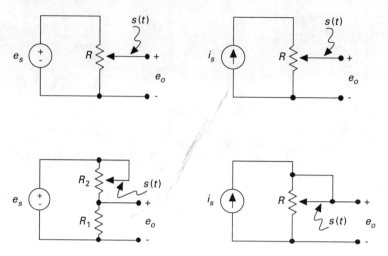

Figure E1.4.3.a

1.4.4 Sensors That Make Direct Physical Variable to Electric Variable Conversion

There are three other classes of sensors that make use of physical phenomena different from the three types discussed in the preceding sections. These three classes are distinct from the earlier ones in that they convert the physical variable directly to an electric variable, frequently without the necessity of an output interface unit. These classes are as follows:

1. Piezoelectric sensors. This type of sensor was referred to earlier in Section 1.2 where an example of a sonar detector was given. These sensors are based on the phenomenon that certain crystalline materials—such as quartz, Rochelle salt, metaniobate, and ammonium dihydrogen phosphate (KDP)—produce a charge or voltage across faces of the material when subjected to a force (pressure). Sensors based on this device exhibit high stiffness and ruggedness. They also function effectively at high frequencies, to hundreds of kilohertz and some to megahertz. These sensors find frequent applications for measuring sound pressure, torque, and acceleration.

2. Thermoelectric sensors. These sensors convert temperature (heat) directly to a voltage. The thermocouple (Seebeck effect) is the most common type of thermoelectric sensor. When two wires of dissimilar materials are connected at a junction, a voltage is produced at the ends of the wires that is a function of the junction temperature. Some thermocouple sensor characteristics are listed in Table 1-1. The table has been adapted from American National Standard C96.2-1973 (ASTM E230-72) [5]. Care must be taken in connecting the output leads to a system, since additional dissimilar metal junctions may be inadvertently produced.

Other physical phenomena in this class of sensor include the Peltier effect, the Thompson effect, and the Volta effect. These will not be described here.

TABLE 1-1 THERMOCOUPLE TYPES AND THEIR CHARACTERISTICS.

VOLTAGE AS A FUNCTION OF TEMPERATURE

Temperature °C	B Output mV	B Tempco µV/°C	E Output mV	E Tempco µV/°C	J Output mV	J Tempco µV/°C	K Output mV	K Tempco µV/°C	R Output mV	R Tempco µV/°C	S Output mV	S Tempco µV/°C	T Output mV	T Tempco µV/°C	Temperature °C
−200	—	—	−8.824	25.1	−7.890	21.8	−5.891	15.2	—		—		−5.603	15.8	−200
−100	—	—	−5.237	45.1	−4.632	41.1	−3.553	30.5	—		—		−3.378	28.4	−100
0	+0.000	—	0.000	58.7	0.000	50.4	0.000	39.5	0.000	5.25	0.000	5.4	0.000	38.8	0
+100	+0.033	0.9	6.317	67.5	5.268	54.4	4.095	41.4	0.647	7.5	0.645	7.3	4.277	46.8	+100
+200	+0.178	2.0	13.419	74.0	10.777	55.5	8.137	39.9	1.468	8.85	1.440	8.45	9.286	53.2	+200
+300	0.431	3.05	21.033	77.9	16.325	55.4	12.207	41.5	2.400	9.75	2.323	9.1	14.860	58.1	+300
+400	0.786	3.95	28.943	80.0	21.846	55.2	16.395	41.9	3.407	10.35	3.260	9.6	20.869	61.8	+400
+500	1.241	5.0	36.999	80.8	27.388	55.9	20.640	42.6	4.471	10.9	5.234	9.9	—		+500
+600	1.791	5.95	45.085	80.7	33.096	58.5	24.902	42.5	5.582	11.3	6.237	10.15	—		+600
+700	2.430	6.8	53.110	79.8	39.130	62.3	29.128	41.9	6.741	11.8	6.274	10.55	—		+700
+800	3.154	7.65	61.022	78.4	45.498	64.6	33.277	41.0	7.949	12.3	7.345	10.8	—		+800
+900	3.957	8.4	68.783	76.7	51.875	62.4	37.325	39.9	9.203	12.7	8.448	11.2	—		+900
+1000	4.833	9.1	76.358	75.0	57.942	59.2	41.269	38.9	10.503	13.2	9.585	11.5	—		+1000
+1100	5.777	9.75	—		63.777	57.8	45.108	37.8	11.846	13.6	10.754	11.9			+1100
+1200	6.783	10.35	—		69.536	57.2	48.828	36.5	13.224	13.9	11.947	12.0			+1200
+1300	7.845	10.9	—		—		52.398	34.9	14.624	14.0	13.155	12.2			+1300
+1400	8.952	11.2	—		—		—		16.035	14.1	14.368	12.2			+1400
+1500	10.094	11.6	—		—		—		17.445	14.1	15.576	12.1			+1500
+1600	11.257	11.7							18.842	13.9	16.771	11.8			+1600
+1700	12.426	11.7							20.215	13.5	17.942	11.5			+1700
+1800	13.585	11.5							—		—				+1800

B: PLATINUM −30% RHODIUM/PLATINUM −6% RHODIUM.
E: CHROMEL/CONSTANTAN.
J: IRON/CONSTANTAN.
K: CHROMEL/ALUMEL.
R: PLATINUM/13%RHODIUM-PLATINUM.
S: PLATINUM/10%RHODIUM-PLATINUM.

TEMPERATURE AS A FUNCTION OF VOLTAGE READING

mV	B °C	B °C/mV	E °C	E °C/mV	J °C	J °C/mV	K °C	K °C/mV	R °C	R °C/mV	S °C	S °C/mV	T °C	T °C/mV	mV
-10.000	—		—		—		—		—		—		—		-10.000
-5.000	—		-94.4	21.70	-109.1	25.10	-153.7	43.48	—		—		-166.5	49.5	-5.000
-2.000	—		-35.3	18.40	-40.8	21.10	-63.1	28.13	—		—		-55.1	30.0	-2.000
-1.000	—		-17.3	17.60	-20.1	20.40	-25.9	26.49	—		—		-26.6	27.6	-1.000
0.000	+42.0	4°/μV	0.0	17.06	0.0	19.84	0.0	25.35	0.0	190.5	0.0	185.2	0.0	25.8	0.000
+1.000	449.6	220	16.8	16.64	19.6	19.25	+25.0	24.69	+145.0	122.7	+146.4	125.8	+25.2	24.6	+1.000
+2.000	634.2	160	33.2	16.21	38.9	19.09	+49.5	24.27	258.2	106.4	264.3	111.7	+49.2	23.4	+2.000
+5.000	1018.2	109	80.3	15.19	95.1	18.43	+122.0	24.42	548.2	90.1	576.6	99.0	115.3	20.9	+5.000
+10.000	1491.8	87	153.0	14.02	186.0	18.03	246.3	24.60	961.7	76.6	1035.8	86.2	213.3	18.6	+10.000
+20.000	—		286.7	12.90	366.5	18.13	485.0	23.50	1684.1	73.5	—		385.9	16.3	+20.000
+30.000	—		413.2	12.47	546.3	17.57	720.8	23.95	—		—		—		+30.000
+40.000	—		537.1	12.36	713.9	15.94	967.5	25.48	—		—		—		+40.000
+50.000	—		661.1	12.47	870.2	15.79	1232.3	27.78	—		—				+50.000
+60.000	—		787.0	12.71	1035.0	16.95									+60.000
+70.000	—		915.9	13.09	—										+70.000

B: PLATINUM – 30% RHODIUM/PLATINUM – 6% RHODIUM
E: CHROMEL/CONSTANTAN.
J: IRON/CONSTANTAN.
K: CHROMEL/ALUMEL.
R: PLATINUM/13% RHODIUM-PLATINUM.
S: PLATINUM/10% RHODIUM-PLATINUM.
Note: Reprinted by permission of Analog Devices, Inc. [5]

Chap. 1   Transducers, Transducer Parameters (Specifications)

3. Photoelectric sensors (including optoelectric sensors). As the name indicates, these sensors generate a voltage when photons impinge on them. The light measurand is usually caused to excite dissimilar materials, such as a silicon photocell sensor constructed from PN junctions. Such PN junctions generate potentials on the order of 500 mV.

Exercises

E1.4.4a Design a transducer that will supply an electrical signal of maximum value $|V_{max}| = 5$ V which is proportional to an acceleration of maximum value

$$|\bar{a}| = 20 \frac{m}{s^2}.$$

This design is not to be just a block diagram; it should show constants of all elements used. Here we will require only one-dimensional capability.

E1.4.4b From a handbook or table select a thermocouple pair and a thermistor and compare the mV/°C factors for similar temperature ranges of operation.

E1.4.4c A copper-constantan thermocouple operates over a −200 to + 400 °C temperature range and has a sensitivity of 0.4 mV/°C. Suppose that the output is unipolar with −200 °C at 0 volts output and an amplifier is used to make the maximum voltage 10 volts. To obtain 1 °C increments, how many bits are required, and what is the value of LSB in volts?

1.5 TRANSDUCER SPECIFICATIONS, AN INTRODUCTION

In order to give the performance characteristics of a transducer, it is necessary to define the various properties of transducer response, that is, the operating parameters that give the input-output relationships and the transducer-generated signals. From this point on in the text the term transducer can be considered to mean sensor as well. Our main interest here is to relate $s(t)$ to $x(t)$, shown in Fig. 0-5. Obviously, if there are no input or output interface elements the transducer is identical to the sensor as we have defined it in Fig. 1-1. In practice one usually obtains the entire transducer as a single unit for which specifications are given.

Transducer specifications can be divided into four performance categories [1]: static performance, dynamic performance, environmental performance, and reliability. Static characteristics relate to transducer performance when the measurand is either non-time-varying or changing very slowly. These specifications are usually determined at a set of standard operating conditions such as room temperature, standard pressure, and specified relative humidity. Examples of static characteristics are linearity and hysteresis (to be defined precisely later).

Dynamic characteristics relate to transducer responses for time-varying measurands and describe the fidelity with which the output can follow the measurand variations. This class of characteristics is exemplified by step response and frequency response.

External factors that affect transducer response are called environmental characteristics. Among these characteristics we find temperature, pressure, humidity, and vibration.

Reliability characteristics include life expectancy, aging properties, and transducer consistency of performance (the latter is sometimes defined by the static hysteresis).

In this text we consider only the static and dynamic characteristics and how they affect the data words ($X(nT)$) from the converter (Fig. 0-5). Another important result of a study of these two characteristics is that an awareness of the ambiguity of the definitions of specifications arises. Frequently, when one looks at specification sheets from two manufacturers of the same transducer, the specification numerical value may differ even though the devices are identical because two different definitions are used. We will give some of the alternative definitions of the specifications so the differences can be appreciated. The art of "specsmanship" often dictates which definition is used, since the manufacturer wants a device to have the best possible specifications.

Many data sheets do not give a complete set of specifications for the transducer. It is frequently up to the designer to determine the specs that are important to a given design and obtain these specs either by direct contact with the vendor or by laboratory measurement.

Some transducers and their specifications obtained from actual manufacturers, illustrating the ideas given above, follow.

1.5.1 An LVDT

This LVDT measures 2.5 inches in length and has an outside diameter of 0.81 inches. The core has a length of 1.65 inches and a diameter of 0.25 inches. The nominal full range of core movement is ±0.2 inches. Within this range, the linearity is specified as ±0.25 %FSO (full-scale output). This figure is based on a best-fit straight line. The sensitivity is specified as 7.5 mV rms output per 0.001 inch movement. This assumes a nominal input voltage of 3.0 V rms. The full-range output voltage is calculated as

$$e_{max} = 0.2 \text{ in.} \times \frac{7.5 \text{ mV}}{0.001 \text{ in.}} = 1.5 \text{ V}$$

This LVDT can be excited with frequencies from 400 Hz to 10 kHz. The friction error is zero, since the core does not touch the coil structure and the resolution is infinite. No specification is given for temperature drift or overall static accuracy.

1.5.2 A Pressure Transducer

This DC-excited capacitance transducer has a full range of 0 to 5000 psia or psig. It has an overall end-point accuracy of 0.11 %FSO. This error figure is based on the RSS of nonlinearity (or linearity), hysteresis, and repeatability. The output range can be either 0 to 5 V or −2.5 V to +2.5 V. The temperature drift is specified in terms of a 2% FSO/100° F maximum sensitivity shift.

If this device were to be used over a 50° F temperature range, the maximum zero shift would be 1% of 5 V, or 50 mV. The maximum error due to sensitivity shift would be 0.75% of 5 V, or 3.75 mV. These errors are seen to be considerably more than the overall static errors.

1.5.3 A Strain Gauge Accelerometer

This transducer measures up to 50 g of acceleration in forward and reverse directions. The sensitivity is 2 mV/g, giving a maximum output of 100 mV peak. The suggested frequency range of the measurand is 0–300 Hz. The resonant frequency of the transducer is 1000 Hz. A DC excitation voltage of 5 V is required. The linearity and hysteresis errors are specified as ±1 %FSO, and these are worst-case, end-point specifications. The zero shift due to temperature is 0.015 %FSO/° F while the thermal sensitivity shift is 0.05 %FSO/° F. The physical dimensions of this transducer are just 6.35 mm by 9.53 mm.

1.5.4 A Piezoelectric Accelerometer

This unit is a triaxial accelerometer consisting of three mutually perpendicular quartz crystals. Acceleration along any of three orthogonal axes can be measured up to peak values of ±500 g. The sensitivity is 4 mV/g. The resolution is 0.01g. The linearity of output is ±1 %FSO. However, the basis for this linearity figure is not mentioned. The frequency response is specified as being within ±5% from 1 to 20 kHz. The thermal sensitivity shift is −0.03 %FSO/° F. A DC power supply of 12 V and 4 mA is required for built-in charge amplifiers that can provide a full-scale output of ±2 V and a current of ±2 mA.

1.5.5 A Capacitance Angle Transducer

This transducer is a capacitive gravity sensor that can measure angles of inclination over a ±45° range. The linearity is ±1 %FSO, however the basis for this specification is not indicated. A 9-V battery is required for this transducer, and a digital LCD readout is provided.

The preceding examples indicate a few of the many available sensors. It is obvious from the specification sheets that standardization of specifications is, unfortunately, not a major concern of the manufacturers. Many do not indicate a basis for their specifications of accuracy for example. Often, the manufacturer will use a least-squares accuracy because this figure is always lower than the end-point value. These will be discussed in detail later.

An ideal transducer would respond to an input quantity and produce an electrical output that bears a simple one-to-one linear relationship with the input quantity. As an equation, using the nomenclature in Fig. 0-5,

$$x(t) = Ks(t) \tag{1-12}$$

would pertain, where $x(t)$ is the output, $s(t)$ is the measurand, and K is a constant. This equation represents a perfectly linear relationship between the sensed or input variable and the output. Each value of output voltage would correspond to some precise value of input measurand with no ambiguity involved.

In the actual case there are several departures from the ideal conditions. Generally, other variables, such as temperature, have an effect on the constant K. The relationship may also depart from a linear one, especially if a large range of the input variable must be converted. Linear relationships lead to less complex mathematics, and many transducers are designed to have an approximately linear characteristic. On the other hand, some sensors are characterized by a highly nonlinear input-to-output relationship. The equation may also depart from the ideal as higher frequencies of input variable are encountered or as aging takes place. These departures from ideal must be characterized in the specifications and considered in the overall system performance if accuracy is an important consideration. The treatment of these nonidealities occupies a substantial portion of this textbook, both for transducers and other components that comprise the data acquisition system.

1.6 TRANSDUCER SPECIFICATION: LINEARITY

Linearity has several definitions among manufacturers, and they will be discussed shortly. Unfortunately, the definitions are not often given on spec sheets, so about all one can do is assume the worst-case definition and go from there. If an engineer designs a transducer (or any other device that has linearity as an important property), the users could be well served by precisely defined terms.

Another term often used to describe the linear operating characteristic of a transducer is nonlinearity. This may seem a bit paradoxical, but often the term refers to the same measure of deviation from "straight line" performance as linearity. Again it is important to determine the manufacturer's definition.

We begin by defining a linear system: A system is said to be linear if it obeys the properties of homogeneity and additivity. These properties are (see Fig. 1-10):

Homogeneity: If x_1 is an input and y_1 is a response (note these must be clearly identified) such that $y_1 = f(x_1)$, then a system has the property of homogeneity if, for a constant A, the input $x_2 = Ax_1$ yields

$$f(x_2) = f(Ax_1) = Af(x_1) = Ay_1 \tag{1-13}$$

In words, this means that the response due to a signal Ax_1 is A times the response due to signal x_1 alone.

Additivity: If x_1 and x_2 are inputs with corresponding outputs $y_1 = f(x_1)$ and $y_2 = f(x_2)$, then a system has the property of additivity if, for an input $x_3 = x_1 + x_2$,

$$y_3 = f(x_3) = f(x_1 + x_2) = f(x_1) + f(x_2) = y_1 + y_2 \tag{1-14}$$

In words, this means that the response due to the sum of two signals is equal to the sum of the responses due to each input acting alone.

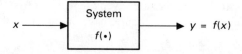

Figure 1-10 Configuration showing the quantities used to define linearity.

Chap. 1   Transducers, Transducer Parameters (Specifications)

The properties of additivity and homogeneity can be shown to be equivalent to superposition, and the superposition theorem states that for a linear system

$$f(Ax_1 + Bx_2) = Af(x_1) + Bf(x_2) = Ay_1 + By_2$$

Example 1-1

Examine the system in Fig. 1-11(a) for linearity. The applied voltage V is the input, and the current I is the response. One is tempted to conclude that since the 5-Ω resistor and 10-V source are constant the system is linear. This is not the case, as we will now show.
First we check for homogeneity. Suppose $V_1 = 1$ V. Then

$$I_1 = \frac{1-10}{5} = -\frac{9}{5} A.$$

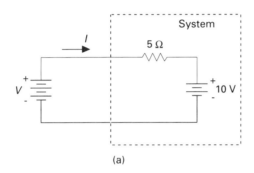

(a)

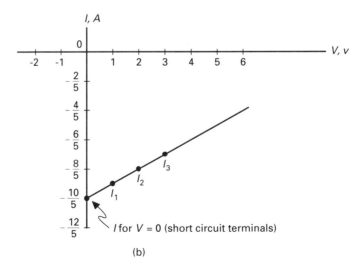

(b)

Figure 1-11 (a) System to be tested for linearity, having input V and response (output) I. (b) Characteristic of system (a) which, while having straight-line variation, is *not* linear.

Now if we form an input $V_2 = KV_1 = K \cdot 1 = KV$, then

$$I_2 = \frac{K-10}{5} = \frac{K}{5} - 2 \ A \neq -K\frac{9}{5} \ A.$$

Therefore, the system is not homogeneous and is thus not linear.
This same conclusion would have also been reached if we had checked for additivity.
Suppose $V_1 = 1$ V. Then

$$I_1 = \frac{1-10}{5} = -\frac{9}{5} \ A.$$

Next let $V_2 = 2$ V for which

$$I_2 = \frac{2-10}{5} = -\frac{8}{5} \ A.$$

Now form $V_3 = V_1 + V_2 = 3$ V. This would result in a current

$$I_3 = \frac{3-10}{5} = -\frac{7}{5} \ A$$

But,

$$I_1 + I_2 = -\frac{17}{5} \ A \text{ so } I_3 \neq I_1 + I_2$$

and the system is nonlinear.

An important fact to be drawn from this example is that a straight-line relationship is not enough to specify linearity. For example, if we plot the three values of current obtained above we would obtain Fig. 1-11. Note that the system as defined has a straight-line variation, but it does not pass through the origin.

Manufacturers often describe the nonlinear properties of their components by using the term linearity, which may seem a bit paradoxical. Most often they give the largest deviation from a straight line, but the placement of the straight line may be specified in one of several ways. Three common ones are shown in Fig. 1-12 for unipolar transducers. These are static characteristics, as defined in the preceding section.

Observe that the maximum error can be quite different for the three cases, possibly by a factor of two or more. Case (a) is called an endpoint-based linearity, since the straight line passes through both extremes of operation of the transducer. Case (b) is called zero-based linearity since the straight line passes through the origin (note (a) is a special case). In terms of deviation, linearity is often simply expressed as the largest departure from the straight line, and it would have the same dimensions as the output x. Thus the linearity in Fig. 1-12(a) would be simply stated as Δx_{max}. Some manufacturers also call this nonlinearity or linearity error. (See exercise 1.6d for transducers having bipolar outputs.)

Chap. 1   Transducers, Transducer Parameters (Specifications)   51

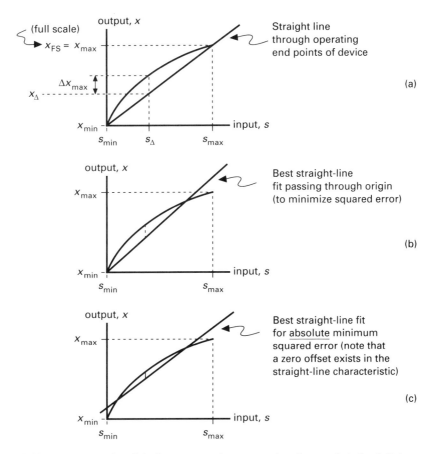

Figure 1-12 Placement of straight lines on transducer operating characteristic for defining linearity.

Most often the linearity (or nonlinearity) is expressed as a percentage. Unless otherwise specified by the manufacturer, it is usual to use one of the two following representations:

$$L = \% \text{ linearity} = \frac{|\Delta x|_{max}}{|x_\Delta|} \times 100\% \quad (1\text{-}15)$$

or

$$L = \% \text{ linearity} = \frac{|\Delta x|_{max}}{|x_{max}|} \times 100\% \quad (1\text{-}16)$$

The second of these is by far the most common in practice, since the first definition would require x_Δ to be given (probably an average value).

These equations can be used regardless of which straight line is used, but the numerical values of L will be different. The actual differences in percentage values may or may not be significant, depending upon the number of bits being used in the digital processing, as will be shown.

One will find that engineering judgement must be applied to whatever term a manufacturer uses to describe the deviation of the transducer from the desired linear input-output relationship.

A comment is appropriate regarding the measurement of the transducer input-output characteristic. For those transducer types that are low-pass in terms of frequency response, the characteristic is most often performed for DC inputs; that is, one constant value of the variable is applied and the output is measured. This yields what is often referred to as the static transfer characteristic. A separate frequency response characteristic is then usually given to account for the ability of the transducer to follow time-varying signals. This will be discussed in Sections 1.8 and 1.9. For those transducers that do not respond to static or DC signals, the measurement of the transfer characteristic is usually made at some optimum or intended operating frequency in the passband of the device.

Next, suppose one knows that the output of the transducer is to be digitized for digital signal processing and that N bits are to be used for the data words. What is the relationship between the number of bits N and the linearity (or nonlinearity) if each data word is to represent the exact value of the input variable and not be in error by the deviation from the ideal straight line input-output relationship?

Let L be the percentage of linearity. If we suppose that the manufacturer of the transducer has used Eq. (1-16) to define L, then

$$maximum\ output\ error = \frac{L}{100} \times |x_{max}|$$

Now if this error is to be negligible it must be below 1/2 LSB in value. Let x_{span} be the total output span of the transducer ($|x_{max}|$ for unipolar output or $|2x_{max}|$ for symmetric bipolar output or $|x_{max} - x_{min}|$ for general case). Then

$$1/2\ \text{LSB} = \frac{1}{2} \times \frac{x_{span}}{2^N} = \frac{x_{span}}{2^{N+1}}$$

We then require (see Eq. (1-3))

$$L\frac{x_{max}}{100} \leq \frac{x_{span}}{2^{N+1}} \tag{1-17}$$

Thus the desired linearity would be

$$L \leq \frac{x_{span}}{x_{max}2^{N+1}} \times 100\%.$$

Conversely, for a given L one could use Eq. (1-17) to determine the number of bits one would specify to guarantee negligible linearity error. Looking at this another way, these results give the number of bits below which no linearity correction is required.

Example 1-2

Suppose a manufacturer of transducers specifies that the linearity is 0.2% based on the best straight-line fit for minimum squared error. The spec is based on full-scale value (zero-based). The output is a voltage covering the range 0 to 10 V nominally. What is the number of bits justified for the signal processing that follows an A/D conversion? First compute the output maximum error:

$$0.2 = \frac{|\Delta x|_{max}}{10} \times 100 = 10 \, |\Delta x|_{max}$$

or

$$|\Delta x|_{max} = 0.02 \text{ V}$$

In order to be negligible, this must be below the 1/2 LSB level of the true value of x. For the voltage range given and using N bits:

$$1/2 \text{ LSB} = \frac{1}{2} \cdot \frac{10}{2^N} = \frac{10}{2^{N+1}} \text{ V}$$

This requires

$$\frac{10}{2^{N+1}} \geq 0.02$$

$$500 \geq 2^{N+1}$$

$$250 \geq 2^N$$

$$N \leq \frac{\log_{10}(250)}{\log_{10} 2} = 7.9$$

$$N \leq 7 \text{ bits}$$

Suppose that one needs to use 10 bits using the transducer in the preceding example. All is not lost, however, for there are ways of compensating for the nonlinearity. The two most popular ways are the following:

1. Determine the actual characteristic of the transducer and then place a nonlinear device at the transducer output to compensate for the nonlinearity, as shown in Fig. 1-13.
2. Determine the actual characteristic of the transducer and then compensate by letting the computer scale the input digital data. This takes time, but in many cases would not impose a difficulty. The time problem can be minimized by programming the correction into a ROM and using the digitized data as the address of its corrected value. This is one of the advantages of digital processing.

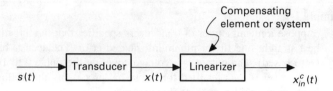

Figure 1-13 Analog linearity compensation.

One final point is worthy of mention. From the three graphs in Fig. 1-12 we see that the values of $\Delta x_{\max}$ are different even for the same device. This means that the values of N (maximum number of bits) will be different for each case. Only in case (a) do we obtain the true value of deviation from true linearity (straight line passing through the end points). Differing values of N for the same device may be disconcerting, but if one wishes to determine the worst case, what could be done is to double the error $|\Delta x|_{\max}$ if case (b) or (c) is known to be applicable.

Another alternative is to perform a calibration on the device one is going to use and determine the absolute deviation. Finally, if one realizes that manufacturers' data are frequently average transducer characteristics, then one is probably justified in accepting whatever error is obtained irrespective of the definition used for the nonlinear effect. In any case, it is doubtful that the computed values of N would differ by more than 1 for any of the definitions.

Example 1-3

Evaluation of the linearity specification for a Wheatstone bridge.

For the circuit of Fig. 1-2, the output voltage can be expressed as

$$V(T) = V(R(T)) = V_s \left[\frac{R(T)}{R_3 + R(T)} - \frac{R_2}{R_1 + R_2} \right] \quad (1\text{-}18)$$

The variation of $V(T)$ as T changes can be found by differentiating $V(T)$ with respect to T. This yields

$$\frac{dV(T)}{dT} = \frac{dV(T)}{dR(T)} \frac{dR(T)}{dT} = \frac{R_3 V_s}{(R_3 + R(T))^2} \cdot \frac{dR(T)}{dT} \quad (1\text{-}19)$$

A linear relationship between $V(T)$ and T would require a constant derivative,

$$\frac{dV}{dT} = \frac{V_{\max} - V_{\min}}{T_{\max} - T_{\min}}$$

Equation (1-19) can approximate a constant value only if $R(T)$ is much smaller than R_3 and dR/dT is a constant. These conditions can often be met, but the output voltage given by Eq. (1-18) is smaller when R_3 is much greater than $R(T)$.

For the bridge, then,

Chap. 1   Transducers, Transducer Parameters (Specifications)   55

$$\left| error \right| = \left| V(T) - \frac{V_{max} - V_{min}}{T_{max} - T_{min}}(T_{max} - T) \right| = deviation$$

so

$$L = \frac{deviation}{|V_{max}|} \times 100\%$$

A two-resistor voltage divider produces only a unipolar output; that is, the output voltage has the same sign as V_s. The bridge circuit allows a single reference voltage to produce a bipolar output at the expense of requiring a different common terminal for the reference, V_s, and output voltages. The common terminal for the reference voltage in Fig. 1-2 appears at the junction of resistors R_2 and $R(T)$. The common terminal for $V(T)$ appears at the junction of resistors R_1 and R_2, or R_3 and $R(T)$.

The bridge circuit produces zero output if

$$\frac{R(T)}{R_3} = \frac{R_2}{R_1}$$

The bridge is said to be balanced for this condition. If $R(T)$ increases, with increasing T (positive temperature coefficient or PTC), the output voltage calculated from Eq. (1-18) becomes positive. A decrease in $R(T)$ results in a negative value of $V(T)$.

Perhaps the most important feature that makes the bridge circuit much more popular than the voltage divider circuit is the fact that the output voltage can be zero for some nominal value of $R(T)$. The voltage divider circuit will have a nonzero output for all finite values of $R(T)$. Although the variation of $V(T)$ with $R(T)$ is the same for both circuits, the bridge circuit can be selected to have zero output for a particular value of $R(T)$. Equation (1.3) shows that R_1, R_2, and R_3 can be used to create this null condition. The output variation can then extend from zero to some upper limit.

The bridge circuit or voltage divider circuit can use an AC source rather than a DC source as long as this source amplitude is precisely controlled. Capacitive bridges can be used in transducers having capacitive element sensing, as shown earlier.

Example 1-4

A transducer output can be expressed as

$$x = 10s + 0.01s^2 - 0.01s^3$$

where s is the measurand and x is the output. If $s = 0$ and $s = 2$ are the measurand end points, calculate the deviation from linearity and the end-point linearity of the transducer.

Solution: In order to calculate the deviation from linearity, the ideal (linear) equation must be found. The slope is found by first determining both end points. Assuming the given equation passes through the end points, we have

$$x_{end} = (10)(2) + (0.01)(4) - (0.01)(8) = 19.96$$

In order to find the end-point linearity, the maximum error occurring anywhere between end points must be found. The equation for the straight line connecting the two end points is

$$x = \frac{19.96 - 0}{2 - 0} s = 9.98s$$

The difference between the straight line and the actual curve is given by

$$\Delta x = 10s + 0.01s^2 - 0.01s^3 - 9.98s = 0.02s + 0.01s^2 - 0.01s^3$$

The maximum error over the range of s can be found by differentiating Δx and equating the result to zero. This results in

$$\frac{d(\Delta x)}{ds} = 0.02 + 0.02s - 0.03s^2 = 0$$

Solving for s gives $s = 1.215$. The maximum error is then

$$\left| \Delta x \right|_{max} = \left| 0.02 \times 1.215 + 0.01 \times 1.215^2 - 0.01 \times 1.215^3 \right| = 0.0211$$

The maximum value of x is 19.96, so the end-point linearity is

$$L_{ep} = \frac{0.0211}{19.96} \times 100\% = 0.106\%$$

Exercises

E1.6a An inductor of constant value 2 mH has applied to it a sine wave of 10-V peak amplitude having variable frequency f. If the frequency is considered the input and the current amplitude the response, is the system linear? Repeat for a capacitor of 2 µF. Do for both general time and sinusoidal steady state variations.

E1.6b A transducer has a maximum linearity error of 0.1% on a zero-based spec passing through the maximum transducer output. The transducer input is temperature and the output is voltage. If the output signal ranges from 0 to 10 V (unipolar device), what is the maximum number of bits that would be justified in a signal processing system? The transducer operates over a temperature range of 80° C to 110° C.

E1.6c A computer used in a data computation system uses 10 bits in a data word. What is the maximum linearity error (as a percentage) that can be accepted based on the zero-based straight line through the origin? The transducer is a bipolar output device having the range $-10 \text{ V} \leq x \leq 10 \text{ V}$.

E1.6d The definitions, figures, and examples in the text were tacitly based on transducers with unipolar input and output signals. Draw the DC transfer characteristics for bipolar input and output signals for a transducer and comment on the defini-

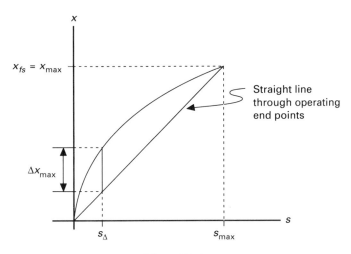

Figure E1.6g

tions as they relate to such transducers. Are any modifications or other definitions appropriate? (Discuss them if they are.)

E1.6e Explain the difference between homogeneity and additivity, and apply these two tests to the following systems:

1. $u \to \boxed{Ku} \to v \qquad K$ constant

2. $v(t) = \int_0^t u(x)\, dx$

3. $v(t) = 4\dfrac{du}{dt}$

E1.6f A transducer has bipolar voltage output of ± 2 V. If the output is 30 mV when the input is zero, what is the number of significant bits in the data words?

E1.6g The response curve of a transducer is shown in Fig. E1.6g. The transducer is manufactured by company A and second sourced by company B. It is known that the transducers are identical whether manufactured by A or B, but company A rates the linearity at 0.2% and B at 0.05%, both based on the same straight line. Is it necessarily true that one company is not being truthful? Explain.

E1.6h If $R_1 = R_3 = 1\ \Omega$, $R_2 = 100\ \Omega$, and $V_s = 12$ V in Fig. 1-2, plot $V(T)$ as $R(T)$ varies from 90 to 110 Ω. On the same set of axes, plot a linear variation of $V(T)$ assuming the slope is taken to be

$$\left.\dfrac{dV(T)}{dR(T)}\right|_{R(T)} = 100\ \Omega$$

1.7 TRANSDUCER SPECIFICATION: HYSTERESIS (FRICTION, BACKLASH)

When the physical input variable to a transducer increases and decreases slowly over its operating range, one would hope that the output would follow the single linear response line. A new device will often trace this curve only on the first application of the range of the input variable. However, when the variable decreases, the output often follows a separate set of values denoted by curve 2 in Fig. 1-14. Furthermore, when the input variable again increases over the design range, it may yield yet another set of output values denoted by curve 3 in the figure. Slow cyclings of the values thereafter generally result in a tracing of the path 2–3. The result of this hysteresis is to yield an error in the output value of x. As usual we try to select a transducer which has Δ_2 max and Δ_3 max below the 1/2 LSB level, or else be aware of how much compensation must be written into any algorithms that use the data. Note that this effect is what would be produced by a set of gears that were not perfectly (tightly) meshed. If the shaft controlling one gear were to change direction of rotation, the second gear would not move until the teeth of the first gear traversed the small space between the gear teeth. This is the familiar backlash phenomenon.

The total band of error described by Δ_2 and Δ_3 is called the error band and is denoted by

$$\text{error band} = \Delta = \Delta_2 + \Delta_3 \qquad (1\text{-}20)$$

From this error band it is easy to see that the effect of hysteresis is to produce nonlinearity, since the hysteresis loop represents the deviation from the ideal linear response. For this reason, many manufacturers include the hysteresis effects with the nonlinear effects, so that hysteresis is automatically accounted for in the linearity parameter. Thus the examples given in the preceding section apply to hysteresis.

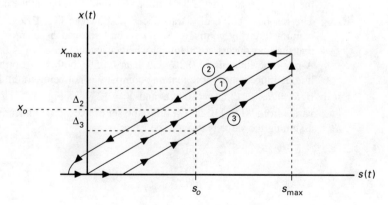

Figure 1-14 Transducer hysteresis.

Figure 1-15 Effect of transducer time response on input waveshape.

1.8 TRANSDUCER SPECIFICATION: TIME RESPONSE

The transducer time response has two effects:

1. The introduction of a delay in the output, since the output cannot change immediately upon the input variable change. If it is important to know the physical parameter value at precise values of time, then the delay time must be determined. Delay correction can be applied to digital data.
2. The distortion of input signal so that the transducer output signal waveshape does not truly represent the input signal waveshape. For example, if the input to a pressure transducer were a sudden step of pressure, the output may have rounded edges, as shown in Fig. 1-15. This second effect can, of course, be interpreted as a frequency response effect wherein high frequencies are not passed. Thus such transducer characteristics yield some initial filtering of the physical variable.

The transducer time response effects are usually quantified by defining a rise time. The standard definition, Fig. 1-16, is the time it takes the output signal to change from 10% of its final value to 90% of its final value, measured in seconds.

Delay time has several possible definitions: one is the time from the start of the pulse to 50% of its final value, and another is the time to get to 90% of final value. The former definition is most common.

For digital data, however, the usual definition of rise time is not really adequate, since the required or operating rise time needs to be defined as the time it takes to get within 1/2 LSB of the final value. This time is obviously considerably longer if many bits are used.

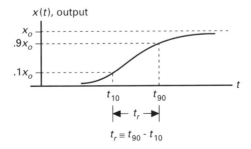

Figure 1-16 Quantities used to define rise time.

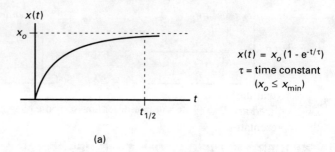

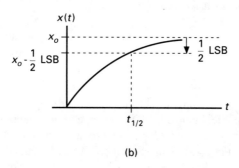

Figure 1-17 (a) First-order transducer unit step response (unipolar). (b) Detail at the 1/2 LSB time, $t_{1/2}$.

Consider the first-order transducer unit step response shown in Fig. 1-17(a). Suppose we are using N bits to represent transducer outputs and that x_o is the final value of a present output being converted. There are initially a total of 2^N different levels (increments) that can be represented for the maximum signal output x_{max}. Thus one-half of an increment (1/2 LSB) would be (for a unipolar system)

$$1/2 \text{ LSB} = \frac{x_{max}}{2^N} \cdot 1/2$$

where x_{max} is the largest possible transducer output assumed positive here. The time it takes to get to within 1/2 LSB ($t_{1/2}$) is then found from the fact that we want the output value at a time $t_{1/2}$ to be greater than x_o minus 1/2 LSB, or,

$$x_o - 1/2 \frac{x_{max}}{2^N} \leq x_o \left(1 - e^{-(t_{1/2}/\tau)}\right)$$

$$-1/2 \frac{x_{max}}{2^N} \leq -x_o e^{-(t_{1/2}/\tau)}$$

Solving for $t_{1/2}$,

Chap. 1 Transducers, Transducer Parameters (Specifications)

$$t_{1/2} \geq \tau \ln\left(2^{N+1} \frac{x_o}{x_{max}}\right) \tag{1-21}$$

(Operating rise time for unipolar output)

Note that the result depends upon how much of the total available transducer range is being utilized and that this "operating rise time" depends upon the number of bits used in the representation and the time constant τ. The worst case (largest $t_{1/2}$) is when $x_o = x_{max}$.

Example 1-5

Suppose we have a transducer with a step response time constant of 1 μs and a signal processing system that uses 10 bits. If the input signal is driving the transducer at 3/4 of its design maximum, what time must be allowed before a conversion to binary form is made? (This is a very respectable transducer response time.)

Using Eq. (1-21):

$$t_{1/2} \geq 10^{-6} \ln\left[(2^{11})(3/4)\right] = 7.3 \times 10^{-6} \text{ s}$$

If the transducer were run at full capability so that $x_o/x_{max} = 1$, then $t_{1/2} \geq 7.6 \times 10^{-6}$ s.

Thus if one is trying to convert data having step values of input, it would not be permissible to convert any faster than 7.3 μs, otherwise the LSB is being wasted. Note also that $t_1/2$ is significantly greater than the usual rule of thumb which states that a first-order response reaches its final value in 5τ seconds. This would be valid for the example only if the transducer were being driven at less than 7% full capability!

As mentioned earlier there is a relationship between the first-order transducer rise time and its frequency response bandwidth (3 dB). This is easily derived using the low-pass RC network shown in Fig. 1-18. We start with the transfer function:

$$\frac{V_2}{V_1} = \frac{\frac{1}{sC}}{R + \frac{1}{sC}} = \frac{1}{RsC + 1} = \frac{1}{RC} \cdot \frac{1}{s + \frac{1}{RC}}$$

Then

$$f_{3db} = \frac{\omega_{3dB}}{2\pi} = \frac{\left(\frac{1}{RC}\right)}{2\pi} = \frac{1}{2\pi RC} \tag{1-22}$$

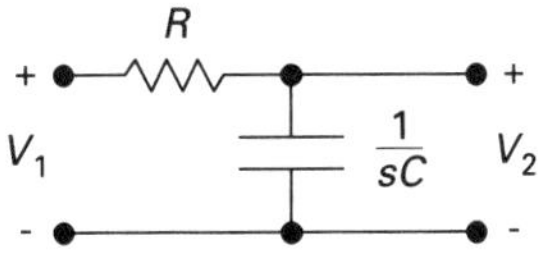

Figure 1-18 First-order low-pass network for relating bandwidth and rise time.

Now for a step input, $V_1 u(t)$, the output can be obtained from circuit theory as

$$v_2(t) = V_1 \left(1 - e^{-t/\tau} \right)$$

where $\tau = RC$.

At the 10% amplitude point

$$v_2(t_{10}) = 0.1 V_1 = V_1 \left(1 - e^{-t_{10}/\tau} \right)$$

$$0.1 = 1 - e^{-t_{10}/\tau} \rightarrow e^{-t_{10}/\tau} = 0.9$$

At the 90% amplitude point

$$v_2(t_{90}) = 0.9 V_1 = V_1 \left(1 - e^{-t_{90}/\tau} \right) \rightarrow e^{-t_{90}/\tau} = 0.1$$

Dividing the two equations

$$e^{(t_{90} - t_{10})/\tau} = 9$$

Thus

$$t_r = t_{90} - t_{10} = -\tau \ln(9) = 2.2\,\tau$$

In summary,

$$t_r = 2.2\,RC = 2.2\,\tau \qquad (1\text{-}23)$$

Eliminating RC from Eqs. (1-22) and (1-23) one obtains

$$t_r = \frac{0.35}{f_{3dB}} \qquad (1\text{-}24)$$

One should remember that the rise time only goes to 90% of final value and that if a system having a large number of bits for data is used, a time significantly longer than t_r must be used.

As an extension of the preceding topic, one may well ask the question: If I wish to pass a pulse of width t_p, what should the bandwidth of the transducer be? Equivalently, we could ask: For a transducer bandwidth f_{3dB}, what pulse width t_p will it handle?

For a temperature transducer this means physically that the input step changes in temperature cannot occur faster than t_p to be accurately resolved, as shown in Fig. 1-19.

What is often done in electronics is to choose (arbitrarily) the bandwidth to be the reciprocal of the pulse width. Thus select

$$f_{3dB} = \frac{1}{t_p} \qquad (1\text{-}25)$$

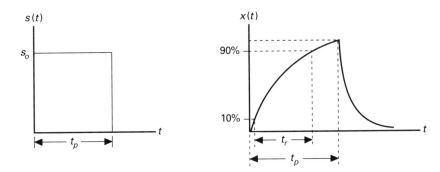

Figure 1-19 Transducer first-order response pulse width determination.

Using the relationship for f_{3dB} developed earlier, this becomes

$$\frac{0.35}{t_r} = \frac{1}{t_p}$$

or

$$t_p = \frac{t_r}{0.35} \tag{1-26}$$

Examining the criterion developed in the preceding equation we ask the question: Using $t_p = t_r/0.35$, how many bits would be justified in a signal processing system? The answer lies in how close the exponential comes to reaching the final value in t_p seconds, for a first-order system. For the rise portion (see Fig. 1-20),

$$x(t) = x_o\left(1 - e^{-\frac{t}{\tau}}\right)$$

Let

$$t = t_p = \frac{t_r}{0.35} = 2.2\frac{\tau}{0.35} = 2\pi\tau$$

(about 6 time constants). Then,

$$x(t = 2\pi\tau) = x_o(1 - e^{-2\pi}) = 0.998x_o$$

Deviation from final value $= x_o - 0.9981x_o = 0.00187x_o$. This must be less than 1/2 LSB to be neglected; therefore,

$$0.00187\, x_o \leq 1/2 \cdot \frac{x_{max}}{2^N}$$

Solving for N,

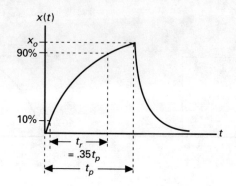

Figure 1-20 Relationship between rise time and pulse width.

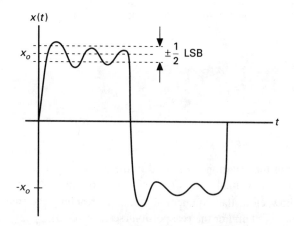

Figure 1-21 Transducer second-order response.

$$N \le \frac{\log_{10}\left(267 \frac{x_{\max}}{x_o}\right)}{\log_{10} 2} \tag{1-27}$$

For the worst case (least upper bound) $x_o = x_{\max}$, which gives $N \le 8$, which is not a bad result for many systems.

If a transducer has a higher-order response, such as that shown in Fig. 1-21, one must ensure that the output has settled to within $\pm 1/2$ LSB.

The real problem involved in allowing for time response can best be seen by referring to the first-order time response plot of Fig. 1-22. It is easy to compute the time $t_{1/2}$, but the question is now shifted to the determination of the time between samples; that is, what is the sample period? If one could be certain that each sample period were measured from the start of the step function input, then $t_{1/2}$ could be used as the sample period. Mother nature and physical processes, however, often produce step changes at

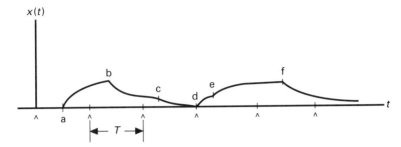

Figure 1-22 First-order response to random step inputs to transducer.

random points, so using $t_{1/2}$ is really not possible in general. Fig. 1-22 is an illustration of this problem.

In Fig. 1-22 the time interval T is the sample interval, and the markers indicate times at which analog to digital conversions are made. Step input changes occur at the points labeled with letters. Note that as long as T is uniform between samples there is really no way to ensure that conversions would be made at the times corresponding to $t_{1/2}$ for each step change. How can we resolve this dilemma? What must be done is to determine the greatest rate of change of the input physical variable one really needs to know about to accomplish tasks or computations on that variable. Perhaps one really does not need to follow every change. What is being suggested here, of course, is that in practice there is an upper frequency limit to which an engineer has any interest. Thus we solve the dilemma by agreeing that we cannot legitimately follow step or other rapid changes by digital processing. The selection of an upper limit is best described by an analysis of the frequency response. This is the subject of the next section. Of course there may be situations for which the time interval T is not constant but for which the input signal is used to start the conversion process. Also, in such systems, the variable steps are usually quite infrequent in time occurrence, so that the equations developed in this section are directly applicable if the signal is used to initiate conversion.

Exercises

E1.8a Suppose the first-order transient response of a transducer is specified by its unit step rise time based on the usual definition of 10%–90% time. Develop a relationship between this rise time and the time to rise to the 1/2 LSB level of an N-bit system.

E1.8b Continuing exercise E1.8a, what is the highest-frequency square wave input signal to the transducer that could be accepted for N-bit conversion to follow in the signal processing? (Hint: A square wave can be thought of as the superposition of step functions, and a certain minimum time must be allowed for successful performance.)

E1.8c Suppose a transducer is subject to square wave pulses of the physical variable being sampled. The output is a voltage with a simple first-order time constant τ. For a 14-bit signal processing system, what is the rise time (expressed in terms of τ) that must be allowed (that is, minimum pulse time) in order to be within 1/2 LSB?

E1.8d A transducer has a second-order step response given by

$$x(t) = x_o - \frac{x_o}{\cos\varphi} e^{-t/\tau} \cos(\omega_o t + \varphi)$$

where τ, ω_o, φ, and x_o are constants ($x_o \leq x_{max}$).

   a. Make a sketch for $f_o = 5/\tau$ and $\varphi = 0$.
   b. What pulse time t_p must be allowed, in terms of τ, ω_o, and φ, if we are using an N-bit processing system?

E1.8e When a transducer is subject to a step change in input, the output may be starting at a nonzero initial value x_i, caused by the previous input. For a first-order response system with an output response to x_o, determine the minimum time from $t = 0$ at which the output can be sampled in terms of the number of bits N. What is the worst-case value of x_i, that is, the x_i value which makes N minimum and t_s maximum? Assume a unipolar transducer starting at zero.

E1.8f Repeat exercise E1.8e for the case where the transducer has an output range that is symmetric and bipolar and for which x_i is negative with $0 > x_i \geq x_{max}$.

E1.8g Suppose that a voltage ramp goes from 0 to $+V_m$ at a slope of

$$A \frac{V}{\mu s}$$

and is to be converted to a digital signal that does not skip a 1-bit level. Find a relationship between V_m, A, N (number of bits), and T (sample interval) for the system. Also compute the serial bit rate for an 8-bit system.

E1.8h An optical transducer is to be used in conjunction with a diode laser to measure velocities very accurately. When an object passes between the laser and transducer it breaks the beam, which initiates a counter that runs until the beam is restored. The velocity is found from the length of the object and the counter reading. The transducer has a first-order time response with a time constant of 1 μs when turning off and 10 ns when turning on. Find the number of bits that could be used to digitize velocity values if the maximum velocity to be measured is 1000 m/s.

E1.8i Derive the equation similar to Eq. (1-21) but for a bipolar, symmetric system having maximum output $\pm x_{max}$. Note that the maximum step size is now $2x_{max}$. Does this bipolar operation offer any improvement on $t_{1/2}$?

1.9 TRANSDUCER SPECIFICATION: FREQUENCY RESPONSE

The frequency response of a transducer is defined as the amplitude and phase of the transducer output (peak values) as functions of the frequency of the input variable, $s(t)$, when $s(t)$ has constant sinusoidal amplitude. In most cases the output $x(t)$ will be a voltage that varies in time the same way that $s(t)$ does. However, as the input variable changes at a faster rate (higher frequency), the transducer output may not be able to follow, so the $x(t)$ amplitude may be in error and a delay (phase shift) of the signal may also result. Graphically, the frequency response is usually only a plot of the amplitude of the output signal as a function of the frequency of a constant amplitude input.

There is, of course, a phase shift effect as described above, but in most cases one is concerned with amplitude. If, however, it is important to determine the exact time of occurrence of an event at the transducer input, then the phase shift (time delay) must also be accounted for.

A generalized transducer frequency response curve is shown in Fig. 1-23(a) with some of the important parameters identified. The parameter x_{o1} is the maximum output response to an input signal of amplitude value s_o, which occurs at ω_1. Notice that the value of x_o will change as the input amplitude at which the test is made is changed. The transducer will not always be run at full output, so some intermediate value is (arbitrarily) selected. A parameter RW, called the *ripple width*, is defined to represent the maximum ripple of the response curve in the passband. The frequency ω_{3dB} is the usual 0.707 response frequency, or half-power point. The new parameter introduced is $\omega_{1/2B}$, which is defined as that frequency for which the transducer output is at the 1/2 LSB level and hence not detected by the following digital processor. This parameter must be used with some care, since it is a function of the amplitude of the input signal and hence dependent upon the value of x_o ($x_o \leq x_{max}$). It is important to notice that if a knowledge of the input amplitude is not of importance in a problem, one can use the output of the transducer far beyond ω_{3dB}. Of course, if the transducer characteristic were specified, one could obtain the true input amplitude by a proper weighting of the transducer output either by an analog scaler or by inserting appropriate amplitude weighting in a digital processing algorithm.

Figure 1-23(b) is another way of looking at the effect of frequency response. For the same input amplitude s_o the output depends upon frequency. The effect is to change the slope of transducer transfer characteristics.

The preceding discussion suggests two important features of the frequency response characteristic of a transducer:

1. A transducer specification should indicate the conditions for which the frequency response is taken, particularly ω_{test} and also what percentage of the total input range is represented by the constant peak value of the test sinusoidal input. ω_{3dB} is often a function of x_o.
2. A transducer specification should indicate the order of the response characteristic and whether the characteristic is low-pass, band-pass or high-pass. If the response order is greater than one, the deviation (ripple) property should be stated in terms of some average or desired response value x_o.

Before going into the details of transducer response, let us examine the ideal situation. There are two basic types of ideal response, depending upon whether or not one wants to use the transducer as a prefilter as well. These two types are shown in Fig. 1-24. If it were desirable to eliminate frequencies above $\omega = \omega_o$, the characteristic illustrated in Fig. 1-24(a) would be the ideal. This shows that the output amplitude is zero for frequencies greater than ω_o. The phase characteristic may not be as obvious, since one might suspect that the ideal phase plot would be a constant, that is, all frequencies delayed the same amount. That this is not so is shown in Fig. 1-25. The input signal $s(t)$ is assumed to

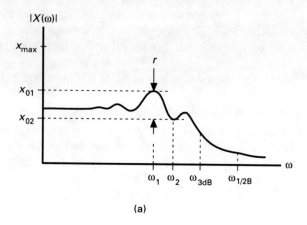

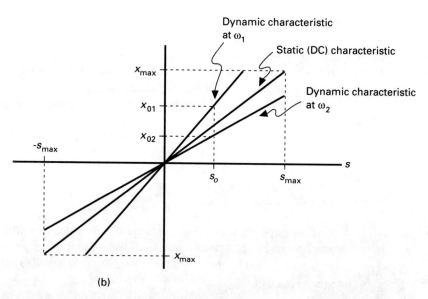

Figure 1-23 Transducer frequency response curves. (a) Magnitude of transducer frequency response. (b) Frequency effects on transducer characteristics.

consist of a waveform containing two frequency components—a fundamental and third harmonic. The curves labeled 1 and 3 in Fig. 1-25(a) combine to produce the actual input indicated. Now if the transducer simply shifts (delays) the total response, the output is merely shifted along the time axis, as indicated in Fig. 1-25(b). We can see from this plot that in order to preserve the waveshape, a 90° phase shift on the fundamental, curve 1, requires a 270° phase shift on the third harmonic, three times as much. Of course, once the frequency exceeds the cutoff, ω_o, it does not matter what the phase shift is.

By similar reasoning applied to Fig. 1-24(b), one would conclude that an ideal all-pass transducer would require an identical negative slope phase characteristic, meaning that the slope is directly proportional to frequency.

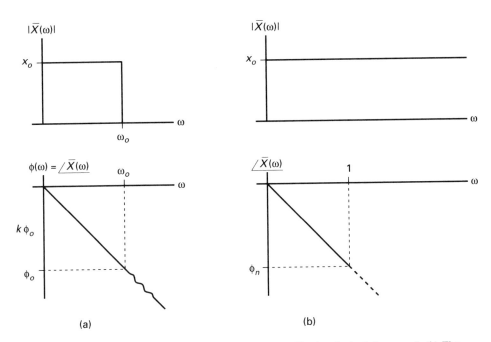

Figure 1-24 Ideal transducer frequency responses. (a) Prefiltering desired (low-pass). (b) Flat response desired (all-pass).

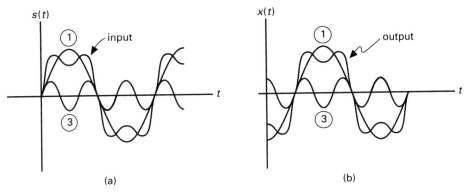

Figure 1-25 Effect of linear phase delay on transducer response: distortionless transfer. (a) Input signal. (b) Transducer output.

Another term used to describe the shifting or delaying of the frequency components is *group delay*. Group delay is defined as

$$\tau(\omega) = -\frac{d\varphi(\omega)}{d\omega} \qquad (1\text{-}28)$$

Thus for an ideal low-pass response, the group delay is a constant in the passband region of operation.

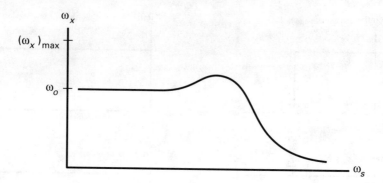

Figure 1-26 Transducer output frequency as a function of the frequency of the constant amplitude input.

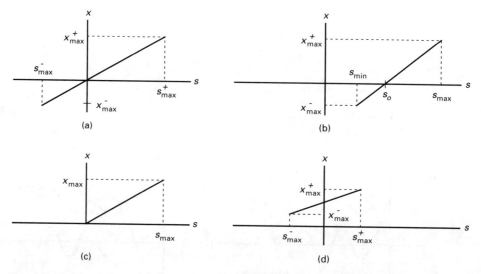

Figure 1-27 Transducer DC transfer characteristics. (a) Bipolar input and output. (b) Unipolar input, bipolar output. (c) Unipolar input and output. (d) Bipolar input, unipolar output.

Although most transducers are designed to produce DC voltages proportional to the instantaneous amplitude of $s(t)$, one should not limit thinking to these types. Some transducers produce output sine waves that have instantaneous frequency values which depend upon the amplitudes of $s(t)$ as shown in Fig. 1-26. Such converters are called amplitude-to-frequency converters, and they have very good low-noise properties. There are other types of transducer outputs, but most are eventually voltage.

The fact that sinusoidal response is being considered should not always be interpreted as an implicit assumption that the transducer is capable of bipolar inputs and outputs. As indicated in Fig. 1-27, the input physical variables may be superimposed on an offset bias of that physical variable. The same is true of the output quantity, so that there are four combinations, as shown. What these plots tell us is that if we are going to test a transducer for frequency response using sinusoidal sources, we must be careful to provide the appropriate signal bias.

We next investigate some properties of the sinusoidal amplitude response curves. The transducer amplitude response in general can be specified by a cutoff frequency and a ripple factor. Such quantities must be used with care when the output is to be processed using digital techniques, as the examples which follow show. For our discussions we assume a low-pass response, realizing that a change of variables can produce the corresponding band-pass or high-pass characteristic. We first examine the ripple factor.

1.9.1 Ripple Factor

Transducers that have ripples in the frequency response curves are often defined or modeled in terms of a parameter ε which appears in a low-pass transfer function of the form

$$|x(\omega)| = \frac{x_o}{\sqrt{1 + \varepsilon^2 \Psi_n^2(\omega)}} \tag{1-29}$$

where x_o is the maximum response of the transducer to an input sinusoid of amplitude s_o (note x_o may not be the DC constant value (0 frequency input)). The function $\Psi_n(\omega)$ is an even or odd polynomial, usually selected to have maximum absolute value of unity in the passband (called a characteristic function), and ε^2 is a constant. In many cases $\varepsilon \leq 1$, but it may take on any value.

Suppose, however, that a transducer characteristic did not have this precise form. What can be done? Usually the value of ripple width (*RW*) (measured by the parameter ε, as will be shown) is very small, so that a small scaling of values or simple small upward shift of x_o can be applied so that an approximate expression can be developed for the actual transducer. Table 1-2 is a list of some of the more common polynomials used in practice. One would examine the available transducer and select a polynomial that matched the transducer characteristic as closely as possible (notice that the polynomials are in terms of a normalized frequency ω_1). The important thing is to be sure that the actual transducer maximum ripple is present in the polynomial fit.

One can easily see from Eq. (1-29) that the maximum value of the function will be x_o for those frequencies which make $\Psi_n^2(\omega) = 0$. The function minimizes when $\Psi_n^2(\omega)$ becomes large. In the so-called passband, $\omega < \omega_1$, $\Psi_n^2(\omega)$ is selected so that it is less than or equal to 1 so $x_{\min}$ occurs when $\Psi_n(\omega) = 1$. These quantities are displayed graphically in Fig. 1-28.

Suppose we let *RW* denote the maximum ripple which occurs when the transducer is being driven to full output capacity, or $x_o = x_{\max}$. The quantity *RW* then represents the largest error that can occur across the frequency spectrum. The total output span of the transducer we define by $x_{span} = |x_{\max}^+ - x_{\max}^-|$ (see Fig. 1-27), assumed to be positive. The maximum possible error must be within 1/2 LSB if amplitude is important, so we require

$$1/2 \frac{x_{span}}{2^n} \geq RW$$

TABLE 1-2 POLYNOMIALS FOR APPROXIMATING TRANSDUCER CHARACTERISTICS IN THE FORM $|\bar{X}(\omega_1)| = \dfrac{\chi_o}{\sqrt{1 + \epsilon^2 \psi_n^2(\omega_1)}}$, $0 \leq \epsilon \leq 1$

Least mean square (LMS) polynomials

n	$\psi_n(\omega_1)$	$\omega_1 = \dfrac{\omega}{\omega_o}$	$|\bar{X}(\omega_1)|$ (for $\epsilon = 1$)
2	$\dfrac{5}{4}\omega_1^2 - \dfrac{1}{4}$,		
3	$\dfrac{7}{4}\omega_1^3 - \dfrac{3}{4}\omega_1$,		
4	$\dfrac{21}{8}\omega_1^4 - \dfrac{7}{4}\omega_1^2 + \dfrac{1}{8}$,		
5	$\dfrac{33}{8}\omega_1^5 - \dfrac{15}{4}\omega_1^3 + \dfrac{5}{8}\omega_1$,		
6	$\dfrac{429}{64}\omega_1^6 - \dfrac{495}{64}\omega_1^4 + \dfrac{135}{64}\omega_1^2 - \dfrac{5}{64}$		
7	$\dfrac{715}{64}\omega_1^7 - \dfrac{1001}{64}\omega_1^5 + \dfrac{385}{64}\omega_1^3 - \dfrac{35}{64}\omega_1$		
8	$\dfrac{243}{128}\omega_1^8 - \dfrac{1001}{128}\omega_1^6 + \dfrac{1001}{64}\omega_1^4 - \dfrac{77}{32}\omega_1^2 + \dfrac{7}{128}$		
9	$\dfrac{4199}{128}\omega_1^9 - \dfrac{1989}{32}\omega_1^7 + \dfrac{2457}{64}\omega_1^5 - \dfrac{273}{32}\omega_1^3 + \dfrac{63}{128}\omega_1$		

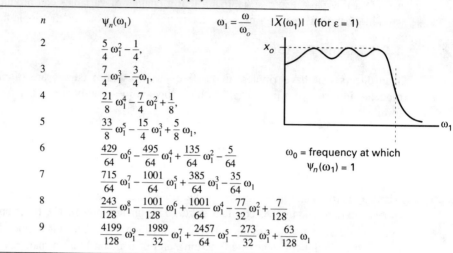

ω_o = frequency at which $\psi_n(\omega_1) = 1$

Butterworth polynomials (maximally flat)

for any n: $\psi_n(\omega_1) = \omega_1^n$, $\omega_1 = \dfrac{\omega}{\omega_o}$ $|\bar{X}(\omega_1)|$
($\epsilon = 1$ always)

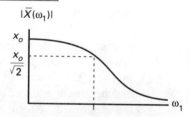

ω_o = frequency at which response is 3 dB down

Chebyshev polynomials (equiripple response)

n	$\psi_n(\omega_1) \equiv C_n(\omega_1)$	$\omega_1 = \dfrac{\omega}{\omega_o}$	$|\bar{X}(\omega_1)|$
0	1,	χ_o	
1	ω_1,	$\dfrac{\chi_o}{\sqrt{1+\epsilon^2}}$	
2	$2\omega_1^2 - 1$,		
3	$4\omega_1^3 - 3\omega_1$,		
any n	$\cos(n \cos^{-1}(\omega_1))$,		

Recursion relation: $\psi_{n+1}(\omega_1) = 2\omega_1 \psi_n(\omega_1) - \psi_{n-1}(\omega_1)$

ω_o = frequency at which $\psi_n(\omega_1) = 1$

Chap. 1 Transducers, Transducer Parameters (Specifications)

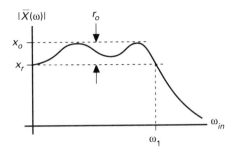

Figure 1-28 Definitions of terms used to define transducer frequency response ripple.

If we denote the ripple factor by

$$r_f = \frac{RW}{x^{span}},$$

the greatest number of bits justified (without requiring any amplitude adjustment correction by the computer) is found from

$$\frac{1}{2^N} \geq 2r_f \qquad (1\text{-}30)$$

or

$$N \leq \frac{\log_{10}\left(\frac{1}{2r_f}\right)}{\log_{10} 2} \qquad (1\text{-}31)$$

As an example, suppose we want to have 8-bit data words. Then we require any ripple be such that

$$r_f < \frac{1}{2^{8+1}} = 0.00195$$

or

$$ripple < 0.2\%.$$

Suppose next that we knew the transducer had a fractional ripple of $r_f = 0.01$ (1%). This would require

$$N \leq \frac{\log_{10}\left(\frac{1}{2 \times 0.01}\right)}{\log_{10} 2} = 5.6$$

or $N \leq 5$ bits.

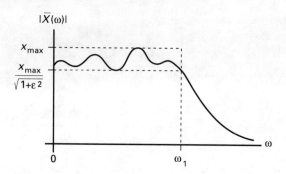

Figure 1-29 Response for the case $\Psi_{max}^2(\omega) \leq 1$ in the passband.

In many instances the transducer ripple is expressed in dB. For functions $\Psi_n(\omega)$ which have maximum passband values of 1, the ripple would be represented in terms of x_r as shown in Fig. 1-29 where

$$x_r = x(\omega) \Big|_{\Psi_n = 1}$$

as follows:

$$RW = x_{max} - \frac{x_{max}}{\sqrt{1+\varepsilon^2}} = x_{max}\left(\frac{\sqrt{1+\varepsilon^2} - 1}{\sqrt{1+\varepsilon^2}}\right)$$

$$\text{Fractional ripple} = \frac{RW}{x_{span}} = r_f = \frac{x_{max}}{x_{span}} \cdot \frac{\sqrt{1+\varepsilon^2} - 1}{\sqrt{1+\varepsilon^2}} \quad (1\text{-}32)$$

This could also be expressed in dB:

$$r_{dB} = 20 \log_{10} r_f = 20 \log_{10}\left[\frac{x_{max}}{x_{span}} \frac{\sqrt{1+\varepsilon^2}-1}{\sqrt{1+\varepsilon^2}}\right]$$

From this relation or Eq. (1-30) we may determine either r_f or ε for the transducer. Note that for this definition of ripple, as $\varepsilon \to 0$ dB, ripple $\to -\infty$. To avoid this problem of having to use large numbers, manufacturers define ripple in dB by another definition which makes 0 dB ripple correspond to a value of $r = 0$ (no ripple). This definition is as follows:

$$\text{dB ripple} = 20 \log_{10} \frac{x_{max}}{x_{min}} = r \quad (1\text{-}33)$$

For the case where $\Psi_{max}(\omega) = 1$ in the passband:

Chap. 1   Transducers, Transducer Parameters (Specifications)   75

$$r = 20 \log_{10}\left(\frac{x_{max}}{x_{max}/\sqrt{1+\varepsilon^2}}\right) = 20 \log_{10}\sqrt{1+\varepsilon^2} \text{ dB} \qquad (1\text{-}34)$$

(Manufacturer's specification definition)

This last equation is the definition that should be assumed unless otherwise stated. Thus $\varepsilon = 0$ corresponds to 0 dB ripple.

In summary, the process used to compute the number of bits on the basis of a manufacturer's dB ripple specification (or if one has matched a curve to a transducer) would be as follows:

1. From the dB specification, use Eq. (1-34) to compute the factor $\sqrt{1+\varepsilon^2}$.
2. Use the factor $\sqrt{1+\varepsilon^2}$ in Eq. (1-32) to obtain r_f. This also requires a knowledge of what fraction of the total transducer capability is being used.
3. Substitute the r_f obtained in the preceding step into Eq. (1-31) to compute N.

Since ε^2 is usually small (≈ 0.1 to 0.01), the equations may be simplified by using the first two terms in the series expansions for $\sqrt{1+x}$ and $\log(1+x)$ for $x \ll 1$ and using these results in Eqs. (1-32) and (1-33). Whether or not this is justified depends upon the number of bits being used, and it must be checked.

1.9.2 3-dB Frequencies

We first define the 3-dB frequencies for the more common frequency responses, which are the first- and second-order responses. These types of responses are most usual for transducers of physical variables, and they are shown in Fig. 1-30. The equations for each response are also given. Note that x_o is defined separately for each response type.

Case I in Fig. 1-30 can be described by considering the resonant hump to be the ripple so that the equations developed for finding ripple can be applied. Two questions are important when amplitude is to be considered important:

1. In order to use the fullest possible range of the transducer, what is the maximum allowable Δ?
2. For a given value of Δ, what is the highest usable frequency?

These questions are, of course, merely different ways of looking at the same response curve.

For question 1, we use the results developed in the discussion of ripple; that is, the number of bits is limited to

$$N \leq \frac{\log_{10}\left(\frac{1}{2r_f}\right)}{\log_{10} 2}$$

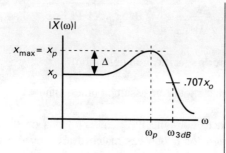
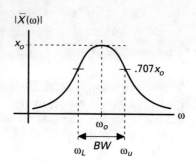

$$|\bar{X}(\omega)| = \frac{\omega_o^2 x_o}{\sqrt{(\omega_o^2 - \omega^2)^2 + (2\zeta\omega_o\omega)^2}}$$

ζ = damping factor, $\quad Q = \dfrac{1}{2\zeta}$

$\omega_p = \omega_o \sqrt{1 - 2\zeta^2} \qquad (\zeta < .707)$

$\omega_p = \dfrac{x_o}{2\zeta\sqrt{1 - \zeta^2}} \qquad (\zeta < .707)$

(a) Case I

BW = bandwidth = $\omega_u - \omega_p$

$$|\bar{X}(\omega)| = \frac{x_o \dfrac{\omega\omega_o}{Q}}{\sqrt{(\omega^2 - \omega_o^2)^2 + \left(\dfrac{\omega\omega_o}{Q}\right)^2}}$$

$Q = \dfrac{\omega_o}{BW}$

(b) Case II

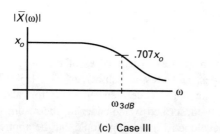

$$|\bar{X}(\omega)| = \frac{x_o \omega_{3dB}}{\sqrt{\omega^2 + \omega_{3dB}^2}}$$

(c) Case III

Figure 1-30 Types of transducer second-order frequency responses. (a) Underdamped, low-pass. (b) Band-pass. (c) Overdamped, low-pass.

where

$$r_f = \frac{\Delta}{x_{span}}$$

This first question implies that N is given and we want to determine the allowable Δ, so we solve for r_f to obtain

$$r_f \leq \frac{1}{2^{N+1}}$$

and then compute Δ as

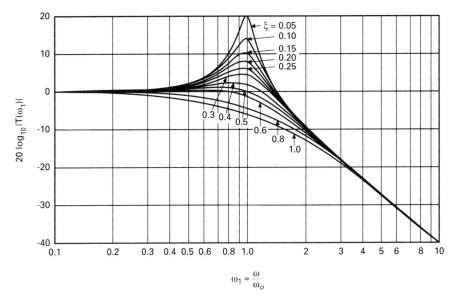

Figure 1-31 Detail of second-order frequency response near resonance, ω_o (magnitude).

$$\Delta \leq \frac{x_{span}}{2^{N+1}} \quad (1\text{-}35)$$

This is reasonable, since if Δ is less than 1/2 LSB we can use the transducer beyond the resonant rise to a point 1/2 LSB below x_o ($\omega > \omega_p$).

The answer to question 2 depends upon the number of bits N and the damping factor ζ (which of course determines Δ for a given x_o). There is a resonant rise Δ only if $\zeta < 0.707$. However, if Δ is less than 1/2 LSB, the highest usable frequency will be greater than ω_p, otherwise the frequency will be less than ω_p. For $\zeta \geq 0.707$ there is no resonant rise and the highest usable frequency will always be greater than ω_o. A plot of a normalized output $|T(\omega)|$ (expressed in dB) as a function of ζ is given in Fig. 1-31.

Let $\omega_{1/2B}$ be the frequency at which the response is within 1/2 LSB of x_o. For Case I transducer response one first takes the given Δ and determines whether it is below 1/2 LSB. If so, one uses the result for $\zeta \geq 0.707$ following. For any values of Δ and ζ we desire to find an $\omega = \omega_{1/2B}$ that yields

$$1/2 \text{ LSB} = \frac{x_{span}}{2^{N+1}} \geq \left| \frac{\omega_o^2 x_o}{\sqrt{(\omega_o^2 - \omega_{1/2B}^2)^2 + (2\zeta\omega_{1/2B}\omega_o)^2}} - x_o \right|$$

Now for $\zeta < 0.707$ and $\Delta > 1/2$ LSB, the quantity inside the absolute value bars will be greater than zero, so we may remove the absolute value bars:

$$\frac{x_{span}}{x_o} \frac{1}{2^{N+1}} \geq \frac{\omega_o^2}{\sqrt{(\omega_o^2 - \omega_{1/2B}^2)^2 + (2\zeta\omega_{1/2B}\omega_o)^2}} - 1$$

for
$$\frac{x_{span}}{x_o \, 2^{N+1}} + 1 \geq \frac{\omega_o^2}{\sqrt{(\omega_o^2 - \omega_{1/2B}^2)^2 + (2\zeta\omega_{1/2B}\omega_o)^2}} \quad (1\text{-}36)$$

$$\zeta < 0.707 \text{ and } \Delta > 1/2 \text{ LSB}$$

For $\zeta \geq 0.707$ or for $\zeta < 0.707$ and $\Delta < 1/2$ LSB the quantity in the absolute value bars will always be negative, so we can negate the difference and remove the bars:

$$\frac{x_{span}}{x_o} \frac{1}{2^{N+1}} \geq 1 - \frac{\omega_o^2}{\sqrt{(\omega_o^2 - \omega_{1/2B}^2)^2 + (2\zeta\omega_{1/2B}\omega_o)^2}}$$

$$-1 + \frac{x_{span}}{x_o \, 2^{N+1}} \geq -\frac{\omega_o^2}{\sqrt{(\omega_o^2 - \omega_{1/2B}^2)^2 + 2\zeta\omega_{1/2B}\omega_o)^2}}$$

$$\frac{-2^{N+1} + \frac{x_{span}}{x_o}}{2^{N+1}} \geq -\frac{\omega_o^2}{\sqrt{(\omega_o^2 - \omega_{1/2B}^2)^2 + (2\zeta\omega_{1/2B}\omega_o)^2}}$$

$$\frac{2^{N+1} - \frac{x_{span}}{x_o}}{2^{N+1}} \leq \frac{\omega_o^2}{\sqrt{(\omega_o^2 - \omega_{1/2B}^2)^2 + (2\zeta\omega_{1/2B}\omega_o^2)}} \quad (1\text{-}37)$$

for
$$\zeta \geq 0.707 \text{ or } \zeta < 0.707 \text{ and } \Delta < 1/2 \text{ LSB}$$

Either of the preceding equations may be used to solve for N or $\omega_{1/2B}$ depending upon which is given, although the direct answer to question 2 would presume Δ to be given.

The derivations of results for Cases II and III follow identical applications of the 1/2 LSB idea and are left as problems.

1.9.3 Concluding Comments on Frequency Response

Some general comments on the application of transducer frequency response curves to determine computer bit level justification are in order.

Comment I. The preceding analyses assumed that the amplitude of the physical variable was the property of primary interest in computation. In many cases some other

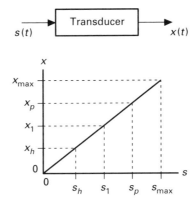

Figure 1-32 Effect of linear transducer on harmonic components of input signal for low frequencies, $f_h \ll f_{3dB}$. s_{max} = maximum transducer capability; s_p = peak value of input signal; s_1 = amplitude of fundamental component of signal; s_h = amplitude of harmonic component of signal.

property of the input signal may be of interest, as, for example, the frequency content or components of $s(t)$. In such cases the amplitude spectrum is not the fundamental bit-limiting factor but rather the required resolution of the signal to recover the desired properties from small signals. Noise limitations often set such bounds as well as the frequency response. Thus it may be in fact possible to operate the transducer well beyond the nominal 3-dB frequency.

As an example suppose we desire to use an N-bit signal processing system in conjunction with a transducer that has a first-order frequency response. The amplitude is not important, but we wish to be able to recover frequency harmonic components in $s(t)$ up to a value f_h whose amplitude is K dB below the maximum amplitude value capability for $s(t)$. What is the ratio of f_h to f_{3dB}?

The transducer characteristic is given by

$$\left| x(\omega) \right| = \frac{\omega_{3dB} x_o}{\sqrt{\omega^2 + \omega_{3dB}^2}}$$

Suppose that $s(t)$ has a peak value of s_p and that the fundamental f_1 has amplitude s_1. Suppose s_{max} is the largest total signal amplitude that the transducer will handle and that this produces an output x_{max}. Let s_h be the amplitude of the input of harmonic f_h. The relations are shown in Fig. 1-32. Then

$$-K = 20 \log_{10} \frac{s_h}{s_{max}}$$

or

$$\frac{s_h}{s_{max}} = 10^{-(K/20)}$$

If the harmonic signal at f_h were at very low frequency, the output would be

$$x_h = \frac{x_{max}}{s_{max}} s_h = 10^{-(K/20)} x_{max}$$

where we have assumed a linear transducer. However, due to the assumed first-order response,

$$\left| x(\omega_h) \right| = \frac{\omega_{3dB} \, x_h}{\sqrt{\omega_{3dB}^2 + \omega_h^2}}$$

Since we must be able to resolve this amplitude, it must represent at least 1/2 bit (greater than 1/2 LSB) at frequency f_h. Therefore

$$\frac{\omega_{3dB} \, 10^{-(K/20)} x_{max}}{\sqrt{\omega_h^2 + \omega_{3dB}^2}} \geq \frac{x_{span}}{2^{N+1}}$$

Solving for ω_h:

$$\omega_h \leq \omega_{3dB} \left(\frac{x_{max}^2}{x_{span}^2} 2^{2N+2} 10^{-(K/10)} - 1 \right)^{1/2}$$

The frequency ratio is then (first-order system)

$$\frac{f_h}{f_{3dB}} \leq \left(\frac{x_{max}^2}{x_{span}^2} 2^{2N+2} \, 10^{-(K/10)} - 1 \right)^{1/2}$$

For a specific application of this result, suppose we use an 8-bit system on a signal for which we want a 30-dB difference between the highest harmonic and the transducer maximum capability. The transducer output is unipolar, so $x_{max} = x_{span}$. Then,

$$\frac{f_h}{f_{3dB}} \leq \left(2^{18} \, 10^{-(30/10)} - 1 \right)^{1/2} = 8.03$$

Thus if amplitude were of no consideration we would operate up to 8 times the nominal "cutoff" frequency. Note, however, that if signals get too small, noise may become a problem.

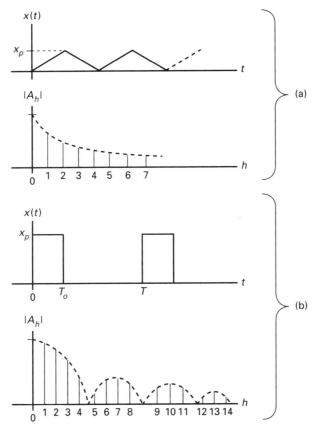

Figure 1-33 Comparison of waveforms having (a) monotonically decreasing harmonic amplitudes, and (b) nonmonotonically decreasing harmonic amplitudes (A_h = amplitude of harmonic h).

Comment II. If frequency components of the input $s(t)$ are of importance, one should not try to recover those frequencies whose amplitudes lie outside the transducer capability. This is often overlooked in some computation systems.

Comment III. It is to be emphasized again that when frequency response curves are given, the amplitude (which should be held constant across the band) of the input test signal should be specified. The frequency response curve is often a function of amplitude. This will also affect the value of x_o on the plot, but for most of the transducers, ranges of operation will result in only a vertical scaling of the response curve.

Comment IV. As a final comment, it is to be noted that the shape of the input function may actually cause a loss of some frequency component lower than the f_h calculated from the preceding equation. This is most evident when one compares the spectrum of a triangular wave with a rectangular wave, for example, as shown in Fig. 1-33. For the triangular waveform the coefficient amplitudes decrease as $1/h^2$ so that if a harmonic limit of, say, $h = 5$ were obtained one would also be assured of obtaining the amplitudes of $h = 0, 1, 2, 3,$ and 4. However, for the rectangular wave, if $h = 11$ were the highest-order harmonic that could be detected, one would miss $h = 4, 5$, and 9 harmonics, since their amplitudes would be below the detection level.

Exercises

E1.9a For a small transducer ripple ($\varepsilon^2 \ll 1$), the usual approximations of the first two terms in the expansions of $\sqrt{1+\varepsilon^2}$ and $\log_{10}(1+\varepsilon^2)$ are used to simplify the equations used to obtain rf and N. For what maximum value of N are these approximations justified?

$$\left[\log_e(1+x) = x - \frac{x^2}{2} + \frac{x^3}{3} \cdots\right]$$

and

$$\left[\sqrt{1+x} = 1 + \frac{x}{2} - \frac{x^2}{4 \cdot 2!} + \frac{3}{8 \cdot 3!} x^3 - \cdots\right]$$

Assume a bipolar transducer output, $2x_{max} = x_{span}$.

E1.9b A transducer has a low-pass first-order frequency response with a given f_{3dB} and maximum output x_{max} (Case III frequency response).

1. What is the relationship between f_{3dB} and the step input rise time for this transducer?
2. Obtain the expression for the frequency $f_{1/2}$ at which the output is down 1/2 LSB from the true value for N bits. Your answer will be in terms of N and f_{3dB} (bipolar response).
3. What is the relationship between $f_{1/2}$ and f_{3dB} for $N = 8$?

Assume symmetric bipolar transducer output ($x_{span} = 2x_{max}$).

E1.9c For the Case I transducer characteristic frequency response, plot

$$\left|\frac{x(\omega)}{x_o}\right|$$

as a function of the variable $\Omega = \omega/\omega_o$, using ζ as a parameter having successive values 0.1, 0.2, 0.5, 0.707, and 1. Make plots for actual ratio and also in dB. For $N = 10$, what are the values of the ratio

$$\frac{f_{1/2}}{f_{3dB}}$$

using $\zeta = 0.5$ and for

$$1 \le \frac{x_{max}}{x_o} \le 10,$$

specifically ratio values of 1, 2, 4, 6, 8, and 10, where we want to preserve amplitude (unipolar)?

Chap. 1   Transducers, Transducer Parameters (Specifications)   83

E1.9d Suppose one has a transducer that has a frequency response characteristic of Case II. For an N-bit computer processing system, develop an equation for the ratio of the usable transducer bandwidth (where output is within 1/2 LSB) to the 3-dB bandwidth. Evaluate this ratio for $N = 8$ and $N = 10$ for the cases where $Q = 1, 5, 10,$ and 50. Amplitude of signal is to be considered important, with bipolar output.

E1.9e For the Case I transducer frequency response, develop an equation for Δ/x_{span} as a function of x_o and ζ. Also find the smallest value of ζ for which there is no resonant rise at $\omega = \omega_p$. What is the value of ω_p at this value of ζ?

E1.9f The input to a transducer is a unipolar triangular wave, and it is desired to compute its frequency components, the amplitudes not being important. The triangular wave has a fundamental frequency that varies from 80 Hz to 500 Hz. For an 8-bit signal processor, what are the highest-frequency values recoverable at the input extreme frequencies? The input is always operated at the maximum transducer input amplitude, and the maximum transducer output is 10 volts. The transducer response is given by

$$\left| x(\omega) \right| = 2500 \frac{x_o}{\sqrt{\omega^2 + 6.25 \times 10^6}}$$

Repeat for operation at $1/2\, x_{max}$. What do you conclude about arbitrary input amplitudes?

E1.9g For Case II transducer response, derive the operating bandwidth as a function of Q, ω_o, N, the peak amplitude of the input signal, and the maximum input that the transducer will accept for a digital computation that does not require absolute signal amplitude; that is, where absolute signal amplitude at the transducer input is not of interest. Use a unipolar output transducer.

E1.9h An 8-bit processing system operates on data from a Case II frequency response transducer. The transducer Q is 20 and has a maximum output of +10 V (unipolar). It is desired to operate the transducer over only 4 times the 3-dB bandwidth. If there are no other frequency-limiting devices in the system, what are the largest amplitude input signals at the edge frequencies? $\Delta\omega_4 = 4BW$.

E1.9i For the low-pass second-order response characteristic (Case I), one can only be certain of not overdriving the transducer if the maximum transducer output is attained only at $\omega = \omega_p$. The transducer DC transfer characteristic may still be linear. Note that if a sine wave signal at frequency ω_p and amplitude s_{max} were applied to the transducer, the transducer would be "overdriven" and saturate at x_{max}. Derive a relationship between ζ and the maximum amplitude a sine wave may have to ensure that the transducer is not overdriven. Consider the cases $\zeta < 0.707$ and $\zeta \geq 0.707$.

E1.9j Using a mathematical equation representation for each of the harmonics shown in Fig. 1-25, prove that the ideal phase characteristic is a linear one. Now suppose the two frequencies are not harmonically related. Show that the same ideal phase characteristic is to be desired.

E1.9k For the Case I frequency response and the condition $\zeta < 0.707$, obtain the expression for the frequency at the resonant peak, ω_p. Next, obtain an expression for Δ. Finally, develop a relationship that would give the number of bits allowed for data words for a given value of Δ.

E1.9l The illustration of distortionless conversion shown in Fig. 1-25 made use of a linear phase delay. Suppose that the phase response of a transducer were a constant 90° delay for all frequencies. Redraw Fig. 1-25 for this case and show that waveform distortion results.

E1.9m A transducer has 0.3-dB ripple in the passband response. For a unipolar output system, how many bits are justified if amplitude is critical?

1.10 TRANSDUCER SPECIFICATION: RESOLUTION (THRESHOLD)

As the measurand, $s(t)$, varies in a smooth, continuous, and monotonic manner, the transducer output may not be perfectly smooth. In fact, the resulting $x(t)$ may have small discontinuities as shown (exaggerated) in Fig. 1-34. The effect is to produce a "graininess" in the input-output characteristic. This leads to an error in the value of x which should correspond to a given s, as indicated at value s_1 in the figure. For input s_1 the output shown would be less than the true value. As long as this increment is less than 1/2 LSB, an A/D converter would not sense the error.

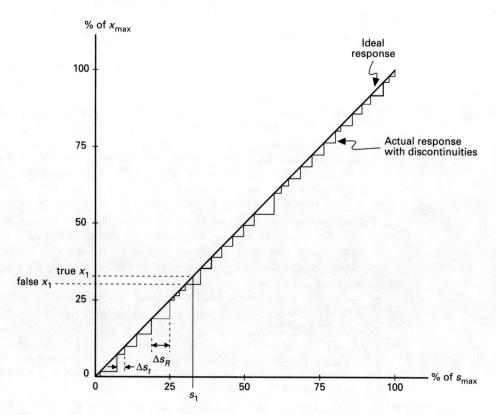

Figure 1-34 Comparison of ideal and actual transducer responses to illustrate the resolution specification.

Chap. 1   Transducers, Transducer Parameters (Specifications)

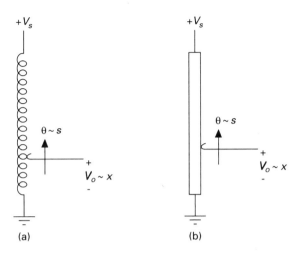

Figure 1-35 (a) Wire-wound potentiometer that has the output discontinuity phenomenon. (b) Film potentiometer that has smooth output.

Example 1-6

Suppose we have a wire-wound potentiometer for which the position of the wiper arm depends upon a shaft rotation Θ—which is s—and the output voltage is x. This is shown in Fig. 1-35(a). As the wiper arm moves up, it will jump from turn to turn so that the resistance increments in steps by the amount of resistance per turn. Thus the output voltage will have finite discontinuities. This could be avoided by constructing the wiper so that it effectively slides around the turns or else constructing the resistor out of a continuous film or composition material (see (b) of the figure). Even in this latter case, however, the least amount of drag or catch in the wiper-resistor interface can produce this equivalent effect. In this type of action the output may even exceed the true value of x momentarily as the spring and recoil of the wiper may overshoot the true setting.

The resolution specification is used to quantify the extent to which a transducer can give an output response for a change in the input. Specifically, the resolution of a transducer is the largest change that s may have before the output responds. This is shown as location (a) in Fig. 1-34. For the step size shown, the resolution would be Δs_R, and amounts to (for this exaggerated case) about 6% of full-scale value. If the full-scale value of s were $50\ N/m^2$ pressure, then the resolution would be $3\ N/m^2$.

Another term used to describe the discontinuity effect is *threshold*. By definition the threshold of a transducer is the smallest input signal change in s which produces a change in the value of the output x. This is shown as Δs_t at point (b) in the figure. Note, however, that if you had to guarantee that a response (output) would result from a given input change, you would need to give Δs_R (largest Δs before response occurs) as the threshold.

There is a second frequently used definition of resolution, which is the reciprocal of the largest change in s that can occur before x responds. On this basis, the resolution would be $1/\Delta s_R$. Suppose one has a transducer for which $\Delta s_R = 0$ (a perfect system).

Then the original definition would give a resolution of 0 (or infinitesimal) whereas the second would give a value of infinite resolution. The definition used by a particular manufacturer is usually clear from the context of the transducer specification sheet.

In order to quantify the relationship between the data word size N and the resolution, we first note that we do not want the system to see these increments. Thus the error introduced must be less than 1/2 LSB. We then have

$$\Delta s_R \leq 1/2 \frac{S_{span}}{2^N} = \frac{S_{span}}{2^{N+1}} \tag{1-38}$$

Note that this equation also helps an engineer to decide upon a particular transducer and can help one avoid overspecifying a unit, since for a given N the system cannot respond to any smaller increments than the right side of the equation.

Solving the preceding equation for N gives the alternate form

$$N \leq \frac{\log_{10}\left(\frac{S_{span}}{2\Delta s_r}\right)}{\log_{10} 2} \tag{1-39}$$

Example 1-7

A pressure transducer consists of a piston and a spring-loaded connecting rod which is attached to the wiper arm of a 1000 Ω, wire-wound potentiometer that has 500 turns (Fig. 1-36). The pressure varies from 0 to 10 Pa. What number of bits is justified for a processing system?

Assuming that the pressure range moves the potentiometer over the full resistance range, or 500 turns, in uniform steps,

$$\Delta s_R = \frac{10\ Pa}{500} = 0.02\ Pa$$

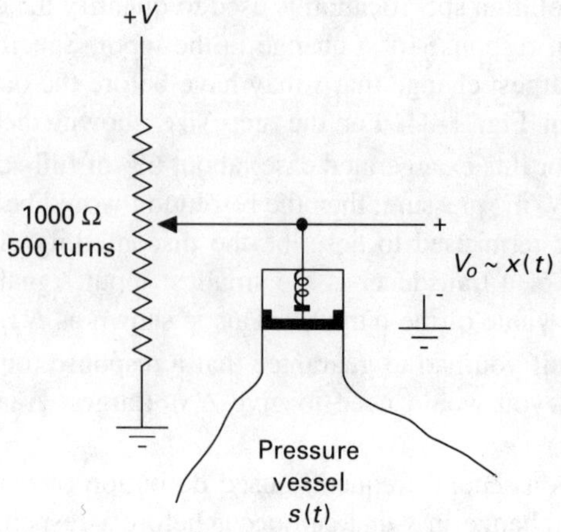

Figure 1-36 System for Example 1-7.

or

$$0.02 \frac{Pa}{turn} \text{ resolution}$$

Using Eq. (1-39) we obtain

$$N \le \frac{\log_{10}\left(\frac{10}{0.04}\right)}{\log_{10} 2} = 7.9$$

Therefore, $N = 7$ would be justified.

There is also in use currently a specification called *resolution error*. This is defined as the error due to the inability of a transducer to respond to changes in the input smaller than a given increment. A moment of thought shows that this is the same as threshold, although in most cases it is assumed to be synonymous with resolution. This term is not widely used, but one needs to be aware of it.

Exercises

E1.10a Suppose you were going to measure the resolution of a temperature transducer whose output is a DC voltage proportional to temperature. Describe the properties of instruments you would use to take measurements in terms of their resolution characteristics.

E1.10b Discuss the resolution of a light-sensitive diode. Is there more than one parameter that may have a resolution specification? List some, if any.

E1.10c A variable voltage is obtained by using a 10-V supply and a wire-wound potentiometer having a value of 1 kΩ, as shown in Fig. 1-35. We neglect any load connected to the wiper arm terminal. A shaft connected to the wiper arm is used to convert the shaft rotation angle to a voltage. The potentiometer has 3000 turns of resistance wire on a form which has a diameter of 2 mm, using 28-gauge wire. Make a plot of a portion of the output voltage near midrange that shows the discontinuity effects as the wiper moves across four turns. Also plot on the graph the output voltage for an infinite resolution resistance element. What number of bits could one justify using for this system?

1.11 TRANSDUCER SPECIFICATION: PRECISION (REPEATABILITY)

When engineers use transducers they hope that every time the physical variable input assumes the same value, the output x will also be the same. Unfortunately this does not happen. The precision specification of a transducer is a measure of how close the values of x will be when the input is reset to the identical value several times (repeatedly). This specification is usually thought of as the number of digits in a value that are significant. For example, a six-place logarithm table is more precise than a four-place one. Another way of describing this parameter is by the terms *coherence* or *repeatability* of a set of measurements.

There are two generally accepted ways for representing precision quantitatively. One is to give a ± error value. The second is to specify the percent that the error value is of full-scale output, ±% FSO; or

$$\frac{\pm error\ value}{full\ scale} \times 100\% \tag{1-40}$$

It is important to understand what this specification does not tell us. Precision does not indicate how close the output value x represents s based on what a linear response curve would give. For example, a transducer output may be $2.436 \pm .0002$ V, whereas the true value, for the given input, might be 2.5 V. This specification is given in the next section.

The quantification of precision is usually obtained by using the standard deviation of a set of output readings obtained when the input is reset to the same value. Thus, for k trials,

$$precision = \sigma_x = \left(\frac{1}{k-1} \sum_{i=1}^{k} (x_i - \bar{x})^2 \right)^{1/2} \tag{1-41}$$

where

$$\bar{x} = mean\ value\ of\ the\ x_i = \frac{1}{k} \sum_{i=1}^{k} x_i$$

Example 1-8

1. A set of voltage output readings is given by

$$4.9764$$
$$4.9769$$
$$4.9762$$
$$4.9768$$

The precision of this set would justify one in stating the value as 4.976. For this example:

$$\bar{x} = \frac{1}{4}(4.9764 + 4.9769 + 4.9762 + 4.9768) = 4.97657$$

Using this mean value we compute a measure of the precision as

$$\sigma_x = \pm \left(\frac{1}{4-1} \left[(-0.000175)^2 + (0.000325)^2 + (-0.000375)^2 + (0.000225)^2 \right] \right)^{1/2}$$

$$= 3.3 \times 10^{-4}$$

If the transducer had a full-scale output (FSO) of 10 V, then the alternate precision measure would be

$$\frac{\pm 3.3 \times 10^{-4}}{10} \times 10046 = \pm 3.3 \times 10^{-3} 46 \text{ FSO}$$

2. A set of voltage output readings is given by

$$4.97644$$
$$4.97649$$
$$4.97648$$
$$4.97642$$

In this case one could use the value $4.9764 = x$ which is more precise than that given in the first part of the example. For these values, using the formulas just given,

$$\bar{x} = 4.9764575$$

and

$$\sigma_x = \pm 3.3 \times 10^{-5}$$

If we compare these two examples we see that the precision measure, or the standard deviation, is smaller in the second case by a factor of 10, which agrees with what one would be justified in using as the number in each case.

The question we now address is the distinction between resolution and precision. *Resolution* is the input change that produces a perceptible change in the output, whereas *precision* represents the number of places in the analog output representation.

As an example of this distinction, let us consider the pressure transducer example of Section 1.10. Suppose that V_s were 5 V. The pressure resolution was shown to be 0.02 N/m^2 which would correspond to an output voltage change of $5 \text{ V}/500\,T = 0.01 \text{ V}/T$. This would be the theoretically smallest value that would be significant. Suppose, however, that the wire in the resistance were not of uniform resistivity, so that the voltage divider would not yield the correct value for a given pressure change, and it might give slightly more (or perhaps less) than the ideal 0.01 V. Thus the total output voltage would be in error. Also, there may be gear and/or wiper arm backlash, which would make the actual reading in error from what it should be for a specified input signal. Now if one were to set the pressure at, say, 2 N/m^2 a number of times and record the transducer output (with an ideal voltmeter), the values would differ, and the standard deviation among the values would be the measure of the precision *independent* of the fact that the smallest significant change is 0.01 or that the value may contain an error from what it should be exactly. Suppose one takes n output readings and they are all identical at 2.6 V for a given input pressure. The average would, of course, be 2.6 V. However, it might be that the ideal output for the pressure specified is 2.5 V.

Another approach from statistics is important to mention at this point. The fact that there is a dispersion of measured values about some average really means that the mea-

TABLE 1-3 CONFIDENCE LIMIT ON A DEVIATION OF A TRANSDUCER OUTPUT READING

Deviation from actual value, ±	Confidence limit, %
$\sigma/2$	38.3
σ	68.3
2σ	95.4
3σ	99.7

sure of precision (standard deviation σ) is only a representation of the probable deviation from the average value. Actually, any one reading could be much smaller or much larger than σ from the average value. The question is more properly phrased as: How confident am I that a given reading from a transducer is within a certain error value so that I can rely on the least significant data bit being at the average value at which a transducer output represents the physical variable? One is then interested in how much confidence can be placed in a given transducer output with respect to how far the output reading is from the value that should be given by the transducer.

To quantify the preceding discussion we use results from statistics based on the normal distribution of data values. These results are summarized in Table 1-3.

To use the table one must first decide how confident one wishes to be in a transducer error value (deviation) and then use the corresponding deviation given in the table. Then this amount of error is made less than 1/2 LSB. Conversely, if the number of bits were already specified, then 1/2 LSB is computed and a transducer is selected that has a σ which, when used in Table 1-3, yields the desired confidence limit. In actual practice a confidence limit corresponding to 2σ is used frequently. The confidence limit in percent is the assurance one has that the reading from a transducer is within the stated deviation from what the transducer would actually indicate for a given input value of the physical variable.

There is another important implication in the precision parameter. The definition given was based on the assumption that we took *one* transducer and reset the input to the same value k times and then measured the outputs. Suppose we took another approach and went to the stockroom and obtained k transducers and then applied the same input to all of them and measured the outputs of all transducers. How would the precision of these readings compare with those from the single transducer? From statistics, the answer would be the *same* if the system for producing the transducers were ergodic. Manufacturing problems, however, will not be pursued here.

Exercises

E1.11a Suppose that the pressure applied to the transducer discussed in this section is 5 N/m^2 (assumed for our purposes to be exact). What is the theoretical (ideal) output voltage of the transducer? Now suppose that the pressure is alternately removed and applied and several readings taken as follows (0-5 V, 0-10 Pa):

> 2.5716 V
> 2.5711 V
> 2.5717 V
> 2.5713 V

What is the standard deviation (measure of the precision) of the transducer? What value would one use to represent 5 N/m^2 on the basis of these data? On the basis of this value, how many bits would one be justified in using in a digital processor? Compare the answer to the result of the example in Section 1.10. Which parameter limits the number of bits, precision or resolution? On the basis of standard deviation, how many bits are justified?

E1.11b For the data given in Table 1-3 obtain an empirical equation for confidence limit (c) as a function of k, where k is the multiplier of σ that one wishes to use as the deviation from the mean; that is, deviation = $k\sigma$. Also obtain an equation for k as a function of c. (Hint: Use a power series. How many terms can you use for four tabular points?)

1.12 TRANSDUCER SPECIFICATION: ACCURACY

A transducer output may not be exactly what it should be to be the analog representation of the physical variable. This parameter is often a function of the units that may be attached to the transducer, such as loading resistors, but not always. It is best, of course, to give such a specification independent of external parameters connected to the transducers. On this basis one might even be justified in defining an *inherent* accuracy (transducer only) and an *apparent* accuracy that is observed to be present when the transducer is part of a system.

Accuracy is a measure of the error in the true value of x regardless of the precision. It is given quantitatively as a percentage defined by

$$A = Accuracy = \frac{error\ in\ x}{true\ value\ x} \times 100\% \tag{1-42}$$

Since the true x cannot be measured, it must either be computed during the transducer design process or calibrated against some standard that is accepted as the determined or calculated value of the highest available accuracy.

Let us first distinguish between the precision and accuracy. This can be most easily demonstrated by considering some possible results of firing at a target, as shown in Fig. 1-37. The figure is nearly self-explanatory. The important point is that the person may fire with precision but still not be very accurate.

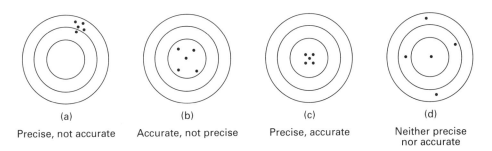

Figure 1-37 Target patterns illustrating the difference between precision and accuracy.

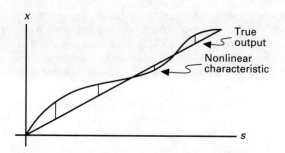

Figure 1-38 Effect of nonlinearity on accuracy.

Several of the parameters discussed earlier obviously have an impact on accuracy. This is one parameter that may be thought of as the combined effect of many of the other parameters. For example, any nonlinearity will cause the output to deviate from its true value. Also, the amount of deviation will depend on the operating region of the transducer, as shown in Fig. 1-38.

Example 1-9

Suppose a transducer converts incoming light intensity into a resistance value (a light-sensitive resistance such as silicon or cadmium sulfide). The resistance is initially calibrated using a standard light source, and measurements indicate an accuracy of 0.1%. What number of bits should be recommended in order to have a good record of light intensity (amplitude)?

Let x_{max} be the maximum output resistance value representing minimum light intensity. Also suppose that at a particular light input results in response x_o and that the minimum resistance is x_{min}. Thus 1 LSB is computed as follows:

$$1 \text{ LSB} = \frac{x_{max} - x_{min}}{2^N}$$

Since the response value may be in error, it will be related to the true value (worst case) by the definition of accuracy:

$$\text{error in } x = 0.001 x_{true}$$

This error must be less than 1/2 LSB to avoid being detected. We then require that

$$0.001 x_{true} \leq \frac{1}{2} \frac{x_{max} - x_{min}}{2^N}$$

$$2^N \leq 500 \frac{x_{max} - x_{min}}{x_{true}}$$

$$N \leq \frac{\log_{10} 500 + \log_{10} \left[\dfrac{x_{max} - x_{min}}{x_{true}} \right]}{\log_{10} 2}$$

Chap. 1   Transducers, Transducer Parameters (Specifications)

The worst case will be when x_{true} has its largest value, since this will minimize the upper limit on N. Let x_{true} be x_{max}, and also use the physical fact that x_{min} and x_{max} are positive. Then

$$N \le \frac{\log_{10} 500 + \log_{10} 1}{\log_{10} 2} = 9 \text{ bits}$$

The development of the general relationship between N and accuracy A in percentage is left as an exercise (1.12a) and is derived following the ideas in Example 1-9.

Exercises

E1.12a Prove that for a transducer accuracy of $A\%$ the number of bits justified to track amplitude variations in a digital processor is given by, for a unipolar system,

$$N \le \frac{\log_{10} \frac{100}{2A}}{\log_{10} 2}$$

(Hint: Generalize the example given in this section.)

E1.12b Describe the relationship of nonlinearity to accuracy, and the relationship of hysteresis to accuracy.

E1.12c A transducer consists of a potentiometer whose wiper arm is controlled by a real-world variable. A fixed voltage is applied to the potentiometer, and the wiper arm output is x. The unit feeds a resistive load R_L. The potentiometer total resistance is R_T, which for purposes here we assume to be ideal. Plot the accuracy as a function of d ($0 \le d \le 1$) for several values of

$$\left(\frac{R_L}{R_T} \right).$$

The variable d is the fractional distance of travel of the wiper arm from ground. R_L is tied between wiper arm and ground.

E1.12d For the example in the text design a method for converting the resistance change into a voltage change. Discuss the parameters in your design that are important if accuracy is to be maintained.

E1.12e Select a particular transducer type (that is, pressure to voltage) and explain the difference between accuracy and precision.

E1.12f You have designed a transducer which has an output of ± 5 V with 0.001% accuracy and 5-mV precision. A potential user calls and tells you he wants to know how many bits he can use without having to make any corrections to the data values. What is your reply?

1.13 TRANSDUCER SPECIFICATION: SENSITIVITY

Sensitivity is a specification that relates to the transfer characteristic of a transducer. Care must be exercised in the use of this parameter so that it is not confused with resolution, accuracy, or precision discussed earlier.

Unfortunately, there are three different definitions of sensitivity in use, and one must examine the specification sheet carefully to determine which is intended. Often the engineer is left to use an intuitive feeling as to what is intended. We will discuss each of these definitions.

Definition 1

The usual definition relates to the transfer characteristic of the device directly, as indicated in Fig. 1-39. This specification is defined by

$$S = \frac{x_{max}}{s_{max}} \qquad (1\text{-}43)$$

from Fig. 1-39(a), or

$$S = \frac{x_{max} - x_{min}}{s_{max} - s_{min}} \qquad (1\text{-}44)$$

from Fig. 1-39(b).

A note is that if the characteristic in (b) of the figure could be extended so it passed through the origin, then the two values of S would be identical in spite of the limited operating range of the latter.

Note that this definition says nothing about the minimum signal or minimum signal change that can be observed. This is the feature that distinguishes sensitivity from resolution.

As an example, suppose a transducer characteristic is given by the curve in Fig. 1-40. Let us investigate the aspects of sensitivity and what it does and does not tell us in terms of predicted outputs. From the curve, according to this definition,

$$S = \frac{10}{4} = 2.5 \; \frac{V}{Nm}$$

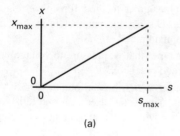
(a)

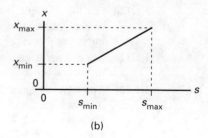
(b)

Figure 1-39 Conventions for defining sensitivity: Definiton One. (a) Full-range device. (b) Limited-range device.

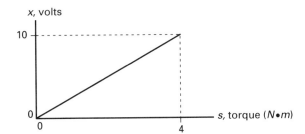

Figure 1-40 DC transfer characteristic of a torque transducer.

Suppose that at a particular time the torque is 0.001 Nm. Then the output would theoretically be computed as

$$x = (2.5)(0.001) = 2.5 \text{ mV}$$

However, it may be that the transducer does not have this resolution capability, for if the output were from a wire-wound potentiometer of 2500 turns, one turn would be 4 mV. Thus whenever one makes a calculation using the sensitivity figure, the resolution must also be checked concurrently.

This same sensitivity factor can be used in the following way. Suppose that the torque *changed* by an amount $\Delta s = 0.0015$ Nm. Then one would generally use

$$\Delta x = (2.5)(0.0015) = 3.75 \text{ mV}$$

but whether or not this is a valid output change depends again upon the resolution, which in this case is not for the potentiometer considered previously.

There are two other terms often used in connection with sensitivity. They are *range* and *span*. The range of a transducer refers to the actual values of the upper and lower limits of operation of the transducer, both input and output. The span of a transducer refers to the *difference* between the upper and lower limit values.

For example, in Fig. 1-39 the ranges would be specified by

$$R_{in} = (s_{min}, s_{max}) \quad \text{for the input and}$$
$$R_{out} = (x_{min}, x_{max}) \quad \text{for the output.}$$

The corresponding spans would be

$$Sp_{in} = s_{max} - s_{min} \quad \text{for the input and}$$
$$Sp_{out} = x_{max} - x_{min} \quad \text{for the output.}$$

For these definitions the sensitivity could be written as

$$S = \frac{Sp_{out}}{Sp_{in}}$$

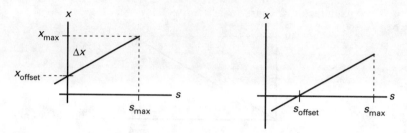

Figure 1-41 Transducer offset.

Some transducers may be constructed so that the response curve is a straight line but does not pass through the origin. The curve could be offset along either axis, as demonstrated in Fig. 1-41. In this case the simple ratio of maximum values will not give the correct slope (although the spans would), so one will frequently specify the offset and then use the ratio of changes in the input and output variables. On this basis, then,

$$\text{Sensitivity} = \frac{\Delta x}{\Delta s}, \quad \text{offset} = D \text{ units (of } x \text{ or } s\text{)}$$

Definition 2

The second definition of sensitivity occasionally used is

$$S = \frac{cause}{response}$$

This is simply the reciprocal of that for definition 1, so all the discussions for definition 1 apply to this case.

Definition 3

The final definition of sensitivity is one given in terms of fractional changes in the input and output variables as:

$$S_s^x \equiv \frac{\left(\dfrac{\Delta x}{x_o}\right)}{\left(\dfrac{\Delta s}{s_o}\right)} \tag{1-45}$$

For this definition it is necessary to specify the reference points x_o and s_o at which S_s^x is calculated.

In more formal mathematical terms we let $\Delta s \to ds$ and $\Delta x \to dx$ and write

$$S_s^x = \frac{\dfrac{dx}{x_o}}{\dfrac{ds}{s_o}} = \frac{d(\ln x)}{d(\ln s)} \tag{1-46}$$

Chap. 1   Transducers, Transducer Parameters (Specifications)

Some manufacturers refer to this sensitivity level in dB as

$$S_s^x \text{ (dB)} = 20 \log_{10} S_s^x \qquad (1\text{-}47)$$

Equation (1-45) may be placed in another convenient form, as follows:

$$S_s^x = \frac{\Delta x}{\Delta s} \frac{s_o}{x_o} \Rightarrow \frac{dx}{ds} \frac{s_o}{x_o} \qquad (1\text{-}48)$$

This allows the reference ratio to be selected as a standard. For instance, for microphones it is usual to specify the reference as

$$\frac{s_o}{x_o} = S_o = 1 \text{ V/N/m}^2 \qquad (1\text{-}49)$$

The applications of these definitions to signal processing are best presented by using examples.

Example 1-10

A transducer that converts electric field intensity ($\overline{E}$) into voltage has a sensitivity of 0.1 V/μV/m, linearly, and has a maximum output voltage of 10 V. If a digital processor utilizing 8 bits follows, what is the smallest change in the electric field that can be indicated digitally? What is the maximum value of electric field intensity allowed for this transducer to operate properly?

$$\overline{E}_{max} = \frac{10 \text{ V}}{0.1 \text{ V/μV/m}} = 100 \frac{\mu V}{m}$$

$$1/2 \text{ LSB} = \frac{10 \text{ V}}{2^{8+1}} = \frac{5}{256} = 0.01953 \text{ V}$$

Therefore,

$$\Delta \overline{E}_{min} = \frac{0.01953 \text{ V}}{0.1 \text{ V/μV/m}} = 0.2 \text{ μV/m}$$

Example 1-11

Suppose that the transducer above has a sensitivity of 0.05 V/μV/m. Answer the questions given in Example 1-10.

$$\overline{E}_{max} = \frac{10 \text{ V}}{0.05 \text{ V/μV}} = 200 \frac{\mu V}{m}$$

$$1/2 \text{ LSB} = \frac{10}{2^9} = 0.0195 \text{ V}$$

Therefore,

$$\Delta E_{min} = \frac{0.0195}{0.05} = 0.39 \frac{\mu V}{m}$$

Note that the device of Example 1-11 has a sensitivity only half that of the one in Example 1-10, which can easily be determined by the fact that the minimum detectable signal is twice that of the first example. However, the total range (assuming linearity extends that far) is twice that of the device in the first example.

In closing it should be pointed out that definition 3 is not a common one. The main application of this definition is to describe the effect of a "variable" within the unit upon the output. For example, if a resistance within the transducer were to change value the output might be affected. This describes how sensitive the output is to a change of component within the transducer.

Exercises

E1.13a For the transducer in exercise 1.10c compute the sensitivity at midrange, using the midrange value as reference, for all three definitions of sensitivity given.

E1.13b The circuit shown in Fig. E1.13b is to operate in a variable temperature environment. Since the resistor will change value, the frequency response curve will change. Suppose that the resistor temperature coefficient is such that the resistor value changes P parts per million / °C. Obtain an equation for the sensitivity of $|V_o|$ on the resistor value at a reference temperature 3-dB frequency, using definition 3 given in the text. For a temperature change of ΔT °C, how many bits N are justified in the digital processing that follows if amplitude is an important parameter and the reference temperature curve is being compensated for in the digital processing algorithm?

E1.13c For a 5-bit temperature data acquisition system, what is the maximum tolerable linearity for a transducer which has an output voltage range of ± 1 mV? The linearity is based upon the best straight-line fit such that at 300 kelvins input the output is 0 mV. The sensitivity of the transducer is 100 $\mu V/K$. Use definition 1. What is the operating temperature of the transducer?

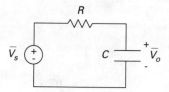

Figure E1.13b

Chap. 1   Transducers, Transducer Parameters (Specifications)   99

1.14 TRANSDUCER SPECIFICATION: SELECTIVITY

When a physical quantity is being monitored, our interest may lie with only one signal out of several that may be incident upon the transducer. The signals may have some property that may be used as a separating parameter. For example, if pressure were being monitored there may be several incident pressure signals from different directions, and the transducer may be selective to one direction. Another example is an antenna that converts a time-varying electric field into a voltage. If there were several electric fields present, selectivity would be a measure of how well the antenna would be able to respond to one of them. For the antenna, designs are available that could select on the basis of frequency or direction of the fields. Polarization of the electric field is yet another property of the electric field that could be used as a separation property.

Selectivity is defined as a measure of the extent to which a transducer is capable of differentiating between several signals of the same physical type on the basis of a common signal parameter. The antenna may use the frequency of the electric field as the separation parameter. As a plot this may be represented as shown in Fig. 1-42. For these types of responses the quality factor Q, defined by ω_o/BW, is used frequently as a measure of selectivity, higher Q being associated with higher selectivity. There is no formal mathematical expression for selectivity in general, but usually some physical characteristic of the transducer response is used, such as angle of incidence of the input $s(t)$, frequency of the input, or polarization of the input. For example, if the frequency responses of two transducers are as shown in Fig. 1-43, we would say that transducer 1 is more selective than transducer 2 since $Q_1 > Q_2$. Note that in this example there is the

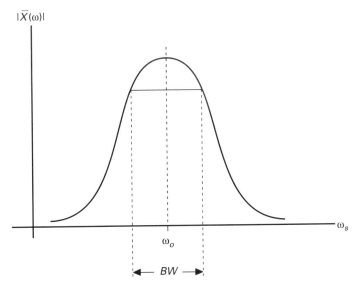

Figure 1-42   Example of the frequency selectivity of a transducer.

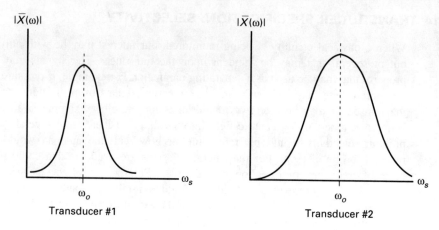

Figure 1-43 Frequency response characteristic as a criterion for selectivity.

obvious relationship to transducer frequency response discussed earlier. It is primarily the point of view one takes (that is, to pass a wide frequency range or to accept only a very restricted range) which determines the specification used.

Example 1-12

A sonar transducer has the angular response characteristic shown in Fig. 1-44, often called the pattern of the transducer. The length of the radius line ρ is proportional to the transducer response for a signal originating from an angle Θ from the reference direction. The equation is

$$\left| x(t) \right| = K \cos \Theta_s$$

Suppose that there are two incoming signals, the one of interest (desired) at angle Θ_d and another (undesired) at Θ_u, both of which are constant angles, and $\Theta_d > \Theta_u$. If we define the selectivity of the transducer as the ratio of the responses of the two signals

$$sel = x_d/x_u$$

determine the required selectivity for N-bit data word signal processing.

$$\left| x_d(t) \right| = K \cos \Theta_{s_d}$$

and

$$\left| x_u(t) \right| = K \cos \Theta_{s_u}$$

For N-bit words,

$$1/2 \text{ LSB} = \frac{x_{span}}{2^{N+1}}$$

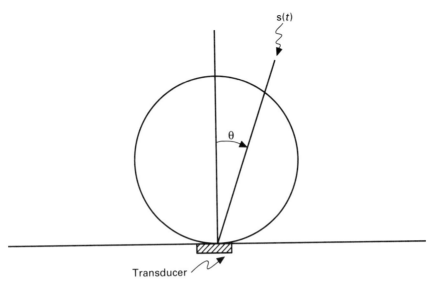

Figure 1-44 Angular displacement as a criterion for selectivity.

In order to reject the undesired signal,

$$\left| x_u(t) \right| \leq 1/2\, \mathrm{LSB} = \frac{x_{span}}{2^{N+1}}$$

therefore,

$$K \cos \Theta_u \leq \frac{x_{span}}{2^{N+1}}$$

Thus

$$sel = \frac{\left| x_d(t) \right|}{\left| x_u(t) \right|} = \frac{\cos \Theta_d}{\cos \Theta_u} \geq \frac{\cos \Theta_d}{\left(\dfrac{x_{span}}{K 2^{N+1}} \right)} = \frac{K\, 2^{N+1} \cos \Theta_d}{x_{span}}$$

Example 1-13

We have a transducer for which the DC input-output characteristic is amplitude linear but for which the frequency response is a low-pass second-order function, as shown in Fig. 1-45. The transducer is subjected to input signals from two different sources, which have frequencies ω_p and ω_u, respectively. The amplitude of source for s_p is always at least K dB above that for s_u. How many bits are justified in the signal processor? We also assume that the transducer is operating amplitude linear so that

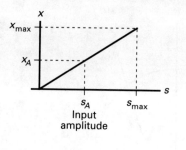
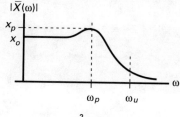

$$|\bar{X}(\omega)| = \frac{\omega_o^2 x_o}{\sqrt{(\omega_o^2 - \omega^2)^2 + (2\zeta\omega_o\omega)^2}}$$

x_o = output due to DC input s_o

Figure 1-45 Sensitivity calculation based on transducer frequency response.

$$x = x_{max} \frac{s}{s_{max}}$$

for the static characteristic (symmetric bipolar). This problem is an example of how selectivity is important as a point of view even though the presentation is in terms of frequency.

At $\omega = \omega_p$,

$$\left| x(\omega_p) \right| = \frac{x_p}{2\zeta\sqrt{1-\zeta^2}} = \frac{\frac{x_{max}}{s_{max}} s_p}{2\zeta\sqrt{1-\zeta^2}} \qquad (1\text{-}50)$$

(See exercise 1.14d.)

At $\omega = \omega_u$,

$$\left| x_u(\omega_u) \right| = \frac{\omega_o^2 x_u}{\sqrt{(\omega_o^2 - \omega_u^2)^2 + (2\zeta\omega_u\omega_o)^2}} = \frac{\omega_o^2 \frac{x_{max}}{s_{max}} s_u}{\sqrt{(\omega_o^2 - \omega_u^2)^2 + (2\zeta\omega_u\omega_o)^2}} \qquad (1\text{-}51)$$

Now in order to reject ω_u this output must be below 1/2 LSB. Then

$$1/2 \text{ LSB} = \frac{x_{span}}{2^{N+1}} \geq \frac{\omega_o^2 \frac{x_{max}}{s_{max}} s_u}{\sqrt{(\omega_o^2 - \omega_u^2)^2 + (2\zeta\omega_o\omega_u)^2}}$$

or,

Chap. 1   Transducers, Transducer Parameters (Specifications)

$$2^{N+1} \leq x_{span} \frac{s_{max}}{x_{max}} s_u \frac{\sqrt{(\omega_o^2 - \omega_u^2)^2 + (2\zeta\omega_o\omega_u)^2}}{\omega_o^2} \tag{1-52}$$

For the linear, symmetric, bipolar case, $x_{span} = 2x_{max}$. This last result can be looked at from two points of view:

1. For an undesired signal s_u it specifies the number of bits allowed.
2. For a given number of bits N it specifies the largest signal s_u that can be rejected at frequency ω_u.

This result is what one would expect just on the basis of frequency response only.

The form of the equation above, however, does not show the selectivity which must account for the desired input x_p. From the problem statement

$$\frac{s_p}{s_u} = 10^{K/20}$$

or,

$$s_u = 10^{-K/20} s_p$$

Thus

$$2^{N+1} \leq \frac{x_{span} s_{max}}{x_{max} 10^{-K/20} s_p} \frac{\sqrt{(\omega_o^2 - \omega_u^2)^2 + (2\zeta\omega_o\omega_u)^2}}{\omega_o^2} \tag{1-53}$$

(This is the Case I response.)

From this form we could determine how far a signal s_u needs to be below s_p for a given number of bits and frequency ω_u, if N and ω_o were fixed along with the transducer damping ζ.

Exercises

E1.14a Suppose we have a pressure transducer that only responds to the component of a longitudinal (compressive-expansive) pressure wave which is normal to the transducer surface. Suppose there are two such waves generated by two different devices and the waves travel at right angles to each other. It is desired to have the transducer respond to only one of these waves, and one wave has its maximum absolute value always less than one-fourth of the minimum magnitude of the other. For an N-bit signal processor, what is the maximum angle between the transducer surface normal and the direction of travel of the larger amplitude wave if we desire to select the larger amplitude wave with the transducer? Assume bipolar output, symmetric.

E1.14b A transducer that has Case I response is to select the signal at ω_p and reject the signal at $\omega_u = 10\omega_o$. For a 10-bit system, what is the allowable input signal amplitude of s_u for a transducer having a damping ratio of 0.65 and being driven by a desired signal s_o equal in amplitude to the maximum allowable input which at frequency ω_p drives the output to x_{max}? Also find the maximum signal in at ω_p (unipolar).

E1.14c Let K dB be the level of s_p above that of s_u for a transducer response at frequencies ω_p and ω_u, respectively. Let the transducer response be given by a general function form

$$\left| X_A(\omega) \right| = F(\omega) s_A m$$

where m is the slope of the transducer DC response, s_A is the amplitude of the input sine wave, and $F(\omega)$ is a real function. For an N-bit system show that the selectivity criterion for x_o is given by

$$2^{N+1} \leq \frac{x_{max}}{10^{-K/20} x_o} \cdot \frac{F(\omega_p)}{F(\omega_u)}$$

E1.14d Derive Eq. (1-50), using the assumption that $\zeta < 0.707$. Hint: Find the ω which maximizes $|X(\omega)|$; this is ω_p.

1.15 TRANSDUCER SPECIFICATION: DRIFT

For a fixed (constant) value of the measurand s_o, the output x_o may change value—a phenomenon called *drift*. Drift most frequently occurs due to either changes in temperature or due to aging of components within the transducer. The latter source of drift is identified most frequently as a time drift. The two parameters we will adopt are as follows:

1. Time drift

$$D_t = \frac{\Delta x}{\Delta t} \bigg|_{x=x_o}, \frac{units}{second} \qquad (1\text{-}54)$$

2. Temperature drift

$$D_T = \frac{\Delta x}{\Delta T} \bigg|_{x=x_o}, \frac{units}{degrees\ C} \qquad (1\text{-}55)$$

Note that the values of drift may depend upon the initial setting of the output, x_o, which will be the value at time t_o, or temperature T_o. Thus the computational forms of the preceding equations are

$$D_t = \frac{x_o(t_1) - x_o(t_o)}{t_1 - t_o}$$

$$D_T = \frac{x_o(T_1) - x_o(T_o)}{T_1 - T_o}$$

Chap. 1 Transducers, Transducer Parameters (Specifications)   105

Although transducers are frequently rated in terms of their temperature range of operation, such specification does not mean that the output will not drift with temperature changes. Thus while a device may function properly over the temperature range, the outputs at two temperatures T_1 and T_o are usually different. For example, a pressure-sensing device rated to operate from -20 °C to $+60$ °C can be set up at any temperature in the range to establish an output, say 0 volts out at 15 °C and 1 N/m^2. Since the device still operates at $+55$ °C, this does not mean that the output of 1 N/m^2 will still be 0 volts. This is because the temperature variation of the transducer components causes the output to drift to, say, 1 millivolt. There are, of course, units that will both operate over the specified temperature range and have "negligible" drift over that range. Manufacturers usually clearly state the latter, since that is a good selling point. The key idea, however, is what "negligible" means. For signal processing it means that the total drift over the desired temperature range produces an error less than 1/2 LSB.

There are two primary classifications of time drift: long-term and short-term. For the long-term drift the obvious limitation is that the final error value at the end of a data gathering period must be below 1/2 LSB. For the short-term drift (which is frequently a plus and minus variation) we require instantaneous fluctuations be less than 1/2 LSB.

Since the basic ideas behind relating bits, N, to the error producing specifications have been repeatedly applied in previous sections, a discussion of the drift specification is left as an exercise for the student to develop.

Exercises

E1.15a Prepare a discussion of one or two pages in length on the drift parameter. Include in your discussion a derivation of the relationship among $D, f_s,$ and N (and any other quantities you may come up with). Suggest ways of minimizing the effects of drift. (Hint: Consider the fact that drift is a low-frequency phenomenon.)

E1.15b Repeat exercise E1.15a for a transducer specification we term output offset. Note that you will first have to define what the specification means.

E1.15c A transducer which converts pH into a voltage level is subject to temperature variation ΔT. The output voltage range of the transducer is from 10 mV to 100 mV as the pH varies from 3 to 7. The transducer is initially calibrated by setting the temperature to T_1 and applying an acid of known pH of 3. The output will then drift away from the 10-mV output as the temperature changes. If the engineer desires to use 10 bits to represent the data and the transducer drift is 50μV/° C, what is the allowable temperature variation ($\pm \Delta T$ to which the transducer may be subjected?

1.16 TRANSDUCER SPECIFICATION: NOISE AND NOISE FIGURE

There are two primary sources of noise present in transducer operation: (1) the noise introduced by the external world and (2) the noise generated within the unit, as shown in Fig. 1-46. Thus even with zero value of $s(t)$ there may be a noise output $x(t)$. If there were a large enough number of bits being used in the data words, this noise would be interpreted as a signal. In this section only the introductory aspects of noise are consid-

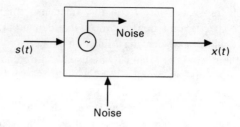

Figure 1-46 Sources of noise that affect a transducer output.

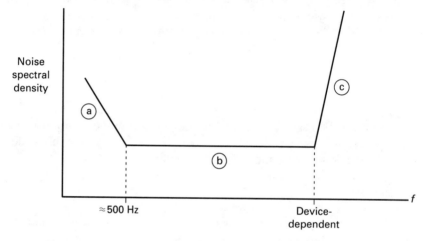

Figure 1-47 General "bathtub" noise spectrum of electronics systems.

ered, but in enough detail that the effect of noise on bit size can be estimated. In Chapter 5 some of the details are given.

Since most noise sources in electronic systems are random phenomena, their characteristics are generally described using the theories of probability and statistics. As a general effect, noise in electronic systems—of which a transducer is an example—can be described by one or more of the portions of the noise spectral density plot of Fig. 1-47. Region (a) of the figure is called by various names: flicker, $1/f$, or pink noise. The slope of the curve is about 3 dB/octave [20]. Region (b) is referred to as the floor or background noise. The higher-frequency region (c) is frequently called the shot noise, Johnson noise, or white noise. This higher-frequency portion begins nominally about 100 kHz (but may be up in the megahertz region, depending upon the system) and has a slope of approximately 6 dB/octave.

The most common source of noise is a single resistor that generates a voltage due to the random motion of electrons in the resistor material. This is called thermal noise. This type of noise has uniform energy distribution over the frequency spectrum and a normal (Gaussian) distribution of levels, and it is also known as Johnson noise or white noise. As a voltage value (RMS):

$$e_n = \sqrt{4kTBR} \qquad (1\text{-}56)$$

TABLE 1-4 RELATIONSHIP BETWEEN NOISE BANDWIDTH AND 3-dB VOLTAGE BANDWIDTH

Asymptotic Voltage order of system	B/f_{3dB}	Voltage high-frequency roll-off	
		dB/octave	dB/decade
1	1.57	6	20
2	1.22	12	40
3	1.15	18	60
4	1.13	24	80
5	1.11	30	100

where

  k = Boltzmann's constant
    = 1.38×10^{-23} joule/kelvin
  T = temperature, kelvins
  B = system noise bandwidth, hertz
  R = resistance, ohms
  e_n = RMS noise voltage, volts

The bandwidth is calculated from the formula

$$B = \frac{\int_0^\infty G(f)\,df}{G} \qquad (1\text{-}57)$$

where $G(f)$ is the system power gain as a function of frequency, f is the frequency in hertz, and G is the system available gain, P_o/P_i, and is the peak value of $G(f)$, with P_o being the power output to a matched load and P_i the input power. A model of the resistor is a noiseless resistor in series with the noise voltage source, e_n, as shown in Fig. 1-48.

Note that the gains used in Eq. (1-57) are power gains, not voltage gains. It is usual to use

$$G(f) = \left| A_v(f) \right|^2 \qquad (1\text{-}58)$$

where $A_v(f)$ is the system voltage frequency response or voltage gain. For many cases B is taken to be the usual voltage 3-dB bandwidth, although the value of e_n thus calculated is smaller than actual. Table 1-4 shows the relationship between the noise bandwidth, B, and the 3-dB voltage bandwidth for various orders of systems. Note that as the order of the system increases, the voltage 3-dB bandwidth approaches B.

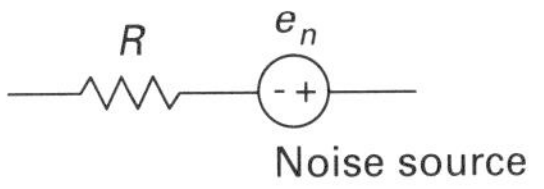

Figure 1-48 Noise model of a resistor.

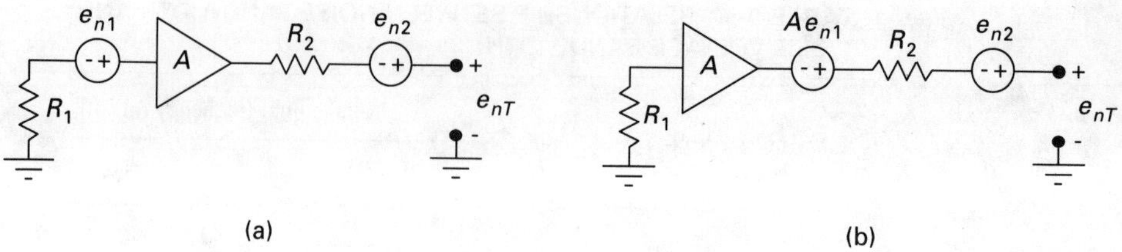

Figure 1-49 Effect of an amplifier placed between two resistive noise sources.

One must also remember that the bandwidth of the transducer itself is not the only consideration in computing B in the noise voltage formula. The proper bandwidth to be used is the total system in which the transducer is embedded, since any amplifiers and filters will affect the operating bandwidth. The first amplifier is usually the dominant factor.

In addition to the Johnson noise, the current flow in the resistor also adds a noise component. This is caused by material granularity and electron bunching processes, and is called shot noise. This can be minimized by using metal film resistors.

If there are several independent noise sources or resistors in series at some branch of the transducer or system, the total RMS noise would be

$$e_{n_T} = \sqrt{e_{n_1}^2 + e_{n_2}^2 + e_{n_3}^2 + \ldots} \tag{1-59}$$

For the Norton equivalent circuit (noise current source in parallel with the resistor), current sources of several parallel resistors would add in similar manner.

The importance of having the noise sources referred to the same branch of the circuit is shown in Fig. 1-49. For this configuration the output noise voltage would be given by

$$e_{n_T} = \sqrt{(Ae_{n_1})^2 + e_{n_2}^2} \tag{1-60}$$

From this we can see that the first stage and its input noise in a system are the most critical, since a high gain will cause the input noise components to dominate the noise level. This is why in many electronic systems a great deal of attention is paid to the design of the first stage. Any noise introduced by the amplifier is also critical, since it will decrease the signal-to-noise ratio which exists as the input prior to R_1.

Since analog-to-digital converters convert total amplitude and not RMS, we need to determine the peak value. Since noise signals are random, probability theory is used to provide the answer. The result is as follows:

$$e_n^P \approx 4e_n \tag{1-61}$$

The factor of 4 assures us that the actual instantaneous peak value will not exceed e_n^P more than 0.01% of the time. This is discussed in more detail in Section 5.2b.

Chap. 1 Transducers, Transducer Parameters (Specifications) 109

Example 1-14

Suppose we have a resistance of 100 k Ω in a circuit that has a noise bandwidth of 15 MHz and is at room temperature of 290 kelvins. If the signal processor is operating on voltages to 5 V maximum (unipolar: 0 to 5 V), what is the maximum number of bits to be recommended?

We begin by computing the noise voltage generated by the resistor:

$$e_n = \sqrt{4 \times 1.38 \times 10^{-23} \times 290 \times 15 \times 10^6 \times 100 \times 10^3} = 0.15 \text{ mV}$$

Thus

$$e_n^P \approx 4e_n = 0.6 \text{ mV}$$

This must be below the 1/2-bit level, or increment, by what is used in current practice:

$$\text{LSB} = \frac{5 \text{ V}}{2^N}, \text{ with } N \text{ to be determined}$$

Therefore,

$$0.6 \times 10^{-3} \leq \frac{1}{2} \cdot \frac{5}{2^N}$$

Solving for N:

$$N \leq 12 bits$$

Other examples of internal noise sources consist of the "graininess" of the transducer and the generation of extraneous signals by moving parts within. Examples of these would be, in the first instance, a wire-wound resistor that might lose contact momentarily with its wiper arm or produce high-frequency components as the turns are added in steps. This also affects resolution, discussed earlier. The generation of signals by moving parts could arise when one part of the transducer is carrying a current and an adjacent moving line would generate an induced emf due to the magnetic field in the region between the two conductors.

Moving parts in contact can produce local "hot spots" and change the transducer signal output. This could, of course, be interpreted as a nonlinearity, but this effect would not be constant, as frictional constants and location of the "spot" may be unpredictable.

Examples of external noise include the following:

1. One particularly troublesome external noise source is radio frequency interference (RFI), a general term used to describe noise introduced by external electric and magnetic fields. Such time-varying fields induce additional signals onto an output. Power line frequencies are often the culprits here, and much effort is spent shielding transducers from this effect. Higher-frequency noise is often called electromagnetic interference (EMI).

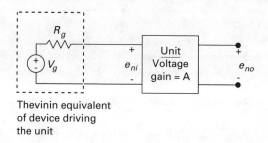

Thevinin equivalent
of device driving
the unit

Figure 1-50 Configuration of system used to derive the noise figure of a unit.

2. Additive noise may arise in two ways. The first is an addition of an external signal of the same type as the physical variable being monitored. For example, if a very precise measurement is being attempted on a magnetic field being produced by current flow, the earth's magnetic field may add to that field already present. Thus either a calibration must be performed to null out the earth's field or the number of bits used in the signal processing must be limited to that precision.

3. An amplifier may introduce additional noise. The output of many transducers is small, so that in order to take advantage of the full range of an analog-to-digital converter an amplifier may be required. These amplifiers are frequently a part of the transducer and would be part of the output interface unit of Fig. 1-1. The amplifier generates its own internal noise, which increases that which may come from a transducer or equivalent resistance. The amount of noise an amplifier generates at its own input terminals was described by Friis and is quantified by a noise factor, F. The total effective input noise is defined in terms of the transducer noise and the noise factor as

$$e_{n_{total}}^2 = F e_n^2 \tag{1-62}$$

The noise effect of a unit is most often expressed by a term called the noise figure of the amplifier, where

$$F_{dB} = 10 \log_{10} F \tag{1-63}$$

Most operational amplifiers have F_{dB} values in the range 1 to 10.

The noise figure of a unit is derived using the system configuration of Fig. 1-50. For simplicity here we use a full voltage in-voltage out model. For a transducer, the Thevenin generator would be the equivalent resistance of the medium or variable driving the system. The results obtained from the following derivation apply to any type of unit: amplifier, antenna, filter, or other electronic transducer.

The Thevenin resistance of the driving device generates a noise voltage of value

$$e_{n_g} = \sqrt{4kTBR_g}$$

The output noise of the unit as connected is denoted by e_{n_o} and consists of the combined effects of the noise of R_g and the noise added by the unit. The equivalent input voltage that would produce this output noise is

Chap. 1 Transducers, Transducer Parameters (Specifications)

$$e_{n_i} = \frac{e_{n_o}}{A}$$

Now the noise factor F is defined in terms of signal-to-noise ratios by

$$F = \frac{\frac{\text{signal power in}}{\text{noise power in}}}{\frac{\text{signal power out}}{\text{noise power out}}} = \frac{\left(\frac{S}{N}\right)_{in}}{\left(\frac{S}{N}\right)_{out}}$$

It is also common in practice to give signal-to-noise ratios in dB as

$$10 \log_{10}\left(\frac{S}{N}\right).$$

Rearranging, we have

$$F = \frac{\text{signal power in}}{\text{signal power out}} \times \frac{\text{noise power out}}{\text{noise power in}} = \frac{1}{A^2} \frac{\text{noise power out}}{\text{noise power in}}$$

$$= \frac{\frac{\text{noise power out}}{A^2}}{\text{noise power in}} = \frac{\text{noise power out referred to input}}{\text{noise power in}}$$

Since power is proportional to the square of the voltage, and if we assume the same proportionality constant for all powers (which is exact if all resistors are equal, or what we call maximum available power condition), then we can write an equivalent definition:

$$F = \frac{(\text{noise output voltage referred to input})^2}{(\text{noise input voltage})^2}$$

using the symbols given on the diagram:

$$F = \frac{e_{n_i}^2}{e_{n_g}^2}$$

or,

$$e_{n_i}^2 = F e_{n_g}^2 \tag{1-64}$$

Now if we draw an equivalent circuit for the system and let e_{n_A} be the equivalent noise source which the unit appears to produce at its own input, we have the configuration shown in Fig. 1-51. The total noise input is then (voltages add as squares):

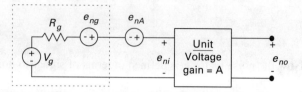

Figure 1-51 System showing an effective input noise voltage, e_{n_A}, generated by the unit only, referred to the input terminals (RTI).

$$e_{n_i}^2 = e_{n_g}^2 + e_{n_A}^2 \qquad (1\text{-}65)$$

Substituting this into Eq. (1-63) we obtain

$$e_{n_g}^2 + e_{n_A}^2 = F e_{n_g}^2$$

or,

$$e_{n_A}^2 = (F - 1) e_{n_g}^2 \qquad (1\text{-}66)$$

Ideally, the unit would add no noise at all, so $e_{n_A} = 0$. This means that the ideal noise factor F is 1, or an ideal noise figure, NF, of 0 dB.

Equation (1-64) has an interesting interpretation. It tells us that the total input noise is a constant (noise factor) times the noise fed in from the driving system.

One should not conclude from this section that if the signal were smaller than the noise is all lost. This is in fact not the case. There are many digital signal processing techniques and algorithms for detecting signals that are embedded in high noise levels. The engineer should work closely with the programmer and others who are interested in recovering small signals, for the desired signal may be recoverable from a noisy environment. At least the 1/2 LSB criterion yields an acceptable noise level below which no special considerations need to be made in system design.

Exercises

E1.16a To obtain an idea of the effect of approximating the noise (power) bandwidth B by the 3-dB bandwidth f_{3dB}, consider a first-order RC low-pass network with the output taken across the capacitor (Fig. E1.13b). Derive the expression for the voltage 3-dB frequency, f_{3dB}, and also the bandwidth B using the formula given in the text. Remember that G is the power gain, which is the square of the voltage gain. What is the percentage error in calculation of the resistor noise voltage e_n when using the f_{3dB} as an approximation for B?

E1.16b For example 1-14, suppose that the 15-MHz bandwidth used in the computation was actually the 3-dB voltage bandwidth rather than the noise bandwidth. Compute the number of bits now justified and compare, assuming a first-order system.

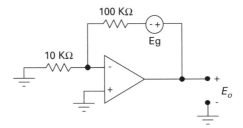

Figure E1.16f

E1.16c A system has a power bandwidth of 50 kHz, and a transducer has an internal impedance of 1000 Ω. The transducer has an output of 3 mV maximum (positive only). What is the permissible number of bits that can be used in the signal processor? Assume room temperature of 290 kelvins. Now determine the number of bits for the following cases:
  a. Signal and noise amplified by 100.
  b. Signal only amplified by 100 and the noise added following amplification.

E1.16d A signal processing system has been designed so that the number of bits being used has the voltage noise level at or below 1/2 LSB level. If the noise voltage is e_n, what is the smallest amplitude signal that the processor can handle in terms of e_n?

E1.16e For the example of the text suppose that the resistor is on the input of an operational amplifier configured so that the amplifier gain is –5. Draw the equivalent amplifier circuit, and determine the number of bits justified for data representation. Suppose that the amplifier noise figure is 3 dB.

E1.16f Determine the output voltage for the Fig. E1.16f amplifier circuit assuming the amplifier and resistors are noiseless. Now move the source to be in series with the input resistance and determine E_o. Now remove all sources, and then compute equivalent sources of noise generated by each resistance (temperature T and bandwidth B) and compute the output noise voltage. Recall that noise voltages are random, so they "add" in root square fashion at output.

E1.16g Prove that the peak noise input (0.01% confidence range) to an amplifier having a noise figure of NF and a grounded input resistor R is given by

$$e_n^p = 4 \times 10^{NF/20} \times \sqrt{4kTBR}$$

E1.16h Suppose a transducer is used on board a satellite that is to pass by the sun and take data. With optimum cooling the transducer temperature will be kept below 800° C. The transducer maximum output is 50 mV (unipolar), and the equivalent (Thevenin) impedance is 1 kΩ. Find the number of bits justified for data words if the system power bandwidth is 20 kHz.

E1.16i Analog master recordings of audio musical performances are to be remastered in digital format for storage. The minimum recommended S/N characteristic (tape hiss, and so forth) is 55 dB. How many bits digitizing would you recommend? Treat audio as a noiselike signal.

1.17 GENERAL DISCUSSION AND CONCLUSIONS

Some comments are appropriate here concerning types of transducers implicitly assumed in the preceding sections. Whenever transducer transfer characteristics were drawn, it may have been presumed by the reader that both input and output were DC levels. To obtain such characteristic curves one would apply a time-invariant input real-world variable and then measure the output time-invariant response (voltage or current). If a time-varying signal were then applied, the output would increase and decrease according to the instantaneous amplitude of the input. The frequency response of the transducer would then modify the output signal variations.

There are, however, other types of output signal forms. For example, the output may be an amplitude-modulated waveform, the amplitude of which would vary with the physical variable. FM techniques illustrate another possibility. The same ideas for relating transducer performance to the digital signal processing can be applied to such output forms.

One should always remember that the parameters discussed are not independent, so it may be a matter of either deciding upon the best way to look at a particular characteristic and/or determining how other parameters impact upon the particular ones of importance in a given application. The art of engineering really manifests itself here, since one must be inventive in order to match transducer and system requirements. Understanding basic physical principles and chemistry is a big necessity here.

Finally, with so many parameters one may be inclined to throw up both hands in dismay and say: Of all the parameters, many of which are not even given on the spec sheet for the transducer, which ones count? The answer to this question depends upon the goal of the computations and the needed information. One needs to decide which parameters impinge most severely upon the intended results and then analyze their effects. If a parameter is not available, one must then make one's own transducer tests for the parameter of interest. Additionally, just having a "feel" for the way things work may provide the answer. Here again is where a sound understanding of physical principles will give great aid. In a moment of panic one could use the "First Approximation" rule for selection of parameters and devices, which states that a useful rule of thumb that usually provides satisfactory results is this: For the critical specs of a multicomponent (multiparameter) system, choose each component (parameter) to perform roughly ten times better than the overall desired performance [24]. Even this will not always work.

Some manufacturers simplify the selection process by giving a specification that combines several parameters whose effects are similar. For instance, hysteresis, nonlinearity, and drift have the similar effect of producing an offset from the ideal straight line through the origin.

For a discussion on the cumulative effects of errors and combining errors see Section 1.3, particularly Eq. (1-4).

GENERAL EXERCISES

1.1 Select a thermocouple pair and design an analog system that can be used to measure temperature. What is the temperature range of your system? Do not be too particular

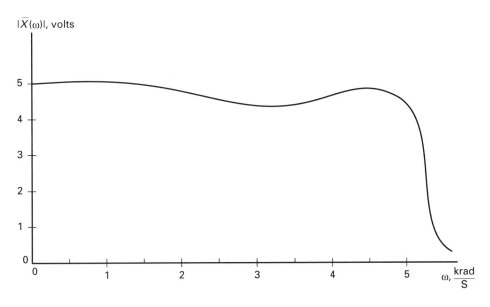

Figure PR1.3

about details at this point, but give enough design so that another person could construct your system and use it.

1.2 Design a system using a strain gauge that can be used to measure length change (in compression) for an iron bar. Specify the strain gauge and the detectable length change and the output of your system over the range.

1.3 The frequency response characteristic of a transducer is shown in Fig. PR1.3. Determine the value of ε and write the equations for $|x(\omega)|$ for Least Mean Square and Chebyshev response approximations. Use the lowest possible orders, n (see Table 1-1). What are the normalizing frequencies ω_o in both cases? Plot the two results superimposed over the actual curve.

1.4 Explain how transducer noise and linearity produce lower limits on transducer resolution.

1.5 Explain the similarities and differences between backlash (hysteresis) and resolution.

1.6 A rotation-to-voltage transducer consists of an antenna shaft connected to the wiper arm of a potentiometer. The potentiometer is a continuous-film potentiometer whose resolution is infinitesimal. A 10-V DC source is used as the reference voltage in a system designed to work at a 14-bit binary level. The resistance value is 100 kΩ, and the ambient temperature is 30° C. The voltage output works into a system that has a bandwidth of 50 kHz. What is the required angular resolution of the transducer in degrees if one antenna revolution corresponds to 10 revolutions of the potentiometer shaft? Note you will have to compare the LSB level of the system with the noise generated by the resistance. The effective resistance will be a function of the wiper arm position. Think in terms of a Thevenin equivalent circuit looking into the wiper terminal.

1.7 For the transducer designed in problem 1.2, comment on your design with respect to each of the transducer parameters discussed in the text, giving values where possible.

1.8 Test any existing transducer with respect to any four of the parameters discussed in the text. Examples include microphone and light-sensitive resistor. These are to be physically conducted tests.

1.9 Describe how you would go about determining the LSB value for a transducer.

1.10 Select a manufacturer's specification sheet for a transducer and explain (if possible) what is meant precisely by each specification.

1.11 A piezoelectric pressure transducer is used as part of an underwater sonar system. In the receive mode the transducer output is to be sampled at a 200-kHz rate. Identify the important transducer parameters and explain why you selected particular ones.

1.12 A pressure transducer is used as part of the altimeter transponder system on a small commercial plane. Identify the important parameters most critical in this application and explain why.

1.13 A moving iron vane galvanometer optical scanner contains transducers that track the galvanometer position. The position transducer operates by detection of capacitance variation between the rotating armature and a set of stationary electrodes. Given the specifications below, determine the maximum number of bits that could be justified by a digital processor that uses the transducer output to determine the position correction signal for the scanner. Assume the processor can convert fast enough. State any assumptions.

Transducer specifications:
position sensitivity (0 Ω load) $2 \frac{mA}{rad}$
linearity $\pm 0.3\%$
resolution 1" of arc
response time 10 μs

Scanner specifications:
output 0.42 V/deg $\pm 20\%$
excursion 0 to 20 degrees
frequency (max) 500 Hz

1.14 An analog voltmeter is said to have a sensitivity of 20,000 Ω/V. Explain exactly what is meant by this term.

1.15 A sales brochure for a model XYZ Absolute Shaft Position Encoder indicates that the digital output will increment from 0 to 1023 during one complete rotation of the transducer shaft. The device is called an absolute encoder because the shaft position is not determined by counts past a reference mark or other similar scheme. It is claimed that for the shaft in a given position, whenever the power is switched off and on the digital output remains the same. Your present application requires high-tolerance shaft positioning that will use all 10 bits available. What are some of the major specifications you would need to request of the manufacturer to evaluate the product?

1.16 A student wants to use a microprocessor in conjunction with an off-the-shelf data acquisition system (DAS). The DAS has on the board an eight-channel analog multiplexer (MUX) so that eight different data sources can be sampled. The student wants to use the eight inputs to sample different audio signals that have frequencies of interest up to 12 KHz, and then use the microprocessor to do digital data compression. The microprocessor has a basic clock rate of 2 MHz. What constraints is the student likely to face in the choice of microprocessors? What problems might arise with the off-the-shelf DAS when it is used with the selected microprocessor?

Chap. 1 Transducers, Transducer Parameters (Specifications) 117

1.17 In the figure for exercise E1.4.1c, replace the inductor by a resistor R. Determine the frequency value for which $|\overline{E}_o|$ is essentially a linear function of C_s for a source of constant frequency ω_c.

REFERENCES

1. Norton, H. N., *Handbook of Transducers for Electronic Measuring Systems*, Prentice Hall, Englewood Cliffs, NJ, 1969.
2. Norton, H. N., *Sensor and Analyzer Handbook*, Prentice Hall, Englewood Cliffs, NJ, 1982.
3. Warring, R. H., and S. Gibilisco, *Fundamentals of Transducers*, TAB Books, Blue Ridge Summit, PA, 1985.
4. Allocca, J. A., and A. Stuart, *Transducers: Theory and Applications,* Reston Publishing Company (Prentice Hall), Reston, VA, 1984.
5. Sheingold, D. H. (ed.), *Transducer Interfacing Handbook*, Analog Devices, Norwood, MA, 1980.
6. Barney, G. C. *Intelligent Instrumentation*, Prentice Hall International, Englewood Cliffs, NJ, 1985.
7. Jordan, E. C. (ed. in chief), *Reference Data for Radio Engineers*, 7th ed., Howard W. Sams and Co. (Macmillan), 1985.
8. Mattiat, O. E. (ed.), *Ultrasonic Transducer Materials*, Plenum Press, New York, 1971.
9. Trietley, H. L., *Transducers in Mechanical and Electronic Design*, Marcel Dekker, New York, 1986.
10. Neubert, H. K., *Instrument Transducers: An Introduction to Their Performance and Design*, 2nd ed., Clarendon Press, Oxford, England, 1975.
11. Cerni, R. H., and L. E. Foster, *Instrumentation for Engineering Measurement*, John Wiley and Sons, New York, 1962.
12. Giles, A. F., *Electronic Sensing Devices*, George Newnes, London, England, 1966.
13. Garrett, P. H., *Analog Systems for Microprocessors and Microcomputers*, Reston Publishing Co. (Prentice Hall), Reston, VA, 1978.
14. Ott, H. W., *Noise Reduction Techniques in Electronic Systems*, John Wiley and Sons, New York, 1976.
15. Rheinfelder, W. A., *Design of Low-Noise Transistor Input Circuits*, Hayden Book Co., New York, 1964.
16. Morrison, R., *Grounding and Shielding Techniques in Instrumentation*, 2nd ed., Wiley-Interscience, New York, 1977.
17. Parratt, L., *Probability and Experimental Errors in Science*, Science Editions, John Wiley and Sons, New York, 1961.
18. Young, H. D., *Statistical Treatment of Experimental Data*, McGraw-Hill, New York, 1962.
19. National Semiconductor Corp., *Pressure Transducer Handbook*, Santa Clara, CA, 1977.
20. Vergers, C. A., *Handbook of Electrical Noise: Measurement and Technology*, TAB Books, Blue Ridge Summit, PA, 1979.
21. Barnes, J. R., *Electronic System Design: Interference and Noise Control Techniques*, Prentice Hall, Englewood Cliffs, NJ, 1987.
22. Ambrozy, A., *Electronic Noise*, McGraw-Hill International, New York, 1982.
23. VanderZiel, A., *Noise in Measurements*, Wiley-Interscience, John Wiley and Sons, New York, 1976.

24. Sheingold, D. H. (ed.), *Analog-Digital Conversion Notes*, Analog Devices, Norwood, MA, 1977.
25. Hordeski, M., *Transducers for Automation*, Van Nostrand Reinhold Company, New York, 1987.

2

Amplifiers and Their Integration into Signal Processing Systems

2.1 INTRODUCTION

Many sensors and transducers produce relatively small outputs, typically in the millivolt and microvolt ranges. These outputs must be amplified before the analog signal can be accurately converted to the digital code. If a transducer output signal is of the same order of magnitude as the 1 LSB value of the converter, conversion is not possible. An amplifier can increase the span of the signal to be converted to take advantage of the total input range of the converter.

Amplifiers are also used to convert the form of the transducer output, that is, from a voltage to a current or a current to a voltage, for example. Most analog-to-digital converters (ADCs) are designed to respond to input voltages rather than currents. A transimpedance amplifier (described later) that produces an output voltage proportional to an input current can be placed between the transducer and ADC to solve this problem.

Another amplifier application is to provide a level shift or bias for a transducer output. A transducer may produce a signal that varies from –2.5 V to +2.5 V, while the ADC may be designed for an input range of 0 to +5 V. A level-shifting amplifier can provide the necessary interface between the two units.

An amplifier may be used to compensate for the transducer specifications discussed in Chapter 1. For example, the amplifier may be designed to compensate the frequency response of the transducer. If an engineer wanted to widen the effective operating bandwidth, it would only be necessary to design an amplifier that could boost the high-frequency components of transducer output. Also, the transducer output can be linearized using an amplifier whose amplitude response compensates for the nonlinear transducer response.

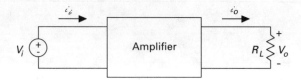

Figure 2-1 Basic amplifier configuration from which various gains are defined.

Some transducers may not be capable of driving the system that follows, that is, the DAS in Fig. 0-5. Additionally, there may be a long cable run between the transducer and the DAS. In these cases the amplifier can provide both a driving function and a matching function. A cable (coax, twisted pair, or other) may require more current than the transducer can deliver, so a driver amplifier becomes necessary. The amplifier can also be used to compensate for cable attenuation. Amplifiers may in fact be placed at both ends of the cable. As is discussed in a later section, optimum performance of a cable (in the sense of preserving transducer output waveshape) is obtained if the cable is driven by a source which has a Thevenin impedance the same as that of the cable (called the characteristic impedance). Also, the cable should be terminated in that same impedance. Amplifiers can be designed to provide this matching function.

Another use of amplifiers is to restore signal size or span after attenuation due to filtering. The Nyquist sampling theorem (described in detail later) requires that when signals are sampled, only frequencies less than one-half the sampling frequency can be present in the analog signal. If a transducer output contains frequencies higher than one-half the sampling frequency, a filter must be used to remove them. When passive filters are used, the input signal is usually attenuated across the desired band. Amplification can then be used to provide the proper signal span to the ADC. If active filters are used, amplifiers are integral parts of the filter.

2.1.1 Amplifier Gain

Amplifier gain is defined as the ratio of the output quantity to the input quantity. The gain may use power, current, and voltage as the quantities of interest. However, most data acquisition systems use either current or voltage quantities, with voltage being the most common of the two. Figure 2-1 shows the basic amplifier configuration from which the various gains are defined. Since the gains may depend upon the value of R_L, any gain values should be accompanied by the amplifier operating conditions which include the value of R_L.

Using the conventions in the figure, the four possible gain functions are defined by the following equations:

1. Voltage gain:

$$A_v = \frac{v_o}{v_i} \qquad (2\text{-}1)$$

For an ideal voltage amplifier, A_v^I is independent of R_L and $i_i = 0$.

Chap. 2   Amplifiers and Their Integration into Signal Processing Systems

2. Current gain:

$$A_i = \frac{i_o}{i_i} \tag{2-2}$$

For an ideal current amplifier, A_i^I is independent of R_L and V_i.

3. Transimpedance:

$$A_{vi} = \frac{v_o}{i_i} \tag{2-3}$$

4. Transadmittance:

$$A_{iv} = \frac{i_o}{v_i} \tag{2-4}$$

The gain expression used in a particular application depends upon the type of input variable and the desired output variable. If the transducer that produces a voltage output is to be connected to a DAS that requires a voltage input, the amplifier that couples these units should be a voltage amplifier. If the transducer generates a current that is proportional to the measurand, the transimpedance amplifier would be appropriate. This latter case is exemplified by a photomultiplier tube monitoring a light source. For most applications the voltage gain is used, so we will confine our discussions to that type.

2.1.2 Amplifier Impedances

The input and output impedances of an amplifier influence the overall gain of a stage because of loading effects. Figure 2-2 shows a general amplifier configuration. The element Z_s represents the source (Thevenin) impedance of the transducer, while Z_L represents the impedance of the circuit to be driven, a cable or a filter for example. The practical amplifier consists of an ideal amplifier component having gain

$$A_v^I = \frac{V_{Ao}}{V_{Ai}} \tag{2-5}$$

and two impedances Z_{in}, the amplifier input impedance, and Z_{out}, the amplifier output impedance. Due to the presence of the input and output impedances and their concomitant voltage divider effects, the overall gain of the system will be less than A_v^I.

The overall gain of the circuit is defined to be

$$A_v^o = \frac{E_o}{E_s} \tag{2-6}$$

Using voltage division,

$$\frac{V_{Ai}}{E_s} = \frac{Z_{in}}{Z_s + Z_{in}}$$

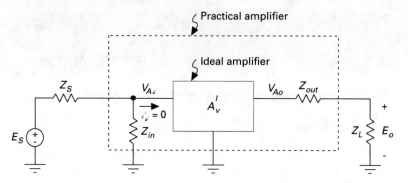

Figure 2-2 A practical amplifier connected between a source and a load.

and

$$\frac{E_o}{V_{Ao}} = \frac{Z_L}{Z_L + Z_{out}}$$

Expanding Eq. (2-6) we have

$$A_v^o = \frac{E_o}{E_s} = \left(\frac{E_o}{V_{Ao}}\right)\left(\frac{V_{Ao}}{V_{Ai}}\right)\left(\frac{V_{Ai}}{E_s}\right)$$

$$= \left(\frac{Z_L}{Z_L + Z_{out}}\right)(A_v^I)\left(\frac{Z_{in}}{Z_s + Z_{in}}\right)$$

$$A_v^o = \frac{Z_L Z_{in}}{(Z_L + Z_{out})(Z_s + Z_{in})} A_v^I \qquad (2\text{-}7)$$

In order to obtain the maximum gain, the ideal situation would be to have $Z_{out} = 0$ and $Z_{in} = \infty$. (Justify this using Eq. (2-7).)

The following example demonstrates the effect of these loading problems.

Example 2-1

A microphone is used as a sound transducer, and the output is to be amplified and applied to a system that has an effective impedance of 600 Ω (resistive). Compute the overall gain and the output voltage. See Fig. 2-3. Equation (2-7) gives

$$A_v^o = (100)\left(\frac{600}{600 + 100}\right)\left(\frac{10^6}{10^6 + 2 \times 10^5}\right) = (100)(0.857)(0.833) = 71$$

In this example the loading effects are quite pronounced, the ideal gain of 100 being reduced to 71. The output voltage is then

$$E_o = (71)(50 \times 10^{-3}) = 3.57 \text{ V}$$

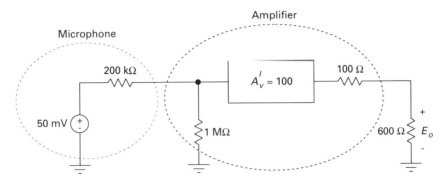

Figure 2-3 System for Example 2-1.

Exercises

E2.1a For Example 2-1, change Z_{in} to 10 MΩ and Z_{out} to 10 Ω and compute A_v^o and E_o. Repeat for Z_{in} = 100 MΩ and Z_{out} = 5 Ω. Also compute the percentage error for each case.

E2.1b For Example 2-1, what value of A_v^i would be required to have an overall gain of 100?

2.2 AMPLIFIERS AND THE TRANSMISSION LINE PROBLEM

In those cases where the computer is located some distance from the transducer, it may be possible or even preferable to make the analog-to-digital conversion prior to sending the data along the cable. In practice the transmission of data in digital form is often preferred, due to the inherently better noise immunity which often provides larger signal-to-noise ratio properties. The connection between the transducer and amplifier or amplifier and converter system may be as simple as a printed circuit trace or as elaborate as a coaxial, twisted pair, or shielded pair cable. In any case a transmission line is involved, regardless of the distance involved or whether the analog or digital transmission mode is selected. Optical transmission systems are also being used which overcome the induced noise problem.

Two possible configurations for data transmission are shown in Fig. 2-4. Both schemes involve amplifiers and transmission lines, and both suffer from signal distortion and extraneous signal generation. There are two types of errors that can be introduced by a transmission line. Both will be briefly discussed in this section. The errors are as follows:

1. Distortion caused by a mismatched transmission line, that is, a transmission line not terminated at one end in its characteristic impedance (to be defined shortly). This type of distortion can also cause extraneous signals to be generated and is caused by signal reflections at the ends of the line. This is of primary concern in both analog and digital transmission over metallic conductor systems.

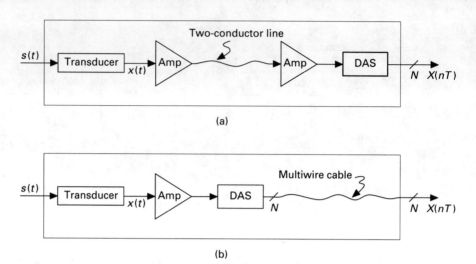

Figure 2-4 Possible configurations for data transmissions. (a) Analog data transmission over transmission line. (b) Digital data transmission over transmission line.

2. Distortion caused by line losses in the transmission line, called dispersion losses, occur in both electrical and optical systems. This effect is caused by the fact that the value of the propagation constant for a line is different for each frequency component of a signal.

2.2.1 Distortion Caused by Transmission Line Mismatch

In this text it will only be possible to give a brief discussion of transmission lines. Detailed discussions are found in references 8–14, 20, and 21. Here we give only those parts of the theory that will allow us to describe the distortion problems.

Any two-conductor system can be modeled by cascading incremental length sections of passive circuit elements. The resulting models are shown in Fig. 2-5. The various terms and symbols are identified as follows. The Δz term is an increment of length along the cable. Line inductances L_1 and L_2 in part (a) of the figure are the inductances per meter of line of each of the conductors of the line. These inductances arise because the currents flowing in the conductors produce magnetic fields which generate the inductive effects. Since the two conductors have the same current flow at each increment, L_1 and L_2 are in series, so it is easier to modify the model by adding L_1 and L_2 to produce the equivalent shown in the (b) part of the figure, $L = L_1 + L_2$. Similarly, the resistances R_1 and R_2 represent the resistances per meter of line of the two conductors. Since these resistors are also in series, it is usual practice to replace them by a single resistance of value $R = R_1 + R_2$. The capacitance C is the capacitance per meter of line between the two conductors, and G is the conductance per meter of line and represents the dielectric leakage current effects. Thus if we multiply each line parameter R, L, G, C by Δz, the result is an incremental electrical component, as shown in Fig. 2-5(b).

There are two important parameters that affect the transmission of signals along a line: the characteristic impedance and the phase velocity. Without derivation we simply

Chap. 2   Amplifiers and Their Integration into Signal Processing Systems   125

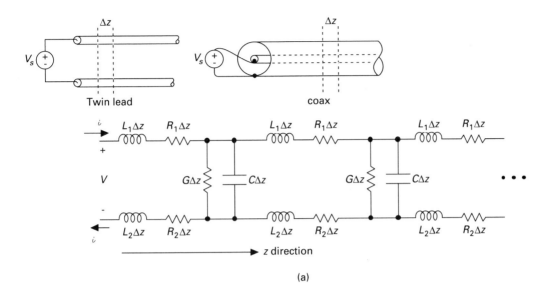

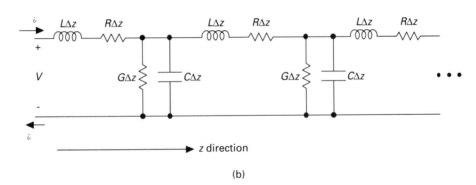

Figure 2-5 Models of two conductor transmission lines. (a) Basic transmission line incremental model. (b) Modified transmission line incremental model.

state the results for these parameters, using the sinusoidal steady state formulation and then show how they affect a signal:

$$\text{Characteristic impedance} \equiv Z_o = \left(\frac{R + j\omega L}{G + j\omega C} \right)^{1/2} \text{ ohms} \quad (2\text{-}8)$$

This is "characteristic" of a given line because it depends only upon the "per meter" line constants in the model. For many cases, R and G may be set to zero with little error so that

$$Z_o = \sqrt{L/C} \text{ ohms}$$

For example, when one speaks of a 50-Ω coaxial cable, this is the approximate Z_o:

$$\text{Phase velocity} \equiv V_p$$

$$= \left[\frac{2}{LC - \dfrac{RG}{\omega^2} + \left(\dfrac{R^2 G^2}{\omega^4} + \dfrac{L^2 G^2 + R^2 C^2}{\omega^2} + L^2 C^2\right)^{1/2}} \right]^{1/2} \frac{\text{meter}}{\text{second}} \quad (2\text{-}9)$$

Physically, the phase velocity is the velocity of propagation along the line of a sinusoid of frequency ω. For completeness we give a third parameter, the attenuation, which is accounted for by multiplying the signal amplitude at location z by the exponential factor

$$e^{-\alpha z} \quad (2\text{-}10)$$

where z is the length of the line and the attenuation constant α is defined by

$$\alpha \equiv \left[\frac{-\omega^2 LC + RG + \left(R^2 G^2 + \omega^2(L^2 G^2 + R^2 C^2) + \omega^4 L^2 C^2\right)^{1/2}}{2} \right]^{1/2} \frac{\text{neper}}{\text{meter}} \quad (2\text{-}11)$$

One should be aware of the fact that the line parameters R, L, G, and C are also functions of frequency, but this will not be pursued here, other than to point out that R and G generally vary as the square root of the frequency.

At any point on the line the current and voltage are given by (where $d = 0$ is at the load and d is a positive value):

$$\overline{I}(d) = \overline{I}_{inc} e^{\gamma d} + \overline{I}_{refl} e^{-\gamma d} + I_{dc} = \frac{\overline{V}_{inc}}{Z_o} e^{\gamma d} - \frac{\overline{V}_{refl}}{Z_o} e^{-\gamma d} + I_{dc}$$

$$\overline{V}(d) = \overline{V}_{inc} e^{\gamma d} + \overline{V}_{refl} e^{-\gamma d} + V_{dc}$$

$$\text{where } \gamma \equiv \sqrt{(R + j\omega L)(G + j\omega C)} \text{ neper/meter} \quad (2\text{-}12)$$

In these results $\overline{V}_{inc}$ and $\overline{I}_{inc}$ are the components traveling to the right, while $\overline{V}_{refl}$ and $\overline{I}_{refl}$ are the components traveling to the left. Note that "moving to the right" means d decreases. We have also provided for a DC component to exist in general.

Of interest specifically are the voltages and currents at the load and at the source, for which the preceding become

$$\overline{I}_L = \overline{I}(0) = \overline{I}_{inc} + \overline{I}_{refl} + I_{dc} = \frac{\overline{V}_{inc}}{Z_o} - \frac{\overline{V}_{refl}}{Z_o} + I_{dc}$$

$$\overline{V}_L = \overline{V}(0) = \overline{V}_{inc} + \overline{V}_{refl} + V_{dc}$$

$$\overline{I}_S = \overline{I}(l) = \overline{I}_{inc} e^{\gamma l} + \overline{I}_{refl} e^{-\gamma l} + I_{dc}$$

$$\overline{V}_S = \overline{V}(l) = \overline{V}_{inc} e^{\gamma l} + \overline{V}_{refl} e^{-\gamma l} + V_{dc}$$

Chap. 2 Amplifiers and Their Integration into Signal Processing Systems

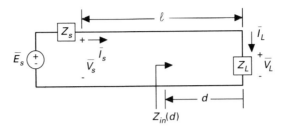

Figure 2-6 General configuration for a terminated transmission line.

Consider a section of line of total length l that is terminated in an impedance Z_L and driven by a generator with a Thevenin impedance $\overline{Z}_s$. This configuration is shown in Fig. 2-6. The transmission line is represented by the lines for simplicity; the R, L, G, C parameters are implied. If a voltage $\overline{E}_S$ is applied to the line, it propagates to the right with a velocity V_p where a signal $\overline{V}_{inc}^L$ reaches the load. At the load, part of the voltage is reflected according to the value of the incident voltage:

$$\overline{V}_{refl}^L = \Gamma_L \overline{V}_{inc}^L \tag{2-13}$$

where

$$\Gamma_L = \frac{\overline{Z}_L - Z_o}{\overline{Z}_l + Z_o} = \frac{\overline{V}_{refl}}{\overline{V}_{inc}} \tag{2-14}$$

The reflected component then propagates back toward the source, where another reflection occurs, given by

$$\overline{V}_{refl}^S = \Gamma_S \overline{V}_{inc}^S \tag{2-15}$$

where

$$\Gamma_S = \frac{\overline{Z}_S - Z_o}{\overline{Z}_S + Z_o} \tag{2-16}$$

Thus multiple reflections occur back and forth on the line, and considerable signal distortion occurs as the signals combine.

Also of interest is the impedance seen along the line looking toward the load. This is an input impedance denoted by the general symbol $\overline{Z}_{in}(d)$, where d is the distance (positive) from the load to the point of interest, as shown in Fig. 2-6. In the sinusoidal steady state (leaving off the DC components)

$$\frac{\overline{V}(d)}{\overline{I}(d)} = \overline{Z}(d) = \frac{e^{\gamma d} + \Gamma_L e^{-\gamma d}}{e^{\gamma d} - \Gamma_L e^{-\gamma d}} \cdot Z_o = \frac{1 + \dfrac{Z_o}{\overline{Z}_L}\tanh \gamma d}{\tanh \gamma d + \dfrac{Z_o}{\overline{Z}_L}} \cdot Z_o \tag{2-17}$$

Although the preceding results have been given for the sinusoidal steady state, the equations still give us a qualitative approach to any waveform, since we may determine what will happen to each individual frequency component of any general waveform. There are important special cases.

Case I

Low frequency or large losses (large R and G). In either instance $R \gg \omega L$ and $G \gg \omega C$. Then

$$Z_o = \sqrt{R/G} \quad \text{ohms (real)}$$

$$V_p \to \infty$$

$$\alpha = \sqrt{RG} \; \frac{neper}{meter}$$

$$\gamma = \sqrt{RG} \; \frac{neper}{meter}$$

$$\Gamma_L = \frac{\overline{Z}_L - \sqrt{R/G}}{\overline{Z}_L + \sqrt{R/G}}$$

$$\Gamma_S = \frac{\overline{Z}_S - \sqrt{R/G}}{\overline{Z}_S + \sqrt{R/G}}$$

$$\overline{Z}_{in}(d) = \frac{e^{\sqrt{RG}\,z} + \Gamma_L e^{-\sqrt{RG}\,z}}{e^{\sqrt{RG}\,z} - \Gamma_L e^{-\sqrt{RG}\,z}} \sqrt{R/G} \tag{2-18}$$

Case II

High frequency or small losses (small R and G). In either instance $R \ll \omega L$ and $G \ll \omega C$. Then

$$Z_o = \sqrt{L/C} \quad \text{ohms (real)}$$

$$V_p = \frac{1}{\sqrt{LC}} \; \frac{meter}{second}$$

$$\alpha = 0$$

$$\gamma = j\omega\sqrt{LC} \; \frac{neper}{meter}$$

$$\Gamma_L = \frac{\overline{Z}_L - \sqrt{L/C}}{\overline{Z}_L + \sqrt{L/C}}$$

$$\Gamma_s = \frac{\overline{Z}_s - \sqrt{L/C}}{\overline{Z}_s + \sqrt{L/C}}$$

$$\overline{Z}_{in}(d) = \frac{\overline{Z}_L + j\sqrt{L/C} \tan \omega\sqrt{LC}\,z}{j\overline{Z}_L \tan \omega\sqrt{LC}\,z + \sqrt{L/C}} \tag{2-19}$$

For this case, expressions for Z_o in terms of physical line dimensions can be found in Chapter 29 of [21] and in [10]. Typical numerical values of Z_o range from 50 to 500 ohms in the lossless case. This case is assumed unless otherwise stated.

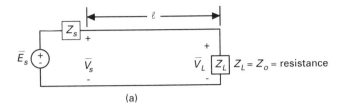

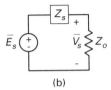

Figure 2-7 Matched lossless transmission line. (a) Basic line configuration. (b) Equivalent circuit.

For many applications the results of Case II can be used because losses are usually small for conductors. Even if there are appreciable losses, the results using these equations give good qualitative insights on line performance. We next consider some examples that are important in practice.

Example 2-2

The obvious way to eliminate distortion caused by reflections is to have Γ_L equal to zero. This is done by setting $\bar{Z}_L = Z_o$ (see Eq. (2-14)). With this condition the system configuration is as shown in Fig. 2-7(a). This is called a matched line. For Case II, Z_o is real (resistive), and since $\Gamma_L = 0$, the impedance for any length is, from Eq. (2-17), a constant equal to Z_o. Thus the source terminals at V_S see a constant load of value Z_o. This leads to the equivalent circuit shown in part (b) of the figure. The voltage V_S is also the load voltage except for a time delay of value

$$T = \frac{l}{V_p} = l\sqrt{LC} \text{ seconds}$$

where l is the line length. Note that these results are independent of frequency for the lossless ($R = G = 0$) case. The amplitude reaching the load, however, is for $\bar{Z}_S = R_S$,

$$|V_L| = |V_S| = |E_S|\frac{Z_o}{Z_o + R_S}$$

Thus if the source is an analog signal, the amplitude is reduced. To compensate for this we must either increase E_S or else have an amplifier at the load with $Z_L = Z_o$. This means that in order to preserve the matched line the amplifier input impedance must be effectively Z_o, usually a small number (50–500 ohms). For digital signals, the voltage division must not reduce $|V_L|$ below the logic threshold level. Also, if a digital system is at the load, the logic input impedance must be Z_o to maintain the matched condition. Networks would be required to provide this value at the logic gate input. It is not practical to make $Z_L = Z_o$ for the large number of gates in a computer; however, making $Z_S = Z_o$, or close to it, often solves the problem.

If the source E_S is an amplifier, the value of $\bar{Z}_S$ (usually resistive) is typically small, say 5 ohms or so, so the division effect is not severe. If the source is a logic gate, $\bar{Z}_S$ is also small, but for Z_o of 100 Ω, a single 5-volt signal must be capable of supplying 50 mA, thus putting a severe load on the logic. There are some techniques for solving this dilemma, but these will not be pursued here [13, 14].

Example 2-3

As an example of problems that arise, consider two logic gates connected by a transmission line (PC board traces or wires) as shown in Fig. 2-8. We consider the lossless case again and assume that the logic gate output impedance is small and the logic gate input impedance is large, but in both cases larger than Z_o. The resulting series of waveforms is shown in the lower part of the figure, and the waveforms are described as follows. Although this is not sinusoidal steady state, by using resistive components we can apply the general ideas.

At $t = 0$ a pulse of amplitude A (less than the source amplitude by the same factor as in the preceding example, it turns out) is sent out as shown in plot (1). The pulse propagates down the line with velocity V_p, and the total time to travel length l is given by

$$T = l/V_p$$

At time $t = t_1 < T$ the pulse has moved to position z_1 as shown in plot (2). Now at time $t = T$ the pulse reaches the input of the gate at the load. Since the gate input resistance is large, Eq. (2-13) yields, for $R_l \gg Z_o$,

$$\Gamma_L = \frac{R_L - Z_o}{R_L + Z_o} \approx 1$$

Thus there is a reflected pulse of amplitude—here we use Eq. (2-12)—giving

$$V_{refl}^L = (1)(A) = A$$

Then $V_L = V_{inc}^L + V_{refl}^L + V_{dc} = A + A + 0 = 2A$. We assume that the initial load voltage (V_{dc}) is zero so that the total load voltage is now essentially $2A$, as shown in plot (3).

The reflected pulse of amplitude A travels to the left as indicated in plot (4), and when the pulse reaches the left gate it sees a reflection coefficient given by Eq. (2-15),

$$\Gamma_S = \frac{R_S - Z_o}{R_S + Z_o}$$

Since we have assumed $R_s > Z_o$, Γ_s will be positive, so that a new pulse is reflected back toward the load end, which gives a reflected component at the source of value

$$V_{refl}^S = \Gamma_S A$$

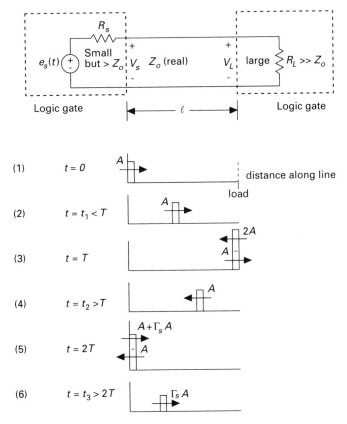

Figure 2-8 Effect of mismatched line on logic pulses.

The voltage at the sending end is now a pulse of amplitude

$$V_S = V_{inc} + V_{refl} + V_{dc} = A + \Gamma_S A + 0 = (1 + \Gamma_S)A$$

However, the pulse reflected from the source end now travels back toward the load end gate and has the effect of being another pulse (plot (6)). The reflected pulse this time will have an amplitude of

$$V_{refl}^L = \Gamma_L(\Gamma_S A) = \Gamma_L \Gamma_S A \approx \Gamma_S A$$

The pulse amplitude seen by the load end gate is now

$$V_L = V_{refl} + V_{inc} + V_{dc} = \Gamma_L \Gamma_S A + \Gamma_S A + 0 = \Gamma_S(1 + \Gamma_L)A \approx 2\Gamma_S A$$

If Γ_S is large enough the gate may see another pulse which was not really sent by the source E_S!

One obvious way for avoiding this problem is to match the transmission line at the input or source end. This can be accomplished by putting a resistance in series with the gate so that

$$R_S' = R_S + R_{add} = Z_o$$

This makes $\Gamma_S = 0$. Even if this is the matching technique used, it must be remembered that the input of the gate at the load must be capable of handling voltages two times the largest pulse sent down the line.

Although both of the examples used digital data for ease of explanation, one can easily imagine what will happen to an analog waveform. At the load end of the line the total voltage is the sum of the incident voltage and the reflected component. Thus if Z_L represents the input of an amplifier, a distorted signal results at the amplifier input. In this instance, matching must be done at the load end.

The examples also assumed that the transmission line was described by Case II conditions. If Case I were examined instead, the only difference is that the signals travel with an attenuation $e^{-\alpha z}$. However, the reflection equations are still valid, since Z_o is real for Case I also.

Exercises

E2.2.1a For the case $Z_o = 200\ \Omega$, $C = 30\ \text{pF/m}$, $|E_S| = 10\ \text{V}$ pulse, plot the values of the seven pulses in Fig. 2-8. Use $R_L = 100\ \text{k}\Omega$ and $R_S = 50\ \Omega$. Assume the pulse width is small compared with T. The seventh pulse is at the right end of the line.

E2.2.1b Suppose a digital system uses level logic rather than pulse logic. Let E_S be a step function of amplitude V_o. Plot the seven graphs shown in Fig. 2-8, using the same assumptions as in Example 2-3. Note that V_{dc} is not zero now, since the step value of total voltage remains on the line. This will be the case at both ends. The seventh step is at the right end.

2.2.2 Distortion Caused by Transmission Line Dispersion

The second source of signal distortion is caused by the loss components of the transmission line, and it is called dispersion. If the values of R and G are not negligible, but are not large enough so that Case I is applicable, the phase velocity V_p is a function of frequency—see Eq. (2-9). When the phase velocity of a line is a function of frequency, the line is said to be dispersive.

The effect this produces on a waveform is easily explained. If a waveform consists of several frequency components and each component propagates at a different velocity, the combination of the components will be different at all points along the line. This is illustrated in Fig. 2-9, where we consider a fundamental and third harmonic as shown in (a), which is at the source end. The two frequencies combine to give the total waveshape indicated. As the two components propagate to the load they do not have the same time relationship, so the total waveform has been changed considerably, as shown in (b).

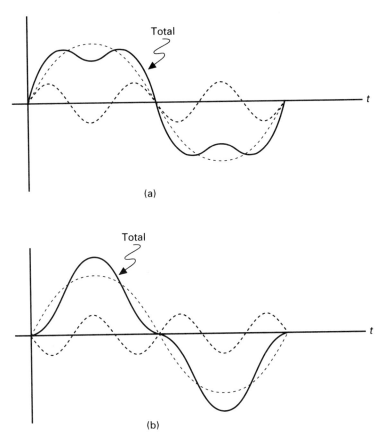

Figure 2-9 Effect of dispersion on an analog signal. (a) Total waveform at source. (b) Total waveform at load.

When pulses are transmitted along a dispersive line, the effect of dispersion is to spread out the pulse in time. The term dispersion has exact meaning here. Suppose we transmit two pulses from a source, as shown in Fig. 2-10(a). As the pulses propagate down the line they begin to disperse, as shown in part (b). Finally, at the load the dispersion phenomenon has merged the two pulses into a single one, as shown in (c). Thus a gate at the load would see only one pulse. From this result it is easy to see that as the data rate is increased the usable transmission distance decreases. This concept is valid for any dispersive data channel.

The conclusion from this subsection is that if an amplifier is inserted in a transmission system, any distortion present is not necessarily due to the amplifier. If exact waveshape is important in a given system, cable selection is worthy of careful consideration.

Exercises

E2.2.2a A table of coaxial cables lists an RG58B type as having $Z_o = 53.5 \, \Omega$ and a capacitance of 93.5 pF per meter. What is the line inductance L per meter?

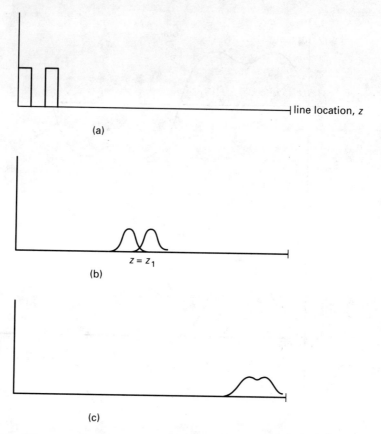

Figure 2-10 Effect of dispersion on digital transmission. (a) Total waveform at source. (b) Pulses dispersing at an intermediate line position. (c) Pulses appearing at the load.

E2.2.2b An RG58B cable operated at 50 MHz has an attenuation of 0.093 dB per meter. What is the attenuation constant α for this cable? (Hint: See Eq. (2-10) for $z = 1$ meter.)

E2.2.2c For the cable of E2.2b, suppose that $G = 0$. Find the value of R. $Z_o = 53.5\,\Omega$, $C = 93.5$ pF/m. See Eq. (2-19) for Z_o and Eq. (2-11) for α. The Z_o given is the high-frequency value.

E2.2.2d For the system shown in Fig. E2.2d, determine the smallest and largest values of R_L such that subsequent reflections back to R_L will produce less than ± 2.5 V total voltage at R_L. (Hints: $\overline{Z}_s = 0$; the 5-V step starts out at the left end of the line and travels to R_L, where a fraction Γ_L is reflected. Assume the transmission line is lossless.)

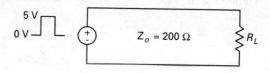

Figure E2.2d

E2.2.2e If you were to put a cable between two amplifiers, what would you do to minimize errors due to
  a. line reflections
  b. dispersion
  c. noise

2.3 AMPLIFIER TYPES AND THEIR GENERAL CHARACTERISTICS

There are three important types of amplifiers used in data gathering systems: the operational amplifier (op amp), the instrumentation amplifier, and the isolation amplifier. The remainder of this chapter is devoted to describing these amplifiers and their characteristics, and to investigating the effects these characteristics have on signal processing. The three symbols for these amplifiers are shown in Fig. 2-11 and will next be described with respect to each amplifier type. There are more abbreviated versions of the symbols shown in the figure that simply remove all the ground lines and symbols and leave only the voltage lines. Ground currents frequently constitute an important part of the system current, so these lines must not be neglected when considering all voltages and currents in the amplifier.

The operational amplifier (op amp) is perhaps the most fundamental amplifier type. In fact, the other amplifiers often consist of three or more op amps in their designs. Active filters (to be discussed in a later chapter) frequently have four op amps, and the op amps make it possible to eliminate inductors in passive network designs. The op amp is thus a basic building block for many signal processor components. Their versatility allows a designer to construct networks that meet many system requirements.

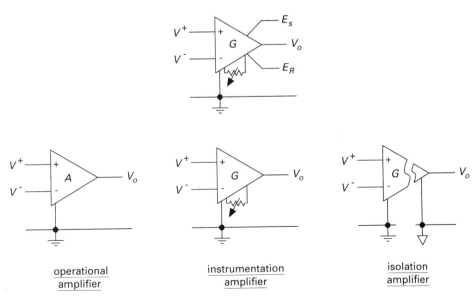

Figure 2-11 Symbols for three basic amplifier types used in data acquisition systems.

The basic specifications of the operational amplifier are very high gain ($A \approx 10^5$), high input impedance ($Z_{in} \approx 10^6$ to 10^{12} Ω), and low output impedance ($Z_{out} \approx 10$ to 100 Ω). For a full operating gain of 10^5 the 3-dB frequency is 10 hertz, and negative feedback—to be described in the following section—is used to obtain more usual gain values of 100, thus increasing the bandwidth to approximately 10 kHz. The wide range of input impedance occurs from the different solid-state implementations. The 10^6 order is obtained from bipolar transistor implementations, whereas the 10^{12} order results from using FET input transistors. This latter op amp implementation is frequently referred to as the BIFET structure, since the output stages are still bipolar to preserve a low output impedance. The fundamental performance equation for the op amp is

$$V_o = A\,(V^+ - V^-) \qquad (2\text{-}20)$$

When it is necessary to have an amplifier with tightly controlled specifications such as shielding for noise, dedicated gain, and high common mode rejection (to be defined later) the instrumentation amplifier is usually selected. This amplifier type has a complex circuit that offers selectable options. Voltage gains are usually adjustable from 1 to 1000 (indicated by the variable resistor on the symbol in Fig. 2-11). These devices have high input impedances on the order of 10^9 to 10^{12} Ω, low output impedance values on the order of 10^{-2} to 10^{-3} Ω, and very good noise specifications at about $1\,\mu V$ to $20\,\mu V$ at the input. In general, the instrumentation amplifier is a more expensive unit, but it is a high-performance amplifier. These are also integrated circuit units. The basic performance equations for the two instrumentation amplifier diagrams shown in the figure are as follows:

1. Bottom diagram:

$$V_o = G\,(V^+ - V^-) \qquad (2\text{-}21)$$

2. Top diagram:

$$E_s - E_R = G\,(V^+ - V^-) \qquad (2\text{-}22)$$

This last equation may seem a little strange, since it does not involve V_o, but in practice the lines E_s (sense line) and E_R (reference line) are connected to other components or circuit points that can be related to V_o. Applications of these two lines will be covered later. A simple example, however, is the case where E_R is connected to ground (0 volts) and E_s to V_o, resulting in operation the same as given by Eq. (2-21).

In many systems it is necessary to have two different references for voltage or to have two parts of a system that are not conductively (metal) connected. For example, digital current pulses in a ground line may couple as noise or common mode voltage signals (defined later) into analog portions of a system. This requirement can be solved using an isolation amplifier (which in reality could be thought of as a subclass of the instrumentation amplifier). One important application of this type of amplifier is in the medical instrumentation field, where it is necessary to isolate a patient from other parts of a system. Such circuit isolation also makes it possible for input signals to have very high DC levels (to thousands of volts) with respect to the input ground and still have safe operation on the output side. Physically, this means that the two ground symbols may in actuality have a significant voltage difference between them. The two primary methods for obtaining this isolation (described in more detail later) are as follows:

1. Transformer coupling, which frequently involves signal choppers and DC-DC converters
2. Optical coupling, which uses photodiodes and phototransistors

Operationally, the isolation amplifier has basic characteristics that are the same as the instrumentation amplifier, with the addition of a high input-output voltage differential (separation). The isolation amplifier diagram shown in Fig. 2-11 is described by

$$V_o = G(V^+ - V^-) \tag{2-23}$$

It is important to remember that the two power supplies applied to the two isolation amplifier sections must also be isolated or separate, otherwise the two grounds would be effectively joined together by the power supply common ground!

Exercises

E2.3 From a manufacturer's product guide select one operational amplifier, one instrumentation amplifier, and one isolation amplifier and compare the values of common parameters listed therein. See Appendix 2.1.

2.4 CHARACTERISTICS AND APPLICATIONS OF IDEAL OPERATIONAL AMPLIFIERS

In order to determine the errors introduced by a network that contains operational amplifiers it is necessary to be able to compute the network response under ideal conditions, such conditions being commonly assumed for initial designs. In practice, the operational amplifier has near ideal performance, so the initial design performance is actually quite close to predictions based on ideal characteristics.

The operational amplifier—or op amp—received its name over forty years ago when analog computers were the dominant electronic computing tool. The name resulted as a natural consequence of the fact that the amplifier configurations used were able to perform most of the algebraic and calculus operations, and to solve differential and integral equations. In the 1950s and early 1960s the op amp was in a package measuring about 12 cm high by 5 cm wide by 15 cm deep, and it consumed some 5 watts of power. The cost of those units was about one hundred dollars. As the integrated circuit came onto the scene, the size and price of the op amp decreased by factors of 10 and 100 respectively, until today four op amps can be purchased in a single package 2 cm by 0.5 cm by 0.3 cm, at a cost of under one dollar. High-performance discrete-component op amps are rarely used at present.

A basic internal structure of the ideal op amp symbol shown in Fig. 2-11 uses two ideal amplifier sections of the type shown in Fig. 2-2 whose performances are governed

138 Amplifiers and Their Integration into Signal Processing Systems Chap. 2

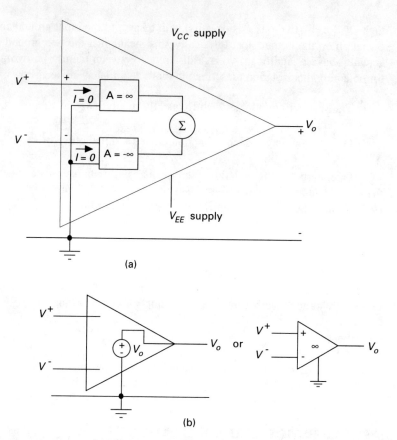

Figure 2-12 Ideal operational amplifier diagrams. (a) Functional block diagram. (b) Simplified equivalent ideal models.

by Eq. (2-5). This internal structure is shown in Fig. 2-12(a). In later sections this ideal structure will be modified to account for nonideal performance. This functional diagram produces the same overall function as Eq. (2-20), since

$$V_o = AV^+ - AV^- = A(V^+ - V^-)$$

except that for $A = \infty$ we seem to have $V_o = \infty$. This paradox can be resolved by setting

$$V^+ - V^- = 0$$

or,

$$V^+ = V^- \tag{2-24}$$

since the product $\infty \cdot 0$ can have any finite limit value. Negative feedback determines the value. Equation (2-24) illustrates the principle of the virtual short, since in this ideal case $V^+ = V^-$ at all times. The adjective virtual means that although $V^+ = V^-$ they are not

Chap. 2   Amplifiers and Their Integration into Signal Processing Systems   139

physically connected, so no current flows between them. One practical consequence of this principle is that if one applies different voltages at V^+ and V^- the output V_o goes to the limit of operation, which is either $V_{CC} - 1$ volt if $V^+ > V^-$ or $V_{EE} + 1$ volt if $V^+ < V^-$ (V_{EE} is usually negative or zero). These are called output saturation voltages.

Figure 2-12(a) also shows that the currents into the plus and minus terminals are zero, since the currents into the internal amplifiers are zero. This of course also means that the input impedances at the input terminals of the op amp are both infinite. Also, the output impedance is zero, since summing two voltages ideally produces a voltage source.

Two simplified equivalent networks are shown in Fig. 2-12(b), which will be used for the work of this section.

Exercises

E2.4a Determine the output voltage of the ideal op amp network shown in Fig. E2.4a. Repeat if the two 5-V supplies are interchanged.

E2.4b Sketch the outputs of the three ideal op amp systems shown in Fig. E2.4b.

2.4.1 Inverting Amplifier Configuration

One of the most important basic applications is shown in Fig. 2-13. Note that for convenience we have placed the negative terminal on top, since negative feedback components are required and easier to draw above the amplifier. This is called the inverting amplifier configuration. From the principle of the virtual short, since V^+ is at 0 volts V^- is also at 0 volts. The physical mechanism that forces V^- to zero is that the amplifier delivers whatever feedback current I_f is required to drive that voltage to zero. This of course means

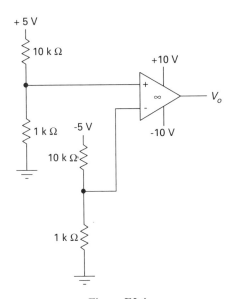

Figure E2.4a

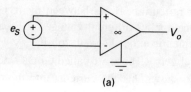

(a)

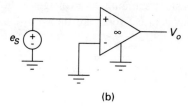

(b)

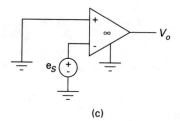

(c)

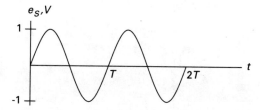

Figure E2.4b

that the voltage across Z_f is V_o. Incidentally, we use capital letters here to represent the Laplace (s-domain) representations of the components and variables. They could also be sinusoidal steady state quantities. Since the ideal op amp has zero input current, KCL at the negative terminal yields (sum of currents leaving node)

$$-I_1 - I_f = 0$$

Using the element volt-ampere characteristics, this may be written

$$-\frac{E_1 - 0}{Z_1} - \frac{V_o - 0}{Z_f} = 0$$

Chap. 2   Amplifiers and Their Integration into Signal Processing Systems   141

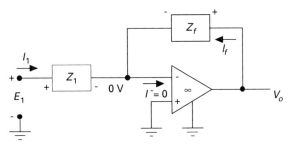

Figure 2-13   Inverting feedback amplifier configuration for an ideal operational amplifier.

Solving for V_o:

$$V_o = -\frac{Z_f}{Z_1} E_1 \tag{2-25}$$

From this the ideal gain is

$$G^I = \frac{V_o}{E_1} = -\frac{Z_f}{Z_1} \tag{2-26}$$

Example 2-4

Suppose Z_f and Z_1 are resistors. Then

$$G^I = -\frac{R_f}{R_1}$$

Thus we have a simple inverting amplifier, for which

$$V_o = -\frac{R_f}{R_1} E_1$$

As a prelude to a later section, note that this result tells us that the output is independent of frequency. In practical op amps the gain A is frequency-dependent and finite, so this simple result is only valid at "low" frequencies. Quantitative values of "low" will be obtained later.

Example 2-5

Assume that Z_f is a capacitor, $\frac{1}{sC_f}$, and Z_1 is a resistor R_1. Equation (2-26) yields

$$G^I = -\frac{1}{sR_1C_f}$$

or,

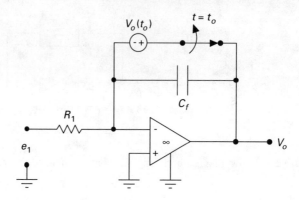

Figure 2-14 Ideal integrator with initial condition.

$$V_o = -\frac{1}{sR_1C_f}E_1 = -\frac{1}{R_1C_f}\frac{E_1}{s} \qquad (2\text{-}27)$$

If we take the inverse Laplace transform we obtain

$$v_o(t) = \frac{-1}{R_1C_f}\int_0^t e_1(\tau)d\tau \qquad (2\text{-}28)$$

This configuration thus performs the integration operation. When it is desirable to start the integration at a time $t = t_0$, we rewrite the preceding equation as

$$v_o(t) = -\frac{1}{R_1C_f}\left[\int_0^{t_0} e_1(\tau)d\tau + \int_{t_0}^t e_1(\tau)d\tau\right]$$

$$= \frac{-1}{R_1C_f}\int_0^{t_0} e_1(\tau)d\tau - \frac{1}{R_1C_f}\int_{t_0}^t e_1(\tau)d\tau = v_o(t_0) - \frac{1}{R_1C_f}\int_{t_0}^t e_1(\tau)d\tau$$

The quantity $v_o(t_0)$ is the initial condition on $v_o(t)$ and can be obtained by placing a voltage source and switch across the capacitor, as shown in Fig. 2-14. The switch opens at $t = t_0$ when integration of $e_1(t)$ is to begin.

While the ideal integrator is an exciting result, it does suffer from practical op amp limitations. The primary effect that one must be aware of is the effect of nonzero amplifier input current. Even with e_1 grounded, some of the input terminal current will flow through the capacitor and produce a change in the output. In most cases the output continues to change until it reaches an op amp saturation value described previously. The initial condition voltage source prevents this from happening until the switch opens.

Example 2-6

Determine the impedance seen by the source E_1 in Fig. 2-13.

Since the negative terminal is at zero volts because of the virtual short principle, we have:

Chap. 2 Amplifiers and Their Integration into Signal Processing Systems

$$Z_{in} = \frac{E_1}{I_1} = \frac{E_1}{\left(\dfrac{E_1 - 0}{Z_1}\right)} = Z_1$$

This result allows us to compute the current that the source must be capable of supplying to the amplifier.

Exercises

E2.4.1a For the network shown in Fig. E2.4.1a, determine the gain in the s-domain and the output as a function of the input in the time domain. Discuss initial conditions if required. Discuss the effect of nonzero amplifier input current when the input voltage source is set to zero (see also exercise E2.4.1f).

E2.4.1b In order to have an amplifier with an ideal gain of −10, the ratio of the feedback resistors is 10. Discuss the amplifier performance and any problems that may result from using the following combinations:

R_1, Ω	R_f, Ω
1	10
100	1000
10^4	10^5
10^6	10^7

E2.4.1c For the integrator described in Example 2-5, discuss the differences in performances for $R_1 = 1$ kΩ, $C_f = 1$ μF and for $R_1 = 100$ kΩ, $C_f = 10$ μF. The supply voltages are $V_{CC} = |V_{EE}| = 10$ volts.

E2.4.1d Derive the ideal gain of the inverting amplifier for the case where Z_1 is a capacitor C_1 and Z_2 is a capacitor C_2. How would the performance of this amplifier compare with the case of two resistors?

E2.4.1e Sketch the ideal frequency response characteristic of the amplifier configuration shown in Fig. E2.4.1e.

E2.4.1f For the network in Fig. E2.4.1e, interchange the input and feedback impedances and plot the ideal frequency response.

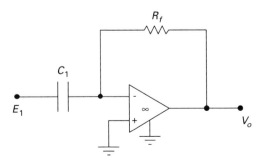

Figure E2.4.1a

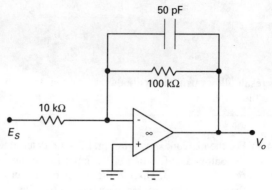

Figure E2.4.1e

2.4.2 Noninverting Amplifier Configuration

Another important basic configuration is the noninverting configuration shown in Fig. 2-15. One advantage of this configuration is that the input impedance is very high, since the source is required to deliver current into the amplifier positive terminal. In the ideal case the terminal current is zero, so the impedance at the terminal is infinite.

The ideal performance of this configuration is obtained by first applying the principle of the virtual short, which yields

$$V_{Z1} = V^- = V^+ = E_1$$

Since $I^- = 0$, Z_1 and Z_F are in series so that we may use the voltage divider formula on V_o to obtain

$$V_{Z1} = E_1 = \frac{Z_1}{Z_1 + Z_F} V_o$$

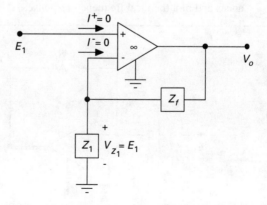

Figure 2-15 Noninverting amplifier configuration.

Chap. 2 Amplifiers and Their Integration into Signal Processing Systems

from which the ideal gain is

$$G^I = \frac{V_o}{E_1} = 1 + \frac{Z_F}{Z_1} = 1 + F_R \tag{2-29}$$

where F_R is the feedback ratio.

Example 2-7

For the case $Z_F = R_F$ and $Z_1 = R_1$,

$$G^I = 1 + \frac{R_F}{R_1} \tag{2-30}$$

Note that the gain is positive and greater than one, so when these circuits are physically constructed, care must be taken to prevent the output terminal at V_o from coupling back into the positive terminal so oscillations do not develop because of nonideal amplifier characteristics.

Example 2-8

A popular application of this circuit is the case where $Z_1 = \infty$ (open circuit). The result is a unity gain amplifier, as shown in Fig. 2-16. This configuration is often called a buffer amplifier because it isolates the source E_1 from whatever network is connected at V_o. (See exercise E2.4.2b.)

One important fact common to both the inverting and noninverting configurations is that the ideal gain formulas are independent of any load connected to the outputs. This means that several stages may be cascaded, as shown in Fig. 2-17, with the result that the overall gain is simply the product of the overall gains, derived as follows:

$$G^I_{total} = \frac{E_o}{E_1} = \frac{E_3 G_3^I}{E_1} = \frac{(E_2 G_2^I) G_3^I}{E_1} = G_1^I G_2^I G_3^I \tag{2-31}$$

This result is valid for any noninteracting stages, whether they are ideal or not.

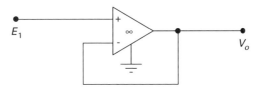

Figure 2-16 Unity gain buffer amplifier.

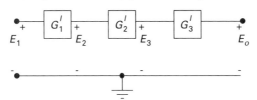

Figure 2-17 Cascaded amplifier stages.

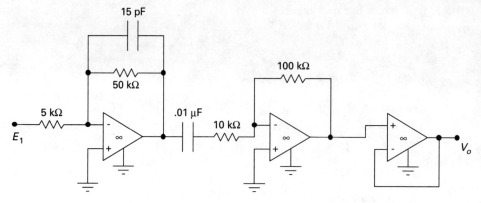

Figure E2.4.2c

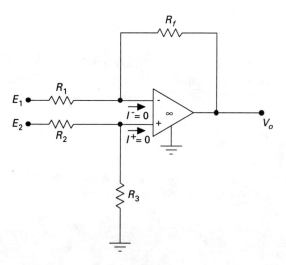

Figure E2.4.2d

Exercises

E2.4.2a For the ideal noninverting amplifier, plot the frequency responses of the systems defined by:

1. $Z_1 = R_1, Z_f = \dfrac{1}{sC_f}$

2. $Z_1 = \dfrac{1}{sC_1}, Z_f = R_f$

E2.4.2b Using the principle of the virtual short, derive the equation for the ideal gain of the system in Fig. 2-16.

E2.4.2c Determine the equation for the overall gain of the system of Fig. E2.4.2c.

Chap. 2   Amplifiers and Their Integration into Signal Processing Systems   147

E2.4.2d Determine the ideal output voltage of the network of Fig. E2.4.2d. Hint: First find the voltage at the positive op amp terminal and then use the principle of the virtual short.

2.5 OPERATIONAL AMPLIFIER SPECIFICATIONS

The op amp may be thought of as a transducer that converts voltage to voltage, so all of the parameters discussed in Chapter 1 apply here as well. However, the nature of the operational amplifier, which consists of many electrical components, makes it susceptible to other problems which require additional specifications. The complexity of the operational amplifier can be appreciated by a quick perusal of Fig. 2-18.

A list of some of the more important parameters and specifications is given in Table 2-1. Along with each quantity, the values for ideal op amps and for most frequently used op amps are given. As was the case with transducers, these parameters are not independent of each other. It is important, however, to understand the effect of each as an independent entity so that an error budget can be constructed that includes the effect of each parameter. The large number of specifications and parameters makes it impossible to give a detailed discussion of each. The method for determining the effect of each specification on a data processing system is exactly the same as for the transducer. Thus it is not necessary to be exhaustive here, and we are content to pick some of the more important ones for study. We will also concentrate on those parameters that are different from those already presented in connection with transducers or whose effects are handled in a slightly different way.

Two of the parameters in the table deserve special mention. First, the extremely narrow op amp bandwidth may be of concern, being only 10 to 100 Hz. Note that this is

TABLE 2-1. IMPORTANT SPECIFICATIONS FOR LOW-POWER OPERATIONAL AMPLIFIERS AND THEIR IDEAL AND PRACTICAL VALUES

Specification	Ideal value	Practical value	Units
Open loop frequency bandwidth	∞	10 to 100	Hz
Offset current (input)	0	5 to 10^6	pA
Offset voltage (input)	0	1 to 20	mV
Common mode rejection	∞	70 to 95	dB
Open loop gain	∞	2×10^4 to 2×10^5	V/V
Bias current	0	10 to 10^6	pA
Input impedance	∞	0.3 to 10^6	MΩ
Open loop output impedance	0	10 to 100	Ω
Settling time	0	0.1 to 100	ns
Slew rate	∞	0.1 to 10	V/μs
Power supply rejection	∞	75 to 100	dB
Overload recovery	0	10 to 100	ns
Temperature drift (of input offset)	0	1 to 10, 10 to 100	μA/°C, pA/°C
Output current (maximum)	∞	10 to 50	mA
Output voltage (maximum)	∞	20	V
Noise figure	0	1 to 5	dB
Ambient temperature range	unlimited	−55 to +125	°C

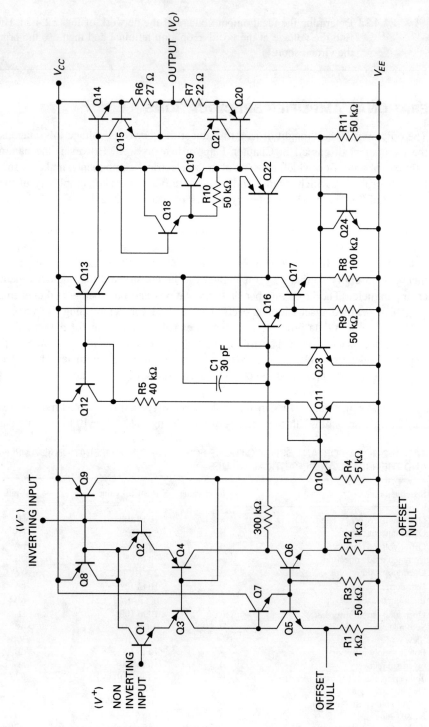

Figure 2-18 Operational buffer schematic diagram.

148

Chap. 2   Amplifiers and Their Integration into Signal Processing Systems   149

the open loop response. Negative feedback which produces reduced gain will widen the bandwidth significantly. For example, an inverting amplifier with a gain of two may operate at hundreds of kilohertz. This will be discussed more fully in the next section. The output resistance is also quite large; but again, the use of negative feedback reduces this considerably (see exercise E2.5a).

Formal definitions of op amp parameters and specifications are given in Table 2-2.

TABLE 2-2.   DEFINITIONS OF AMPLIFIER PARAMETERS AND SPECIFICATIONS

Average Temperature Coefficient of Input Offset Current - The change in input offset current over the operating temperature range divided by the operating temperature range. Similar definitions for power supply and time variations.

Average Temperature Coefficient of Input Offset Voltage - The change in input offset voltage over the operating temperature range divided by the operating temperature range. Similar definitions for power supply and time variations.

Bandwidth - The frequency at which the open loop voltage gain is reduced to 0.707 of its DC value.

Broadband Noise Factor - The base-10 logarithm of the ratio of input signal-to-noise ratio to the output signal-to-noise ratio over the frequency range for which this parameter is nominally flat.

Channel Separation - The level of output signal from an undriven amplifier with respect to the output from an adjacent driven amplifier.

Common Mode Input Resistance - The resistance seen looking into either one of the input terminals and common (ground).

Common Mode Input Voltage - The voltage applied to both terminals tied together.

Common Mode Output Voltage - The output voltage resulting from the application of a specified voltage common to both inputs.

Common Mode Gain - The ratio of common mode output voltage to common mode input voltage.

Common Mode Rejection Ratio - The ratio of the change of input offset voltage to the change in common mode voltage producing it. Also defined as the ratio of open loop gain divided by common mode gain. Both definitions result in the same value.

Common Mode Input Voltage Swing - The peak value of the common mode input voltage which can be applied for linear operation.

Differential Input Capacitance - The effective capacitance between the two inputs, operating open loop.

Differential Input Resistance - The effective resistance between the two inputs, operating open loop.

Differential Load Rejection - The ratio of the change in input offset voltage to the change in differential load current producing it.

Differential Voltage Gain - The ratio of the change in output voltage to the change in differential input voltage producing it.

Equivalent Input Noise Voltage - The equivalent input noise voltage that would reproduce the noise seen at the output if all amplifier noise sources and the source resistances were set to zero.

Equivalent Input Noise Current - The equivalent input noise current that would reproduce the noise seen at the output if all amplifier noise sources were set to zero and the source impedances were large compared to the optimum source impedance.

Input Bias Current - The *average* of the two input currents with no signal applied.

Input Bias Current Drift - The rate of change of input bias current with temperature, supply voltage, or time.

Input Capacitance - The capacitance seen looking into either input terminal with the other grounded. This capacitance is in parallel with the input resistance.

Input Offset Current - The difference in input bias currents into the two input terminals with the output voltage at zero.

Input Offset Voltage - That voltage which must be applied between the input terminals, through two equal resistances, to obtain zero output voltage.

Input Resistance - The resistance seen looking into either input with the other grounded.

Input Voltage Range - The range of voltage on either input terminal over which the amplifier will operate as specified. Exceeding the input voltage range may cause the amplifier to function improperly.

TABLE 2-2. DEFINITIONS OF AMPLIFIER PARAMETERS AND SPECIFICATIONS (cont'd)

Large Signal Voltage Gain - The ratio of the output voltage swing to the change in input voltage required to produce it.

Open Loop Voltage Gain - The ratio of the output signal voltage to the differential input signal voltage producing it, with no feedback applied.

Output Resistance - The small-signal AC resistance seen looking into the output with no feedback applied and the output DC voltage near zero.

Output Short-Circuit Current - The maximum output current obtainable with the output shorted to ground or to either supply.

Output Voltage Swing - The peak output voltage swing that can be obtained without clipping of the output voltage waveform.

Power Bandwidth - The maximum frequency at which the maximum output can be maintained without significant distortion.

Power Supply Rejection Ratio - The ratio of the change in input offset voltage to the change in supply voltage producing it.

Power Supply Sensitivity - The ratio of the change of a specified parameter to the change in supply voltage causing it.

Settling Time - The time from a step change of input to the time the corresponding output settles to within a specified percentage of the final value.

Slew Rate - The maximum rate of change of output under large signal conditions.

Transient Response - The closed loop step function response of the circuit under small signal conditions.

Unity Gain Bandwidth - The frequency at which the open loop gain is reduced to unity.

Unity Gain Crossover Frequency - See Unity Gain Bandwidth.

Exercises

E2.5a Determine the output impedance of the configuration shown in Fig. E2.5a. Note that the amplifier is not an ideal op amp, except that the input currents at the amplifier terminals are zero. (Hint: The Thevenin resistance of a source is $V_{open}/I_{short\ circuit}$.) Compute the system output resistance for $A = 50,000$, $R_1 = 12\ \text{k}\Omega$, $R_f = 120\ \text{k}\Omega$, and $R_o = 100\ \Omega$.

$$\text{Partial answer: } R_{Thev} = \frac{R_o\left(1 + \dfrac{R_f}{R_1}\right)}{\left(1 + A + \dfrac{R_f}{R_1} + \dfrac{R_o}{R_1}\right)}$$

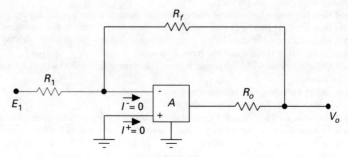

Figure E2.5a

Chap. 2   Amplifiers and Their Integration into Signal Processing Systems

E2.5b Select any two parameters from Table 2-2 and, without reading the definitions given, state what you think they mean and then compare with formal definitions. Are there any significant differences?

2.6 OPERATIONAL AMPLIFIER OPEN LOOP GAIN VALUE

The open loop gain actually has a finite value and causes the input-output relationship to be different from the nice results obtained earlier assuming infinite gain. We next investigate the effects of finite open loop gain but still neglect the frequency variation (frequency response).

Since the gain effects depend upon the circuit configuration we present the ideas by way of examples.

Example 2-9

First let us consider the simple "unity" gain buffer shown in Fig. 2-19. We must remember that the circuits using finite gain op amps do not satisfy the principle of virtual short; that is, $V^+ \neq V^-$ (unless, of course, they are physically tied together—not a practical connection).

From the figure:

$$V^+ = E_1$$

$$V^- = V_o$$

Using Eq. (2-20), with $A = A_o$,

$$V_o = A_o(V^+ - V^-) = A_o(E_1 - V_o)$$

Solving for V_o,

$$V_o = \frac{V_2}{1 + \frac{1}{A_o}}$$

The gain is then

$$G = \frac{V_o}{V_2} = \frac{1}{1 + \frac{1}{A_o}}$$

The crucial question is: What is the relationship between the amplifier open loop gain A_o and the number of bits N in the digital data words? This is answered using the same criterion used with transducer errors; that is, errors should be below 1/2 LSB. To compute the error we first need to know the ideal output from the circuit. This was obtained in Example 2-8 (Eq. (2-31)) for which

$$V_o^I = E_1$$

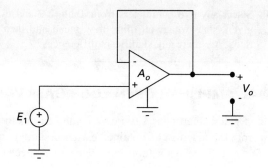

Figure 2-19 Unity gain buffer with finite op amp gain.

Then

$$\text{Error} = V_o - V_o^I = \frac{E_1}{1 + \frac{1}{A_o}} - E_1 = -\frac{E_1}{1 + A_o}$$

The amplifier output limits are essentially V_{CC} and V_{EE} (actually 1 V from these values), so the output span is

$$V_o^{span} = V_{CC} + |V_{EE}| \approx 2V_{CC}$$

for symmetric supplies. Thus

$$|\text{error}| \leq 1/2 \, \text{LSB} = \frac{V_o^{span}}{2^{N+1}} = \frac{2V_{CC}}{2^{N+1}} = \frac{V_{CC}}{2^N}$$

or,

$$\frac{E_1}{1 + A_o} \leq \frac{V_{CC}}{2^N}$$

From this equation we may compute the required gain A_o for a given N or vice versa. The results are as follows:

$$A_o \leq \frac{E_1}{V_{CC}} 2^N - 1$$

$$N \leq \frac{\log_{10}\left[\frac{V_{CC}}{E_1}(1 + A_o)\right]}{\log_{10} 2} \qquad (2\text{-}32)$$

Remember that this result is valid for the unity gain buffer only.

Chap. 2   Amplifiers and Their Integration into Signal Processing Systems   153

TABLE 2-3 REQUIRED MINIMUM OP AMP OPEN LOOP GAIN AND MAXIMUM NUMBER OF DATA WORD BITS FOR UNITY GAIN BUFFER (NEGLECTING FREQUENCY RESPONSE CHARACTERISTIC)

N	A_o
2	3
4	15
6	63
8	255
10	1023
12	4095
14	16383
16	65535

The worst case is for $E_1 = V_{CC}$ for either A_o or N. The corresponding values of N and A_o are given in Table 2-3. These op amp gains are readily available today. For reference, the 741 op amp has a guaranteed minimum A_o of 50,000. Remember also that this example neglects frequency response, which is covered in the next section.

Example 2-10

For the next example, consider the noninverting amplifier shown in Fig. 2-20, which is another commonly used configuration. Again, we neglect any input terminal currents or frequency effects. Using voltage division to obtain V^-, since R_1 and R_f are in series:

$$V_o = A_o(V^+ - V^-) = A_o\left(E_1 - V_o\frac{R_1}{R_1 + R_f}\right) \quad (V^+ = E_1 \text{ here})$$

$$V_o\left(\frac{1}{A_o} + \frac{R_1}{R_1 + R_f}\right) = E_1$$

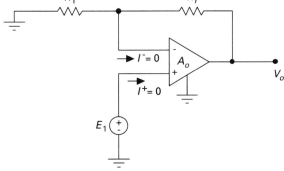

Figure 2-20 Noninverting amplifier with finite op amp gain.

Thus

$$G = \frac{V_o}{E_1} = \frac{1}{\frac{1}{A_o} + \frac{R_1}{R_1 + R_f}} = \frac{R_1 + R_f}{R_1 + \frac{R_1 + R_f}{A_o}}$$

$$= \frac{1 + \frac{R_f}{R_1}}{1 + \frac{1 + \frac{R_f}{R_1}}{A_o}} = \frac{1 + F_R}{1 + \frac{1 + F_R}{A_o}} \qquad (2\text{-}33)$$

where $F_R = \frac{R_f}{R_1}$.

For ideal operation, as $A_o \to \infty$,

$$\frac{V_o^I}{E_1} \to 1 + F_R.$$

This could have been obtained from Eq. (2-30).

$$\text{Error in output} = V_o - V_o^I = E_1 \left[\frac{1 + F_R}{1 + \frac{1 + F_R}{A_o}} - (1 + F_R) \right] = -E_1 \left[\frac{(1 + F_R)^2}{1 + A_o + F_R} \right]$$

As before, let V_0^{span} be the total span of V_0. Then

$$1/2 \text{ LSB} = 1/2 \frac{V_o^{span}}{2^N} = \frac{V_o^{span}}{2^{N+1}}$$

Using magnitude of the error we then require

$$E_1 \frac{(1 + F_R)^2}{A_o + 1 + F_R} \le \frac{V_o^{span}}{2^{N+1}}$$

Solving for N:

$$N \le \frac{\log_{10}\left[\frac{V_o^{span}}{2E_1} \frac{A_o + 1 + F_R}{(1 + F_R)^2} \right]}{\log_{10} 2} \qquad (2\text{-}34)$$

Solving for A_o:

Chap. 2   Amplifiers and Their Integration into Signal Processing Systems

$$A_o \geq \frac{2^{N+1}(1+F_R)^2 E_1}{V_o^{span}} - (1+F_R) \qquad (2\text{-}35)$$

Solving for the resistance ratio (gain factor) $F_R = \frac{R_f}{R_1}$:

$$F_R \leq \frac{1 - 2K + \sqrt{(2K-1)^2 + 4A_o K(K-1)}}{2K} \qquad (2\text{-}36)$$

where

$$K = \frac{E_1 2^{N+1}}{V_o^{span}}$$

Thus given any two of the system parameters we may compute the third.

As a final note as to whether or not the finite gain is of any actual consequence, one must examine the type of information to be extracted from the output signal. If absolute value of signals (or frequency components) is critical, then the effects of finite gain must be accounted for. On the other hand, if only the relative values of, say, frequency components in a signal are of concern then there need be no correction for the finite gain effect, since a gain change will affect all frequency components equally so that amplitude ratios of the components remain the same. (The effects of finite frequency response are of course a separate problem, for then ratios of amplitudes will be changed. This is described in a later section.)

Exercises

E2.6a   Make a plot of overall amplifier gain G of Example 2-10 versus the resistor feedback ratio F_R using the amplifier open loop gain as a parameter having values $A_o = \infty$ (ideal), 10^3, 10^2, 10, 0.

Now for each of the cases above plot the allowable value of

$$\frac{E_1}{V_o^{span}}$$

versus F_R if the input E_1 just drives the amplifier output to $1/2 V_o^{span}$, which is the limiting case for operation allowing equal positive and negative output values. (Note that the value of E_1 is not $1/2 V_o^{span}$.) Using the largest permissible

$$\frac{E_1}{V_o^{span}}$$

determined above, plot A_o as a function of F_R for various N (reasonable values). (Note: The largest permissible

$$\frac{E_1}{V_o^{span}}$$

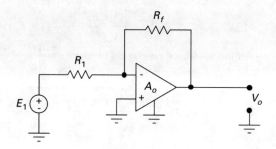

Figure E2.6b

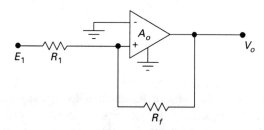

Figure E2.6c

depends upon F_R.) Using the largest permissible

$$\frac{E_1}{V_o^{span}}$$

determined above, plot F_R as a function of N for various values of A_0 as a parameter.

E2.6b Explain the effect of the value of the open loop op amp gain on the performance of the single-input signal inverting amplifier of Fig. E2.6b. Deriving the equation is a good way to show this (assume $Z_{in} = \infty$, $Z_o = 0$). Suppose that the value of the gain error with respect to ideal theoretical value (for $A_o = \infty$) is to be -60 dB. What minimum value of op amp open loop gain will satisfy the requirement? (Your answer will be in terms of the resistors.) Give required values of A_o for $R_f/R_1 = 0.1, 1, 10, 100$. How many bits for data representation are justified for these values (-60 dB level)?

E2.6c Analyze the circuit of Fig. E2.6c and plot V_o/E_1 as a function of A_o for the case $R_f = R_1$. Watch the network polarities carefully. Comment on the result. Would the circuit work this way in practice? Explain.

E2.6d Determine the relationship between the amplifier open loop gain A_o, the number of data word bits N, the resistor values, and the input signals for the summing amplifier of Fig. E2.6d. Assume V_o^{span} is the output voltage span. Amplifier input impedance is assumed infinite.

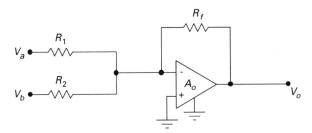

Figure E2.6d

2.7 OPERATIONAL AMPLIFIER FREQUENCY RESPONSE

In the preceding section it was noted that the op amp open loop gain is not a constant A_o. From Table 2-1 we see that the practical values of open loop frequency bandwidth are very low, 10 Hz being the most common value. A typical open loop frequency response curve is shown in Fig. 2-21, using the Bode plot presentation. The subject of this section is the determination of the effect of this frequency response characteristic on signal processing systems. The frequency f_2 is the open loop 3-dB frequency. The frequency f_T is called the unity gain (0-dB) bandwidth.

2.7.1 Effect of Op Amp Frequency Response on Amplifier Designs

The frequency response characteristic of Fig. 2-21 can be described conveniently in terms of the slope of the asymptote, which is

$$slope = \frac{\Delta \text{ dB}}{\# \text{ decades}} = \frac{-100 \text{dB}}{5 \text{ decades}} = -20 \frac{\text{dB}}{\text{decade}}$$

This is called a first-order roll-off (see exercise E2.7a) and is also the same as -6 dB/octave. Octave is a term borrowed from music which means a doubling of frequency. It is usual to express the equation for the open loop gain as

$$A(f) = \frac{A_o}{1 + j\frac{f}{f_2}} \qquad (2\text{-}37)$$

where A_o is the DC open loop gain and f_2 is the corner or 3-dB frequency of the open loop gain—around 10 Hz in the example given. The corresponding magnitude is

$$|A(f)| = \frac{A_o}{\left(1 + \frac{f^2}{f_2^2}\right)^{1/2}} \qquad (2\text{-}38)$$

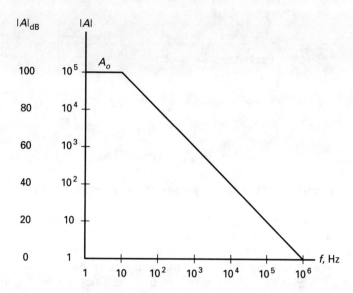

Figure 2-21 Typical small-signal operatnional amplifier open loop frequency response.

For frequencies f much greater than f_2, we may neglect the 1 in the denominator and write

$$f \,|\, A(f) \,| \approx A_o f_2 \tag{2-39}$$

which states that the product of the operating frequency and gain magnitude at that frequency is essentially constant.

One of the consequences of Eq. (2-38) can be obtained by considering the inverting feedback amplifier of Fig. 2-22. Since $I^- = 0$, Kirchhoff's current law at the op amp minus terminal yields

$$I_1 + I_f = 0$$

Using the element volt-ampere characteristics:

$$\frac{E_1 - V^-}{R_1} + \frac{V_o - V^-}{R_f} = 0$$

Solving for V^-:

$$V^- = \frac{R_f E_1 + R_1 V_o}{R_1 + R_f}$$

Since $V^+ = 0$,

$$V_o = A(f)\,(V^+ - V^-) = A(f)\,\frac{R_f E_1 + R_1 V_o}{R_1 + R_f}$$

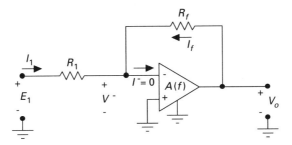

Figure 2-22 Inverting amplifier with a frequency-dependent op amp gain.

Solving for V_o:

$$V_o = -E_1 \frac{R_f}{R_1 + \dfrac{(R_1 + R_f)}{A(f)}} \tag{2-40}$$

The overall gain is

$$G(f) = -\frac{R_f}{R_1 + \dfrac{(R_1 + R_f)}{A(f)}} \tag{2-41}$$

Substituting for $A(f)$ using Eq. (2-36) and rearranging, the preceding equation becomes

$$G(f) = -\frac{\dfrac{R_f}{R_1} A_o}{A_o + 1 + \dfrac{R_f}{R_1} + j \dfrac{\left(1 + \dfrac{R_f}{R_1}\right) f}{f_2}} \tag{2-42}$$

For this particular type of network function the 3-dB frequency occurs when the real and imaginary parts of the denominator are equal. This will yield the 3-dB frequency, $f = f_{2F}$, of the network with feedback. The result is

$$f_{2F} = \frac{1 + A_o + \dfrac{R_f}{R_1}}{1 + \dfrac{R_f}{R_1}} f_2 \tag{2-43}$$

For practical cases, $A_o \approx 10^5$, so we may neglect any terms added to A_o in these last two equations (for the practical cases $R_f/R_1 < 10^4$) with the following results:

$$\lim_{A_o \to \infty} G(f) = -\frac{R_f}{R_1} = G^I = F_R \tag{2-44}$$

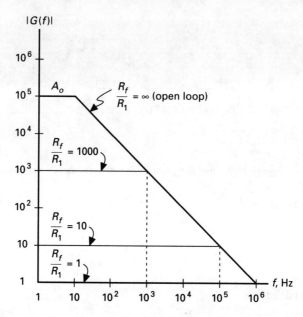

Figure 2-23 Frequency responses of inverting amplifiers having various gains, R_f/R_1.

$$\lim_{A_o \to large} f_{2F} = \frac{A_o}{1 + \frac{R_f}{R_1}} f_2 \qquad (2\text{-}45)$$

From these results we can obtain the gain-bandwidth product of a practical amplifier as

$$\left| G^I \right| f_{2F} \approx \frac{R_f}{R_1} \cdot \frac{A_o}{1 + \frac{R_f}{R_1}} f_2 \qquad (2\text{-}46)$$

For most cases $\left| G^I \right| = R_f/R_1 > 10$, so we may neglect the 1 in the denominator to obtain

$$\left| G^I \right| f_{2F} \approx A_o f_2 \qquad (2\text{-}47)$$

which is a constant. What this result tells us is that when we design an amplifier we can increase the operating bandwidth by reducing the operating gain; or, in other words, there is a trade-off between gain and bandwidth. Although we have restricted the gain magnitude to be greater than 10, this last result gives a nice interpretation of the term unity gain bandwidth; that is, for $\left| G^I \right| = 1$ we have a very wide operating bandwidth of $A_o f_2$ hertz. Thus what originally seemed like a troublesome op amp low bandwidth has been improved with the feedback. In Fig. 2-23 there are plots of the circuit performances as a function of R_f/R_1.

Chap. 2   Amplifiers and Their Integration into Signal Processing Systems   161

One might be tempted to bypass the bandwidth problem by cascading lower gain stages which have wider bandwidths. Unfortunately this does not work, since when one stage is at its 3-dB frequency, n stages will be down $3n$ dB. One is still limited by the gain-bandwidth product (see exercise E2.7.1b and problem 2.1).

A second example of the effect of op amp frequency response comes from the noninverting configuration discussed in Section 2.6 (Fig. 2-20). If we replace A_o by $A(f)$ in Eq. (2-33) we obtain

$$G(f) = \frac{1+F_R}{1+\frac{1+F_R}{A_o}\left(1+j\frac{f}{f_2}\right)} = \frac{1+F_R}{\left[1+\frac{1+F_R}{A_o}\right]+j\frac{1+F_R}{A_o f_2}f} \qquad (2\text{-}48)$$

where

$$F_R = \frac{R_f}{R_1}$$

This function is also of the form that allows us to compute the operating 3-dB frequency by equating the real and imaginary parts in the denominator. Doing this and letting the 3-dB frequency be denoted by f_{2F} yields

$$f_{2F} = \left(1+\frac{A_o}{1+F_R}\right)f_2 \qquad (2\text{-}49)$$

The interpretation of this equation is simplified by putting it into a form that has the ideal network gain given by Eq. (2-29). This substitution yields

$$f_{2F} = \left(1+\frac{A_o}{G^I}\right)f_2 \qquad (2\text{-}50)$$

This shows that as the noninverting amplifier gain increases, the operating bandwidth decreases.

A practical approximation can be applied to the preceding equation by noting that $A_o \gg G^I$ in most cases. This allows us to neglect the number 1 and write

$$G^I f_{2F} = A_o f_2 = constant \qquad (2\text{-}51)$$

We can see here that system gain and bandwidth are again subject to trade-off, since their product is a constant for a given op amp.

Exercises

E2.7.1a Derive the equation for the gain (transfer function) of the first-order RC circuit shown in Fig. E2.7.1a. Plot the gain in magnitude and dB versus normalized frequency, ωRC. From your equation, what is the approximate dB expression

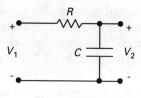

Figure E2.7.1a

when $\omega > 1/RC$? From your approximate solution, how many dB does the output change when the frequency ω is increased by a factor of 10? How much when the frequency is doubled?

E2.7.1b Suppose we cascade three identical stages whose responses are given by Eq. (2-42). By cubing the magnitude of that equation to obtain the overall gain, prove that for large A_o the gain magnitude is cubed but that the 3-dB frequency is reduced.

E2.7.1c Determine the 3-dB frequency for the unity gain buffer amplifier that has a frequency-dependent op amp gain. You may make any practical approximations. (See Fig. 2-19.)

E2.7.1d Write the expression of $A(f)$ for an op amp that has a DC gain of 200,000 and a bandwidth of 10 Hz. What is the gain magnitude of 5 kHz? If the op amp is used in an amplifier having an overall gain of 10, estimate the bandwidth. If an engineer wants to design an amplifier with a minimum bandwidth of 1 MHz, what is the approximate largest gain possible?

E2.7.1e A 741 op amp having $A_o = 10^5$ and $f_2 = 6$ Hz is used in a noninverting amplifier. If $R_f = 120$ kΩ and $R_1 = 10$ kΩ, what is the DC gain?

E2.7.1f An inverting amplifier uses an op amp that has a DC gain of 2×10^5 and a unity gain bandwidth product of 10^6. If $R_f = 47$ kΩ and $R_1 = 2.2$ kΩ, determine the DC gain and operating bandwidth.

2.7.2 Effect of Op Amp Frequency Response on the Bit Size *N* of the Data Words

In this subsection we investigate the effects of the finite op amp frequency response on the allowable number of bits N in the data words. Before results can be derived, the engineer must decide whether

1. Amplitudes of the frequencies in the transducer waveform are critical

or

2. Only the presence or absence of a given frequency is critical.

The number of bits N one uses to represent a data word will be quite different depending upon that decision.

Let us first suppose that the amplitudes of the various frequency components in the transducer waveform are critical. This means that if the frequency response of the ampli-

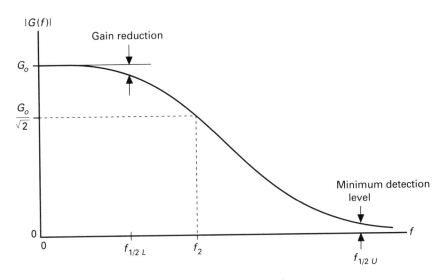

Figure 2-24 Effect of op amp network frequency response on lower 1/2 LSB frequency.

fier reduced the highest desired frequency component by more than 1/2 LSB, then a distortion has occurred. This is the same problem as was discussed for transducers, and it is illustrated in Fig. 2-24.

With respect to Fig. 2-24, we assume that the input amplitude of a given frequency component is E_h, where E_h may be either a single frequency sinusoid or a desired harmonic frequency in a complex waveform. The output voltage would then be

$$V_{oh} = G(f)E_h \qquad (2\text{-}52)$$

The ideal output would be

$$V_{oh}^I = G_o E_h$$

If we let the output voltage span be V_o^{span}, we then require, to preserve amplitudes,

$$|\text{error}| = |V_{oh} - V_{of}^I| = ||G(f)| - |G^I|| \, |E_h \le 1/2 \frac{V_o^{span}}{2^N} \qquad (2\text{-}53)$$

Thus all we need to do is substitute the designed amplifier gain function and solve for the $f_{1/2L}$ (lower 1/2 LSB frequency) if N is given, or solve for N if $f_{1/2L}$ is chosen to be the desired harmonic frequency f_h. Note this also requires the engineer to make an estimate of the largest value expected for E_h, because that value provides an estimate of the largest error. In practice the usable frequency on this basis is well below f_2. Equation (2-53) can also be used to help one determine the required gain function and op amp properties if $f_{1/2L}$ (f_h) and N are given.

Example 2-11

A digital processing system uses 10-bit data words, and the required range of the analog voltage corresponding to the transducer input is ± 5 V. It is desired to determine the amplitudes of frequencies to 6 kHz, for which the peak value at 6 kHz is estimated to be 20 mV. The largest amplitude of any frequency component (including DC) is 400 mV. Design an inverting amplifier that will meet these specifications.

We note in passing that in many engineering design cases the amount of information available may exceed what is needed or may be missing in some particulars. Such is the case with this example.

In order to cover the ± 5 V range across the frequencies of interest, we compute the required amplifier gain in the ideal case. Using the given information and Eq. (2-44), we have for the inverting amplifier:

$$G^I = -\frac{\text{maximum amplifier output}}{\text{maximum amplifier input}} = -\frac{5\text{ V}}{400 \times 10^{-3}\text{ V}} = -12.5 = -\frac{R_f}{R_1}$$

Since amplitude is important, Eq. (2-53) applies with the function $G(f)$ from Eq. (2-42). Using the resistance ratio above, the result is (at $f_{1/2L} = f_h = 6000$ Hz):

$$\left| -\frac{12.5 A_o}{A_o + 1 + 12.5 + j\dfrac{1+12.5}{f_2}6000} - |(-12.5)| \right| 20 \times 10^{-3}$$

$$\leq \frac{V_o^{span}}{2^{10+1}} = \frac{5-(-5)}{2^{11}} = \frac{10}{2^{11}}$$

Take the 12.5 and 20×10^{-3} constants to the right side of the inequality:

$$\left| -1 + \frac{A_o}{A_o + 13.5 + j\dfrac{81000}{f_2}} \right| \leq 0.0195$$

Next, combine terms, take the magnitude, and then square both sides:

$$\frac{13.5^2 + \left(\dfrac{81000}{f_2}\right)^2}{(A_o + 13.5)^2 + \left(\dfrac{81000}{f_2}\right)^2} \leq 3.8 \times 10^{-4}$$

Now we know that f_2 is on the order of 10 Hz, so we may use this in the preceding and solve for A_o, which results in

$$A_o \geq 415{,}430$$

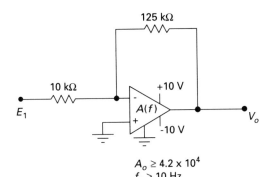

$A_o \geq 4.2 \times 10^4$
$f_2 \geq 10$ Hz

Figure 2-25 Final amplifier design for Example 2-11.

The engineer must select an operational amplifier having an open loop gain in excess of 420,000 and an open loop cutoff frequency of 10 Hz. Another way of saying this is that the op amp unity gain bandwidth f_T must exceed 4.2 MHz.

To complete the design we need to select the power supplies. Since the output must swing ± 5 V, the supplies must exceed ±6 V to avoid clipping and distortion. Since reasonable resistances in circuits are in the kilohm range, we select $R_1 = 10$ kΩ and $R_f = 125$ kΩ. The final network is shown in Fig. 2-25.

We next suppose that the actual amplitudes of the frequencies are not of interest, but we simply want to determine if a given frequency is present. In this case we can tolerate an amplitude reduction as long as the voltage is larger than 1/2 LSB. We can use the amplifier well above the operating 3-dB frequency now, at a frequency denoted in Fig. 2-24 as $f_{1/2u}$. Since we want to detect this signal, the output voltage must exceed 1/2 LSB so that Eq. (2-52) yields

$$| \text{output voltage} | = | V_{oh} | = | G(f)E_h | = | G(f) | E_h \geq 1/2 \frac{V_o^{span}}{2^N} \quad (2\text{-}54)$$

In this case the engineer must estimate the smallest expected amplitude of the signal to be detected, to obtain $f_{1/2u}$ if N is given, or to determine N if $f_{1/2u}$ is given. Also, if N and $f_{1/2u}$ are given, the minimum detectable signal (assuming no noise) can be determined.

Example 2-12

An inverting amplifier with a gain at DC of −200 is to be used to detect the presence of a 10-kHz frequency component in a transducer signal. If the op amp has a DC open loop gain of 10^5 and a gain bandwidth product of 10^6 and operates from ±10 V supplies, what is the smallest detectable signal if 10 bits are used in the data words?

For the nominal gain of −200, Eq. (2-44) gives the required feedback ratio F_R as 200. For the op amp we are given $A_o = 10^5$, from which we can derive

$$f_2 = \frac{\text{GBW product}}{A_o} = \frac{10^6}{10^5} = 10 \text{ Hz}$$

Using these results in Eq. (2-42) we have

$$G(f) = -\frac{(200)(10^5)}{10^5 + 1 + 200 + j\frac{200f}{10}} \approx \frac{-2 \times 10^7}{10^5 + j\frac{200f}{10}} = \frac{-2 \times 10^8}{10^6 + j200f}$$

$$|G(f)| = \frac{2 \times 10^8}{\sqrt{10^{12} + 4 \times 10^4 f^2}}$$

Estimate V_o^{span} as $2V_{CC}$ or 20 V. (We could subtract 2 volts in practice, since the output cannot really get to V_{CC}.)

Putting the preceding results into Eq. (2-54) using $f = 10$ kHz and $N = 10$, we have

$$\frac{2 \times 10^8}{\sqrt{10^{12} + 4 \times 10^{12}}} \times E_h \geq \frac{20}{2^{11}}$$

Solving for E_h:

$$E_h \geq 0.11 \text{mV}$$

Thus any 10-kHz signal having an amplitude greater than this will be detected.

Exercises

E2.7.2a A noninverting amplifier has been designed to have a gain of 100 and uses an op amp with $A_o = 10^5$ and $f_2 = 5$ Hz. The supply voltages are ±15 V. For 8-bit data words, what is the highest frequency that can be used if frequency amplitudes are important? What is the operating bandwidth of the amplifier?

E2.7.2b Repeat exercise E2.7.2a for an inverting amplifier having gain $= -1000$.

E2.7.2c For the amplifier of exercise E2.7.2a, suppose that the largest expected amplitude of E_h is 0.05 V at 1 kHz. What is the allowed value of N? Suppose E_h is of such size as to drive the amplifier to saturation at low frequencies. Now what is N? Amplitude of the frequency at 1 kHz is important.

E2.7.2d Suppose an inverting amplifier is designed to have a DC gain of -50 using an op amp with a DC gain of 2×10^5 and a bandwidth of 10 Hz. The amplifier is to be used in a 10-bit signal processing system for which signal amplitudes are important. What is the highest usable frequency if the largest input signal amplitude is 100 mV? The supply voltages are ±10 V.

E2.7.2e Repeat Example 2-12 for $N = 8$.

E2.7.2f In order to illustrate the effect of the choice of power supplies, suppose $V_{CC} = |V_{EE}| = 20$V, and repeat Example 2-12, using $N = 10$.

E2.7.2g An engineer has designed an amplifier with a gain of -250 using dual supplies of ±10V. The op amp has an open loop gain of 8×10^4 and an open loop bandwidth of 15 Hz. A digital signal processor uses 8-bit data words, and the engineer is asked what the highest frequency of amplitude 400 μV is that can be detected, and what the highest frequency of amplitude 10 mV is that can be passed without noticeable reduction. What are the answers to these queries?

Chap. 2 Amplifiers and Their Integration into Signal Processing Systems 167

E2.7.2h Repeat Example 2-11 for a noninverting amplifier.
E2.7.2i Repeat Example 2-12 for a noninverting amplifier.
E2.7.2j In Example 2-11, change f_2 to 5 Hz and compute A_o.
E2.7.2k Suppose an engineer selects an operational amplifier that has a guaranteed minimum A_o of 10^5 and a minimum f_2 of 10 Hz to use in the amplifier of Example 2-11. What is the new $f_{1/2L}$?

2.8 OPERATIONAL AMPLIFIER SLEW RATE

In the preceding section on amplifier frequency response we saw that the effect on a sine wave was an amplitude reduction. The waveshape was not distorted in that the sine wave remained a sine wave. Of course if a given waveform contains several frequency components, the overall waveform is distorted because each frequency component amplitude is affected differently. For frequency response effects the amplitude of the input signal did not play an important role (other than saturation, of course). If an input signal is reduced by a factor of two, for example, the output waveform would also be reduced by a factor of two but still have the same distorted shape. The subject of this section is an investigation of the ability of an amplifier to follow changes in signal amplitude as a function of time. As one might expect, if a signal is changing very rapidly (large dv/dt not necessarily from a sine wave) the amplifier might not be able to track in time. The parameter used to describe the ability of an amplifier to track a signal is called the *slew rate*, which we will denote by SR in volts/second. The SR is not completely independent of frequency response as we shall see, since both effects are dependent upon the capacitances present inside and outside the op amp.

The source of slew rate limiting can be explained by examining an equivalent circuit representation of an op amp, as shown in Fig. 2-26. The input transconductance amplifier, denoted by g_m, is a differential amplifier that converts the two input voltages into a current according to the equation

$$i_g = -g_m(V^+ - V^-). \tag{2-55}$$

The second voltage amplifier, having gain $-G$, has an output voltage (due to the unity gain buffer) of

$$v_o = -Gv_G \tag{2-56}$$

The feedback capacitor C_C is an internal capacitor responsible for the low-frequency (≈ 10 Hz) open loop cutoff frequency of the op amp. Without going into detail we indicate that the reason for deliberately introducing the low-frequency cutoff is to ensure that the amplifier will be stable over the entire operating range from DC to the open loop unity gain frequency. What this capacitor does is compensate the op amp performance by producing a phase margin sufficient to preclude oscillations within the amplifier for any feedback configuration. For this reason the capacitor is called a compensating capacitor, which explains the choice of the symbol C_C. This is the capacitor C_1 in Fig. 2-18.

Since the capacitor C_C is in the feedback of an amplifier, we may replace it by its Miller effect equivalent, as shown in Fig. 2-27. We have also combined the output and

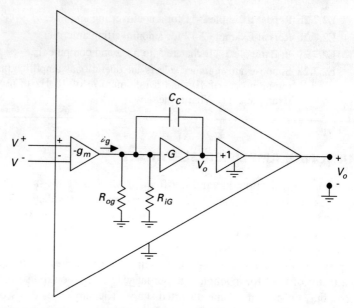

Figure 2-26 Equivalent network of an operational amplifier in terms of a transconductance amplifier, $-g_m$, a voltage amplifier, $-G$, and a unity gain amplifier.

input resistors in parallel to produce R as shown. From this reduced model we can compute the open loop op amp gain as

$$A(s) = \frac{V_o(s)}{V^+ - V^-} = \frac{-GV_G(s)}{V^+ - V^-} = \frac{-GI_g(s)Z_g(s)}{V^+ - V^-}$$

$$= \frac{(-G)\left(-g_m(V^+ - V^-)\right)\left(\dfrac{1}{\dfrac{1}{R} + s(1+G)C_C}\right)}{V^+ - V^-}$$

$$= g_m G \left(\frac{1}{\dfrac{1}{R} + s(1+G)C_C}\right) = \frac{g_m G R}{1 + \dfrac{s}{\left(\dfrac{1}{(1+G)RC_C}\right)}} \qquad (2\text{-}57)$$

Comparing this to Eq. (2-37), using $s = j\omega$, we have

$$f_2 = \frac{1}{2\pi(1+G)RC_C} \qquad (2\text{-}58)$$

and

$$A_o = g_m G R \qquad (2\text{-}59)$$

Chap. 2 Amplifiers and Their Integration into Signal Processing Systems

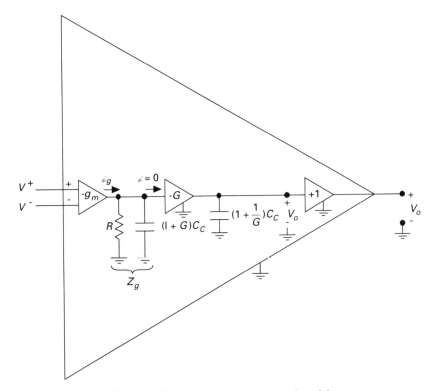

Figure 2-27 Reduced op amp internal model.

The compensating capacitor introduces another effect, which is the slew rate referred to earlier. Although in a typical data acquisition environment the input signals are usually band-limited, let us consider an operational amplifier in the unity gain configuration with a step occurring in the input waveform, as shown in Fig. 2-28. As the input reaches V_1 we assume that the output has been able to follow so that $v_o = V_1$, which is also the voltage on the right side of C_C. For large G the input voltage will be nearly zero and all the current i_g flows through C_C, since zero volts across R means zero current in R. Thus the voltage across C_C is $v_o = V_1$ at $t = t_o$. The equation for the voltage on C_C is

$$i_g(t) = C_C \frac{dv_o}{dt} \tag{2-60}$$

When the input voltage step occurs at t_o the capacitor voltage remains at V_1, since it cannot change voltage instantaneously. This also means that the terminal V^- has V_1 on it and V^+ has V_2 on it, creating a large difference voltage at the input. Thus the amplifier is no longer in the linear mode which saturates the transconductance amplifier output to its limit current, so

$$i_g = I_{\lim}$$

With this current value, Eq. (2-60) becomes

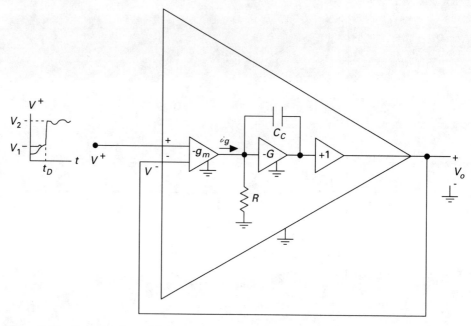

Figure 2-28 Operational amplifier model connected for unity gain configuration with input voltage discontinuity.

$$C_C \frac{dv_o}{dt} = I_{\lim}$$

Thus the maximum possible output slope is

$$\left.\frac{dv_o}{dt}\right|_{\max} = \frac{I_{\lim}}{C_C} \equiv SR \tag{2-61}$$

If any input waveform has a voltage slope greater than this, the output will slope limit (slew rate limit) to the value SR. The output will change linearly at this rate until the input voltage value attains a value that equals v_o or until v_o changes to equal V^+. At this point the amplifier then reverts to the linear tracking mode.

Next consider a sine wave input given by

$$v^+(t) = V_m^+ \sin \omega t$$

For this signal

$$\frac{dv^+}{dt} = \omega V_m^+ \cos \omega t$$

for which, since the amplifier is unity gain,

Chap. 2 Amplifiers and Their Integration into Signal Processing Systems

$$\left.\frac{dv_o}{dt}\right|_{max} = \left.\frac{dv^+}{dt}\right|_{max} = \omega V_m^+ = \omega V_{om} \qquad (2\text{-}62)$$

Thus there will be no slew rate limiting as long as

$$\omega V_{om} \leq SR \qquad (2\text{-}63)$$

There is also maximum voltage level $V_{\lim}$ that can occur at the op amp output, which is given on the op amp specification sheet. Thus the highest frequency that will not be distorted due to slew rate limiting is

$$\omega_{fp} V_{\lim} = SR$$

or

$$\omega_{fp} = \frac{SR}{V_{\lim}} \qquad (2\text{-}64)$$

This frequency ω_{fp} is called the full-power bandwidth, and it is given on the specification sheet or can be computed from the other two given parameters. Suppose that in a particular application the supply voltage is V_{CC}. For this condition the maximum output is $V_{CC} - 1$ and Eq. (2-63) yields

$$\omega_{max}(V_{CC} - 1) = SR \qquad (2\text{-}65)$$

Then the maximum frequency for given power supplies is

$$\omega_{max} = \frac{SR}{V_{CC} - 1} \qquad (2\text{-}66)$$

Note that this will be larger than ω_{fp}.

Example 2-13

For a 741 op amp, the maximum transconductance amplifier output is 20 µA and the compensating capacitor is 30 pF. Estimate the slew rate. Determine the frequency of the largest amplitude sine wave that can be applied without slew rate limiting if the supply voltages are $V_{CC} = +10$ V and $V_{EE} = -10$ V.

Using Eq. (2-61)

$$SR = \frac{10 \times 10^{-6}}{30 \times 10^{-12}} = 0.67 \times 10^6 \, \frac{V}{s}$$

or

$$0.67 \, \frac{V}{\mu s}$$

From Eq. (2-66):

$$\omega_{max} = \frac{0.67 \times 10^6}{10 - 1} = 74 \, \frac{\text{krad}}{\text{s}}$$

or 11.8 kHz.

One important fact to note from the equivalent model of Fig. 2-26 is that the voltage at v_o is independent of any external components connected there. This is because we have assumed zero output impedance of the unity gain buffer. Thus if we have a configuration other than unity gain, the input amplitude is smaller than the output by the design gain, whereas in the unity gain connection the input is the same as the output. External elements, however, must not cause the unity gain output buffer in the model to go into current limit.

Another kind of slope limiting occurs due to the frequency response of the op amp. If we let the op amp characteristic be denoted by (details are left for exercise E2.8b)

$$A(s) = \frac{A_o}{1 + \dfrac{s}{\omega_2}} \tag{2-67}$$

we can determine the transfer function of a unity gain configuration as

$$G(s) = \frac{V_o}{E_1} = \frac{1}{1 + \dfrac{\left(1 + \dfrac{s}{\omega_2}\right)}{A_o}} \tag{2-68}$$

$$= \frac{1}{1 + \dfrac{1}{A_o} + \dfrac{s}{A_o \omega_2}} \tag{2-69}$$

For $A_o \gg 1$ (the usual case) this is

$$G(s) = \frac{1}{1 + \dfrac{s}{A_o \omega_2}} \tag{2-70}$$

Applying a unit step

$$e_1(t) = E_1 u(t)$$

the output response is

$$v_o(t) = E_1(1 - e^{-A_o \omega_2 t}). \tag{2-71}$$

Chap. 2   Amplifiers and Their Integration into Signal Processing Systems

Thus

$$\left.\frac{dv_o}{dt}\right|_{max} = E_1 A_o \omega_2 \qquad (2\text{-}72)$$

As long as this is less than the slew rate, the step slope will be frequency response limited rather than slew rate limited. (The details of the preceding derivation are reserved for exercise E2.8b.)

Since slew rate and frequency response limiting both involve the capacitance C, it is obvious that the two effects are not completely independent. Thus as C decreases one would expect that bandwidth and slew rate would increase. To justify this expectation some commercially available amplifiers were selected and the unity gain frequency (gain-bandwidth product) was plotted against slew rate. The results are given in Fig. 2-29. As expected there is a general increase in both values. The dispersion of values can be explained several ways, the easiest being given by considering the wide range of slew rates at a given unity gain frequency, say at 1 MHz. At that value the slew rates differ by more than two orders of magnitude. This can be explained by simply observing that increasing the saturation current, i_g (see Fig. 2-26), could theoretically increase the slew rate without affecting the bandwidth.

Example 2-14

A noninverting amplifier has an ideal gain of +50 and the supply voltages are ±10 V. The op amp has an open loop gain of 10^5 and an open loop bandwidth of 10 Hz. The amplifier slew rate is 0.4 V/μs. The amplifier is to pass frequencies to 10 kHz. Plot the output waveform that corresponds to the input that would theoretically produce maximum output voltage at 10 kHz. What is the equation for the input voltage?

The largest amplitude is approximately 10 V − 1 V = 9 V, so the theoretical maximum output would be

$$v_o(t) = 9 \sin(2\pi 10^4 t)$$

For a gain of 50 the input is

$$e_1(t) = 0.18 \sin(2\pi 10^4 t)$$

The maximum slope that the output tries to produce is

$$\left.\frac{dv_o}{dt}\right|_{max} = 2 \times 10^4 \, \pi \times 9 = 0.565 \, \frac{V}{\mu s}$$

Since this exceeds SR, limiting will occur. The output then changes at 0.4 V/μs until the output value equals what the ideal waveform would want to produce, which puts the amplifier back into the linear amplifying domain. The time at which this occurs is when

$$9 \sin 2\pi 10^4 t_o = 0.4 \times 10^6 t_o$$

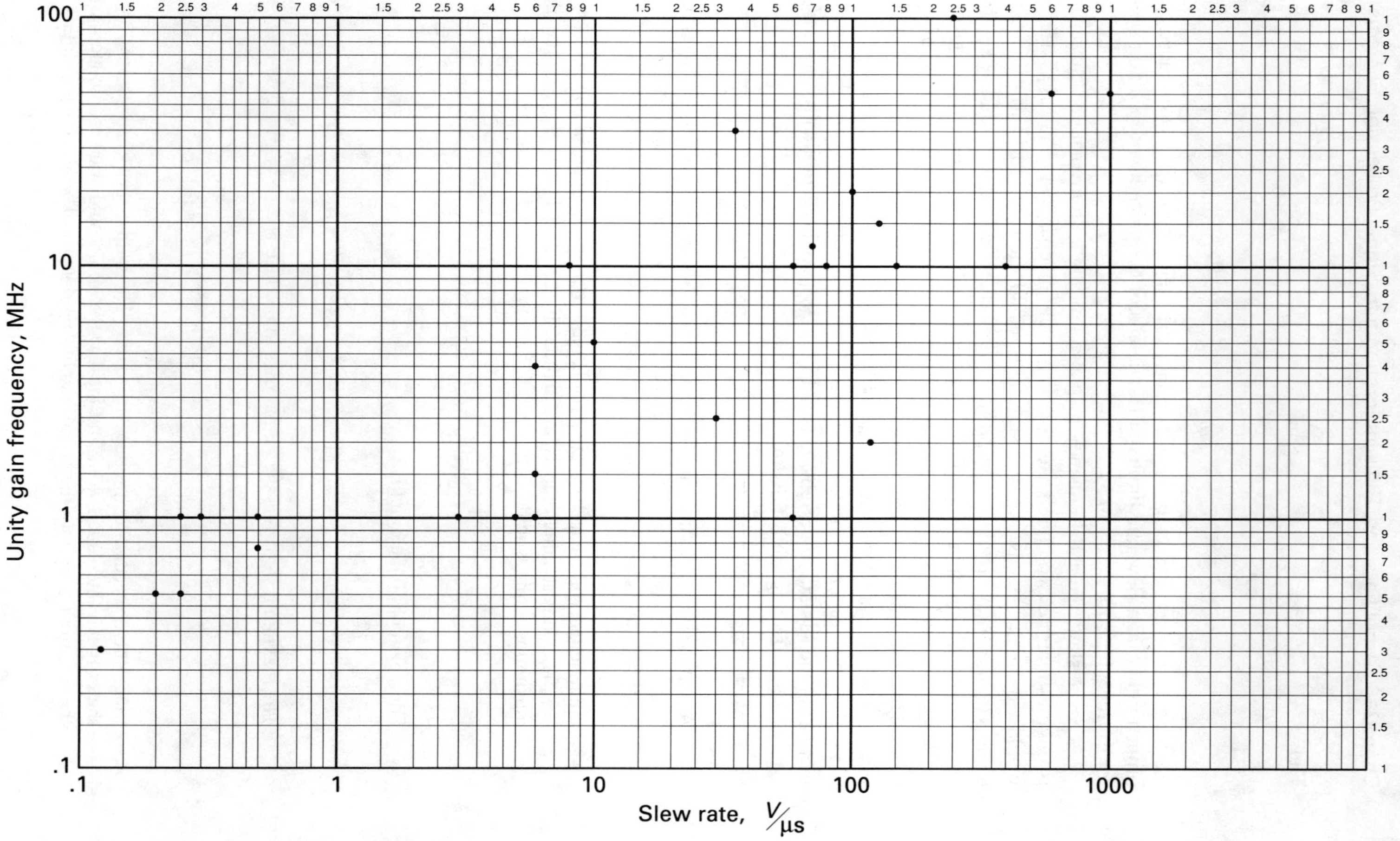

Figure 2-29 Unity gain frequency versus slew rate. Taken from manufacturers' data sheets by sampling the devices randomly. Note positive correlation.

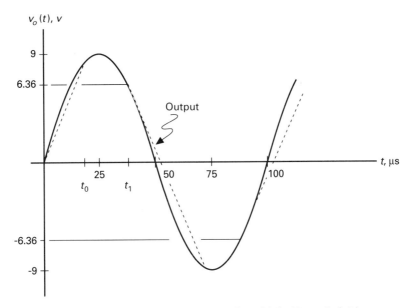

Figure 2-30 Slew rate limiting of a sinusoid, for Example 2-14.

which gives, by numerical solution,

$$t_o = 22.17 \ \mu s$$

At this time the amplitude is 8.86 V.

Thus if we assume that the waveform starts at $t = 0$, the output will rise at the slew rate until it catches up with the ideal output, as shown in Fig. 2-30. The output then follows the ideal curve until the slope again exceeds -0.4 V/μs which occurs at

$$2\pi \times 10^4 \times 9 \cos 2\pi \times 10^4 t_1 = -0.4 \times 10^6$$

or,

$$t_1 = 37.5 \ \mu s$$

The output now decreases 0.4 $V/\mu s$ until it again reaches the ideal output response.

Since the output of a transducer is seldom a single-frequency sinusoid, the question of slew rate limiting of practical signals must be addressed. In most cases the input signals are band-limited, so some knowledge of the highest frequency present is often known. By analyzing typical or expected input signals the maximum amplitudes of the frequencies within the band can often be estimated. It may also be that some frequencies never appear in the signal even though they may lie in the band. Using this information the engineer can make an estimate of worst-case (highest) slope combinations and select a slew rate on that basis. See problem 2.14 at the end of the chapter.

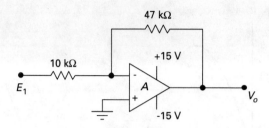

Figure E2.8a

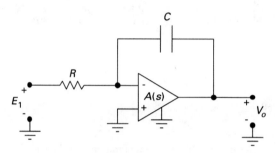

Figure E2.8f

Exercises

E2.8a The op amp in Fig. E2.8a has a slew rate of 3 V/μs and an open loop gain of 10^5. Determine the largest input amplitude and highest input frequency that will not show slew rate limiting.

E2.8b A unity gain amplifier configuration is constructed using an op amp having an open loop gain characteristic

$$A(f) = \frac{A_o}{1 + j\dfrac{f}{f_2}}$$

Convert this characteristic to the Laplace notation $(s, A(s))$ and determine the largest input step that can be placed at the input so that the output slope is just equal to a given SR value; that is, assume SR is known as a number. (If the step slope is less than SR the amplifier is said to be frequency response, FR, limited.)

E2.8c Repeat Example 2-14 for two separate cases:
  a. Frequency of input is 20 kHz.
  b. Slew rate is 1 V/μs.

E2.8d An op amp has a slew rate of 0.3 V/μs and a unity gain bandwidth of 1 MHz. The op amp DC gain is 200,000. It is used in an inverting amplifier configuration having a gain of −10. A step of 1 V occurs in the input voltage. Is the output slew rate or frequency response limited?

E2.8e Make a laboratory test for slew rate and frequency response limiting on an available amplifier. Use both step and sinusoidal inputs for both tests.

E2.8f Analyze the integrator of Fig. E2.8f for slew rate and frequency response limiting. $A(s)$ is first-order response. Input is $Ku(t)$.

E2.8g An operational amplifier has a 0.75 V/μs slew rate and a gain-bandwidth product of 10^6 rad/s. It is connected in a noninverting amplifier configuration for a gain of 10. $A_o = 10^5$.

    a. For a 1-V input step, which will limit the output, slew rate or frequency response?

    b. The input is a 0.5-V peak sine wave. Which will limit the upper frequency, slew or bandwidth?

E2.8h An engineer has available an op amp that has a slew rate of 6 V/μs. If this must be used to amplify a 100-kHz signal, what is the maximum amplitude that the signal can have and still be assured of being in amplifier rating? Amplifier operating gain is 5.

2.9 OPERATIONAL AMPLIFIER INPUT SIGNALS: DIFFERENTIAL AND COMMON MODE VOLTAGES

To this point only differential signal effects have been considered in analyzing the performance of networks containing op amps. This is the configuration shown in Fig. 2-31(a) for which

$$V_o = A(V^+ - V^-) = AV_D \qquad (2\text{-}73)$$

where V_D is the differential or difference input signal at the terminals and the gain A is as described in earlier sections.

There is a second input signal type, illustrated in Fig. 2-31(b), called a common mode voltage (CMV). This is defined as the voltage seen by the amplifier when the input terminals are tied together. If the amplifier were ideal then Eq. (2-73) would predict the desirable zero output value for V_o^{cm}. However, due to practical matters within the op amp (imbalance in transistor stages), there is an output. The source of this error can be

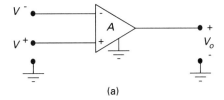

(a)

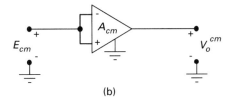

(b)

Figure 2-31 Conventions for describing op amp input signals. (a) Differential mode voltages. (b) Common mode voltage.

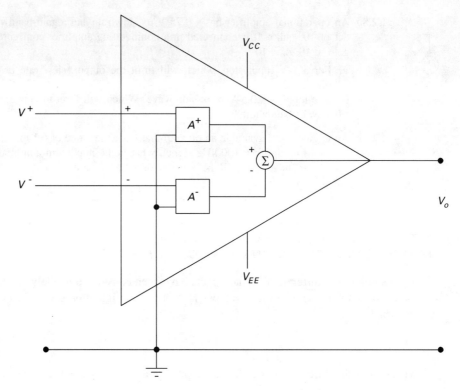

Figure 2-32 Functional op amp diagram that accounts for different gains, to the input signals.

explained by modifying Fig. 2-12(a) to show finite gains in the two inner amplifiers, as shown in Fig. 2-32. For this configuration

$$V_o = A^+ V^+ - A^- V^-$$

When the two input terminals are connected together this reduces to, by definition and using $E_{cm} = V^+ = V^-$,

$$V_o^{cm} = E_{cm}(A^+ - A^-) \equiv E_{cm} A_{cm} \qquad (2\text{-}74)$$

where A_{cm} is the common mode gain. Ideally A_{cm} is zero, but it may be a small positive or negative value. Thus any output produced by a common mode voltage can be attributed to a gain imbalance between the two inputs. Notice that V_o^{cm} would be interpreted as a signal by the processing that follows. The parameter used as a measure of the amplifier immunity to common mode voltages, or rejection of E_{cm}, is the common mode rejection ratio, CMRR. Before defining this parameter we take a closer look at the sources of common mode voltage.

The usual configurations used to represent sources of common mode input signals are shown in Fig. 2-33. In the left part of the figure separate unwanted sources, E_{cm}^+ and E_{cm}^-, are shown for each of the input signals E_1 and E_2. These common mode signals may

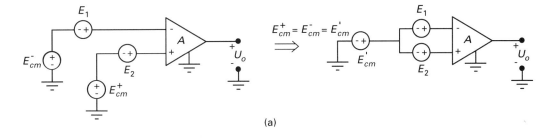

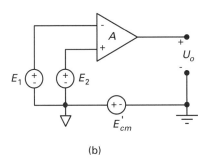

Figure 2-33 Representations of sources of common mode voltages.

be noise or extraneous signals present on the desired signals or introduced by electromagnetic or radio frequency interference (EMI, RFI) induced into the lines. If these sources are of equal magnitude, E_{cm}, they may be replaced by a single source, as the right half of the figure shows, using source multiplication.

A second source of common mode voltage is caused by noise or signals induced in the ground line. Digital current pulses frequently generate these signals. Such signals produce two different "grounds" by raising the levels of the op amp input signals as indicated by the different ground symbols. We denote this by E_{cm2}.

There is a third source of common mode voltage produced by the signals E_1 and E_2 even if other sources shown in the figure are not present. This is not immediately evident from the figure and requires a derivation. In order to make the result as general as possible we work with V^+ and V^-, which are to be interpreted as the total terminal voltages with respect to the output ground reference.

Since the two voltages V^- and V^+ can be varied independent of each other, one could imagine these voltages as two independent variables on orthogonal coordinate axes, as shown in Fig. 2-34(a). Changing V^+ from a value of V_a^+ to V_b^+ leaves the value of V^- unchanged. On this basis we say that the two signals are orthogonal or linearly independent. As a counterexample, if the axes were not at right angles, as shown in part (b) of the figure, then changing V^+ would result in a change of V^- also. One could, of course, move along an oblique line to hold V^- constant, but this process is not as convenient to represent directly. Since the two signals are independent and the amplifier is assumed to be linear, we can apply each of the signals one at a time and add the results at the output (superposition theorem).

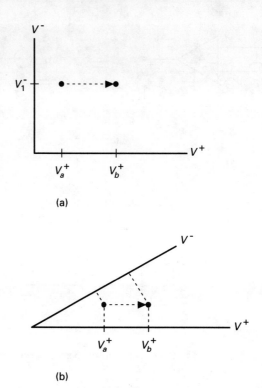

Figure 2-34 Representations of op amp input terminal voltages on coordinate axes.

What we want to do now is obtain a total common mode voltage, E_{cm}, and a differential mode voltage, V_D, in terms of V^- and V^+ such that E_{cm} and V_D are also orthogonal. This will allow us to use superposition in linear amplifier circuits and determine separately the effects of V_D (the desired input) and E_{cm} (an error input). To convert from two orthogonal variables to two other orthogonal variables we use a rotation of axes, as shown in Fig. 2-35. This is called an orthogonal transformation, and a set of simultaneous equations suffices to relate the voltage values in the two systems. In general we write the relationships as

$$V_D = a_{11}V^+ + a_{12}V^-$$
$$E_{cm} = a_{21}V^+ + a_{22}V^-$$

However, from the definition of the difference voltage V_D, which is $V_D = V^+ - V^-$, we would have $a_{11} = +1$ and $a_{12} = -1$ so that

$$V_D = V^+ - V^-$$
$$E_{cm} = a_{21}V^+ + a_{22}V^-$$

For an orthogonal transformation the coefficient determinant must be ± 1. From the preceding equations we obtain

Chap. 2 Amplifiers and Their Integration into Signal Processing Systems

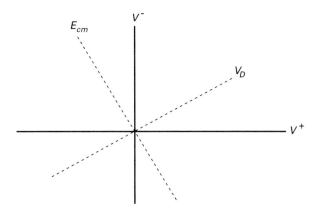

Figure 2-35 Rotation of axes (orthogonal transformation) used to convert op amp terminal voltages into differential and common mode components.

$$\begin{vmatrix} 1 & -1 \\ a_{21} & a_{22} \end{vmatrix} = a_{22} + a_{21} = \pm 1$$

This requires

$$a_{21} = \pm 1 - a_{22}$$

The set of simultaneous equations becomes

$$V_D = V^+ - V^-$$
$$E_{cm} = (\pm 1 - a_{22})V^+ + a_{22}V^-$$

Since E_{cm} and V_D are orthogonal their inner (scalar) product must be zero. Thus

$$(1)(\pm 1 - a_{22}) + (-1)(a_{22}) = 0 \rightarrow a_{22} = \pm 1/2$$

Then we have two valid possibilities:

$$V_D = V^+ - V^-$$
$$E_{cm} = 1/2 V^+ + 1/2 V^-$$

or

$$V_D = V^+ - V^-$$
$$E_{cm} = -1/2 V^+ - 1/2 V^-$$

We select the case that makes E_{cm} the positive function of V^+ and V^- to obtain finally

$$V_D = V^+ - V^- \tag{2-75}$$

$$E_{cm} = \frac{V^+ + V^-}{2} \tag{2-76}$$

Two important ideas come from these results:

1. Any sources or voltages applied to V^+ and V^- generate a common mode component,

$$E_{cmg} = \frac{E_1 + E_2}{2}$$

2. Any external sources of common mode voltage, such as E_{cm1} and E_{cm2} in Fig. 2-33, may be added to the E_{cmg} without affecting V_D, since they are orthogonal. Thus we may write the total common mode voltage using superposition as

$$E_{cm} = E_{cm1} + E_{cm2} + E_{cmg} = E_{cm1} + E_{cm2} + \frac{E_1 + E_2}{2} \tag{2-77}$$

Using the preceding results we can express the total output voltage using superposition and Eqs. (2-73) and (2-74) as follows:

$$V_o = A(V^+ - V^-) + A_{cm}\left(\frac{E_1 + E_2}{2} + E_{cm1}\right) = AV_D + A_{cm}E_{cm} \tag{2-78}$$

In subsequent sections we examine the practical cases where feedback is used.

Example 2-15

For two voltage sources v_x and v_y applied directly to the amplifier terminals as shown in Fig. 2-36(a), develop the equivalent circuit of the right side of Fig. 2-33 with $E_{cm2} = 0$.

Using Eqs. (2-75) and (2-76)

$$E_{cmg} = \frac{V^+ + V^-}{2} = \frac{v_y + v_x}{2}$$

$$V_D = V^+ - V^- = v_y - v_x = \frac{v_y - v_x}{2} + \frac{v_y - v_x}{2}$$

The expression for V_D has been divided into two equal parts in order to provide a junction point for tying in the common mode voltage. The solution is shown in Fig. 2-36(b). To verify this equivalency, note that

V_D = voltage drop from the positive terminal to the negative terminal

$$= V_{ac} = V_{ab} + V_{bc} = \frac{v_y - v_x}{2} + \frac{v_y - v_x}{2} = v_y - v_x$$

Exercises

E2.9a An operational amplifier has a signal $V(t) = 2\sin 100t$ applied to the negative (inverting) input terminal and a square wave of 2 V peak value (a symmetric wave) of frequency 15.92 Hertz applied to the positive (noninverting) terminal

Chap. 2 Amplifiers and Their Integration into Signal Processing Systems

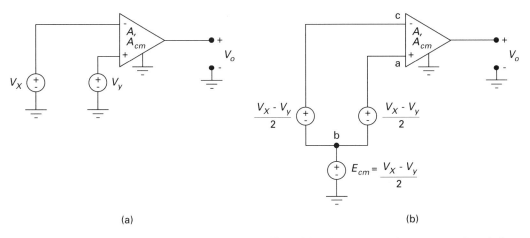

Figure 2-36 Modeling applied sources as differential and common mode sources (a) Actual circuit with applied surces. (b) Equivalent circuit with explicit differential and common mode sources.

($v_2(t)$). Plot the signal-generated common mode voltages and the difference voltages for the cases:
1. Square wave has a negative first half cycle.
2. Square wave has a positive first half cycle.

E2.9b For Example 2-15, give the equation for the output voltage in terms of v_x, v_y, A, and A_{cm}. (See Eq. 2-78.)

E2.9c For the left circuit of Fig. 2-33(a) obtain the expressions for the common mode and difference voltages in terms of E_1, E_2, E_{cm}^+, and E_{cm}^- using $E_{cm2} = 0$. Redraw the circuit showing a single E_{cm} source and two other sources all appropriately labeled with values (expressions). Show that for $E_{cm}^+ = E_{cm}^-$ the right circuit of Fig. 2-33(a) results.

E2.9d Prove the equivalency shown in Fig. E2.9d and from this compute the differential gain output. Do this by obtaining an expression for V_D for the right circuit.

E2.9e Write the expressions for the differential input voltage and common mode voltage for the unity gain buffer (isolation) amplifier.

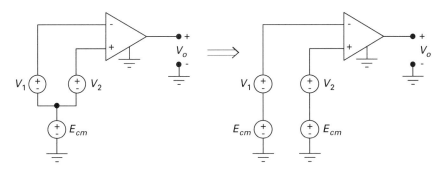

Figure E2.9d

2.10 OPERATIONAL AMPLIFIER COMMON MODE REJECTION

In this section we present the parameters and specifications used by manufacturers to describe the common mode responses of their amplifiers. We also investigate the effect of the common mode error on the signal processing system and specifically relate the specifications to the number of bits used in the digital words.

There are two terms used to describe the ability of an amplifier to reject common mode effects. These are the common mode rejection ratio (*CMRR*) and common mode rejection (*CMR*). Although these are technically different quantities, we find that among various manufacturers and textbooks the terms are used interchangeably; that is, what one calls *CMR* another calls *CMRR*. A further compounding of the problem is that the definitions used are not (or appear not) even consistent.

One should learn from the preceding that whenever a manufacturer specification sheet is used, one should check the definition of the parameter. This also holds for several other parameters. Obviously, it may develop that "specsmanship" is involved, so that the manufacturer will select the definition that will give the product the best apparent specification when in fact it may be equivalent to a competitive unit. This, of course, may not be the case, but rather a manufacturer may use the definition that is felt most useful. Finally, when one begins to perform calculations, it is important that when numerical values are at stake one examines each source if values and/or equations are taken from different sources.

The next few paragraphs are intended to present the various definitions so one can apply appropriate equations. Equivalencies are also developed where possible. The results apply equally well to isolation amplifiers and instrumentation amplifiers presented later.

We now give the basic definitions of common mode rejection ratio (*CMRR*) and common mode rejection (*CMR*) to be used in this text. We have followed the conventions adopted by the majority of the manufacturers, writers, and standards.

For *CMRR* we use the value of the ratio of the quantities involved; that is, as the ratio of voltages, or currents, or gains, as the case may be. The definition is (see Fig. 2-37 for configuration)

$$CMRR \equiv \frac{\text{Desired Output Voltage for Differential Input Gain}}{\text{Common Mode Output Voltage for } E_{cm} \text{ of Same Amplitude as Differential Input}}$$

$$= \frac{V_o}{V_o^{cm}} = \frac{AV}{A_{cm}E_{cm}} = \frac{AV}{A_{cm}V} = \frac{A}{A_{cm}} = \frac{\text{Differential Gain}}{\text{Common Mode Gain}} \qquad (2\text{-}79)$$

Since we want any common mode signals or gains to be as small as possible, a large *CMRR* implies a better amplifier.

From the preceding definition we define the *CMR* as

$$CMR \equiv 20 \log_{10} CMRR \text{ dB} \qquad (2\text{-}80)$$

We assume that positive values are used for A and A_{cm}. Thus based on our definition a larger positive *CMR* means a better amplifier. For practical amplifiers (including the instrumentation amplifier studied later) *CMR* values range from 50 to 120 dB.

Chap. 2 Amplifiers and Their Integration into Signal Processing Systems

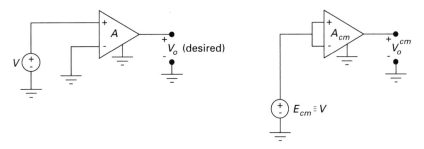

Figure 2-37 Configurations used to define common mode rejection ratio (*CMRR*).

Next we examine some other definitions in common use and show how they relate to the preceding.

1. Recommended by IEEE standards

$$CMRR \equiv \frac{E_{cm}(\text{at input})}{\dfrac{V_o^{cm}(\text{effect at output})}{\text{amplifier gain}}} = \frac{E_{cm}(\text{at input})}{\dfrac{V_o^{cm}(\text{at output})}{A}}$$

This is equivalent to the definition given earlier, since we may write this in the forms:

$$CMRR = \dfrac{\dfrac{A}{V_o^{cm}(\text{at output})}}{E_{cm}(\text{at input})} = \dfrac{\text{Amplifier Gain}}{\text{Common Mode Gain}} = \dfrac{A}{A_{cm}}$$

2. Some manufacturers and texts specify the quantity *CMRR* in dB. All this means is that they have used the name *CMRR* in place of *CMR*. Since the specification sheets give the unit dB, no problem should arise.
3. A third common definition in use simply ignores the differential gain property of the amplifier and uses the direct ratio of common mode voltages as

$$CMRR = \frac{V_o^{cm}}{E_{cm}} = |A^+ - A^-| = \text{Common mode gain} = A_{cm}$$

where

E_{cm} = common mode input voltage
V_o^{cm} = common mode output voltage due to E_{cm}
A^+ = gain of positive input port
A^- = magnitude of gain of negative input port

From this one would have

$$CMR = 20 \log_{10}(CMRR)$$

as before, except that since *CMRR* is very small. This definition yields a negative number. Thus if a specification sheet gives a negative number for dB, one must use this definition.

From the preceding we see that, aside from the interchanging of the terms *CMR* and *CMRR*, there are really only two different definitions in spite of the forms. These are

1.
$$CMRR_1 = \frac{E_{cm}(at\ input)}{\frac{V_o^{cm}(at\ output)}{A}}$$

2.
$$CMRR_2 = \frac{V_o^{cm}(at\ output)}{E_{cm}(at\ input)} \qquad (2\text{-}81)$$

Note that these differ only by the amplifier differential (desired) gain. Therefore, whenever definition 1 is used, the amplifier gain must be specified. This is the most common practice. $CMRR_1$ is a large value; $CMRR_2$ is a small value.

In dB then,

$$CMR_1 = 20\ \log_{10} CMRR_1 = 20\ \log_{10}\left[\frac{E_{cm}(at\ input)}{V_o^{cm}(at\ output)} \times A\right]$$

$$= 20\ \log_{10}\left[\frac{E_{cm}(at\ input)}{V_o^{cm}(at\ output)}\right] + 20\ \log_{10}(A)$$

$$= -20\ \log_{10}\frac{V_o^{cm}(at\ output)}{E_{cm}(at\ input)} + 20\ \log_{10}(A)$$

$$CMR_1 = -CMR_2 + 20\ \log_{10}(A) \qquad (2\text{-}82)$$

Example 2-16

Suppose that an amplifier has a gain of 1000 and that an input common mode signal of 10 V produces an output voltage of 1 mV. Then:

$$CMR_2 = 20\ \log_{10}\frac{1\ mV}{10V} = 20\ \log_{10}(10^{-4}) = -80\ dB$$

$$CMR_1 = -(-80) + 20\ \log_{10}(1000) = 80 + 60 = 140\ dB$$

Thus, the same amplifier has quite different specifications for its *CMR*. Also note that if a spec sheet gives a negative dB we cannot simply ignore the minus sign when comparing with an amplifier from another manufacturer giving *CMR* in positive dB. We must check the definition that a manufacturer uses.

In this book we use the definitions adopted by the majority of users and the IEEE:

Chap. 2   Amplifiers and Their Integration into Signal Processing Systems   187

$$CMRR \equiv \frac{E_{cm}}{\frac{V_o^{cm}}{A}} = \frac{E_{cm}}{V_o^{cm}} A = \frac{A}{A_{cm}}$$

and

$$CMR \equiv 20 \log_{10} CMRR \text{ dB} \qquad (2\text{-}83)$$

We now need to relate the values of differential and common mode gains to the number of bits N used in the data words. By doing this we can determine whether or not the common mode output will affect the data words; or, if the number of bits N is given, we can select an amplifier having appropriate gains.

Suppose that we let V_o^{span} represent the total output voltage span that is to be converted to digital form. For example, if $V_{omin} = -10V$ and $V_{omax} = +10$ V, then $V_o^{span} = 10 - (-10) = 20$ V, and 1/2 LSB would then be

$$1/2 \text{ LSB} = 1/2 \frac{V_o^{span}}{2^N} = \frac{V_o^{span}}{2^{N+1}}$$

Now suppose that with external components connected to the amplifier the ideal output is to be V_o^I. The error is then obtained using Eq. (2-78) for the actual output from which

$$error = \left| V_o - V_o^I \right| = \left| A(V^+ - V^-) + A_{cm} \frac{(V^+ + V^-)}{2} - V_o^I \right| \qquad (2\text{-}84)$$

This result includes the effects of finite gain and common mode gain. Note we have, for simplicity, neglected any induced common modes E_{cm1} and E_{cm2}. If any were present they would simply be added to $(V^+ + V^-)/2$.

From the definition of *CMR* Eq. (2-83) we have

$$A_{cm} = A 10^{-CMR/20} \qquad (2\text{-}85)$$

The error expression becomes:

$$error = \left| A(V^+ - V^-) + A 10^{-CMR/20} \left(\frac{V^+ + V^-}{2} \right) - V_o^I \right| \leq 1/2 \text{ LSB} = \frac{V_o^{span}}{2^{N+1}} \qquad (2\text{-}86)$$

Note that this result is valid whatever the external configuration may be that produces V^- and V^+. Also, we have assumed that A_{cm} is positive. If it were negative—this is usually not known in practice, however—we would negate the common mode component.

We now consider a few examples that illustrate the uses of the general formula developed above.

Example 2-17

Amplifier with resistive feedback.

Suppose we have an amplifier of gain A (usually very large) and we wish to amplify the signals on lines 1 and 2. The configuration is shown in Fig. 2-38. The

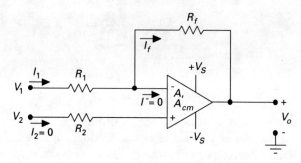

Figure 2-38 Two-input amplifier configuration.

output is to be converted by an N-bit A/D converter. What is the required (minimum) CMR of the amplifier? (Neglect stray common mode, E'_{cm}.)

The largest possible outputs would be nearly $\pm V_s$. If we assume that the total swing is $2V_s$, then the voltage span could be approximated by

$$V_o^{span} = 2V_s$$

$$\text{Input common mode voltage (CMV)} = E_{cm} = \frac{V^+ + V^-}{2}$$

If we neglect the amplifier input current, $I_2 = 0$ so $V^+ = V_2$, then

$$E_{cm} = \frac{V^+ + V^-}{2} = \frac{V_2 + V^-}{2} \qquad (2\text{-}87)$$

To find V^- we proceed as follows. Neglecting the current into the amplifier negative terminal,

$$I_1 = I_f = \frac{V_1 - V_o}{R_1 + R_f}$$

Then

$$V^- = V_1 - I_1 R_1 = V_1 - \frac{R_1}{R_1 + R_f}(V_1 - V_o) = \frac{R_f}{R_1 + R_f} V_1 + \frac{R_1}{R_1 + R_f} V_o \qquad (2\text{-}88)$$

From the properties of the amplifier we have from Eq. (2-78) with $E'_{cm} = 0$:

$$V_o = A(V^+ - V^-) + A_{cm}\left(\frac{V^+ + V^-}{2}\right)$$

from which

$$V_o = A\left(V_2 - \frac{R_f}{R_1 + R_f}V_1 - \frac{R_1}{R_1 + R_f}V_o\right) + \frac{A_{cm}}{2}\left(V_2 + \frac{R_f}{R_1 + R_f}V_1 + \frac{R_1}{R_1 + R_f}V_o\right)$$

Solving for V_o:

$$V_o = \frac{A + \frac{A_{cm}}{2}}{1 + \left(A - \frac{A_{cm}}{2}\right)\left(\frac{R_1}{R_1 + R_f}\right)} V_2 - \frac{\left(A - \frac{A_{cm}}{2}\right)\left(\frac{R_f}{R_1 + R_f}\right)}{1 + \left(A - \frac{A_{cm}}{2}\right)\left(\frac{R_1}{R_1 + R_f}\right)} V_1 \qquad (2\text{-}89)$$

Substituting this into Eq. (2-88) we have

$$V^- = \frac{R_f}{R_1 + R_f}\left(1 - \frac{R_1\left(A - \frac{A_{cm}}{2}\right)}{\left(1 + A - \frac{A_{cm}}{2}\right)R_1 + R_f}\right) V_1$$

$$+ \frac{R_1\left(A + \frac{A_{cm}}{2}\right)}{\left(1 + A - \frac{A_{cm}}{2}\right)R_1 + R_f} V_2 \qquad (2\text{-}90)$$

(As an aside, note that $\lim_{A \to \infty} V^- = V_2$ as it should for the principle of the virtual short.)

If V_1 and/or V_2 has a DC level, then E_{cm} could be appreciable. Levels of several volts are not uncommon.

From the expressions for V^+ and V^- we obtain, after some algebra,

$$V^+ + V^- = \frac{(1 + 2A)R_1 + R_f}{\left(1 + A - \frac{A_{cm}}{2}\right)R_1 + R_f} V_2 + \frac{R_f}{\left(1 + A - \frac{A_{cm}}{2}\right)R_1 + R_f} V_1$$

Using Eq. (2-85) to eliminate A_{cm} this becomes

$$V^+ + V^- = \frac{(1 + 2A)R_1 + R_f}{\left(1 + A - \frac{A}{2}10^{-CMR/20}\right)R_1 + R_f} V_2$$

$$+ \frac{R_f}{\left(1 + A - \frac{A}{2}10^{-CMR/20}\right)R_1 + R_f} V_1 \qquad (2\text{-}91)$$

Similarly we compute

$$V^+ - V^- = \frac{\left(1 - A10^{-CMR/20}\right)R_1 + R_f}{\left(1 + A - \frac{A}{2}10^{-CMR/20}\right)R_1 + R_f} V_2$$

$$- \frac{R_f}{\left(1 + A - \frac{A}{2}10^{-CMR/20}\right)R_1 + R_f} V_1 \tag{2-92}$$

The ideal output can be obtained from Eq. (2-89) by letting $A \to \infty$ and $A_{cm} \to 0$. The result is, as expected,

$$V_o^I = \left(1 + \frac{R_f}{R_1}\right) V_2 - \frac{R_f}{R_1} V_1 \tag{2-93}$$

(Note that for $V_2 = 0$ we obtain the usual ideal inverter with gain and for $V_1 = 0$ the noninverter with gain. These are the usual equations one uses to design amplifiers.)

Now using Eqs. (2-91), (2-92), and (2-93) in the general formula for error, Eq. (2-86), we obtain (again after some algebra)

$$\left| \frac{R_1\left(A10^{-CMR/20} - 1\right) + R_f\left(A10^{-CMR/20} - 2 - \frac{R_f}{R_1}\right)}{\left(1 + A - \frac{A}{2}10^{-CMR/20}\right)R_1 + R_f} V_2 + \frac{R_f\left(1 + \frac{R_f}{R_1}\right)}{\left(1 + A - \frac{A}{2}10^{-CMR/20}\right)R_1 + R_f} V_1 \right|$$

$$\leq \frac{V_o^{span}}{2^{N+1}} \tag{2-94}$$

With regard to this result see exercise E2.10f.

Do not forget that this result holds only for the one amplifier configuration analyzed. Each different circuit must be separately analyzed unless it can be obtained as a special case of this one. Typical special cases are $V_1 = 0$ or $V_2 = 0$.

Example 2-18

Some numerical results for two special case configurations.

Case I: $V_2 = 0$. Suppose we wish to use a 10-bit data word and that the usual amplifier open loop gain is $A = 2 \times 10^5$. If we want to have an ideal gain of -20 on V_1, then $R_f = 20R_1$. The amplifier being used has a voltage supply of ± 11 V. (This would allow nominal full-scale output swings of ± 10 V.)

For $V_2 = 0$ the circuit becomes that of Fig. 2-39. From Eq. 2-94 we obtain (for $V_2 = 0$):

Chap. 2 Amplifiers and Their Integration into Signal Processing Systems

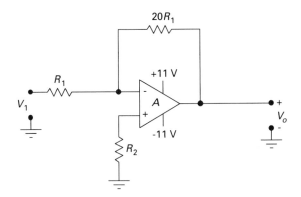

Figure 2-39 Inverting amplifier for Example 2-18 (case (a)).

$$\left| \frac{20R_1(1+20)}{R_1 + 20R_1 + 2 \times 10^5 R_1 - 10^5 R_1 \times 10^{-CMR/20}} V_1 \right| \le \frac{2 \times 10}{2^{10+1}} = \frac{10}{2^{10}}$$

Now for a gain of –20 the largest value of V_1 is ±0.5 V. Also, since the denominator is always positive, the absolute value bars may be removed ($CMR \ge 0$). The preceding reduces to

$$\frac{(20)(21)(0.5)}{\left[21 + 2 \times 10^5 - 10^{(5-CMR/20)} \right]} \le \frac{10}{2^{10}}$$

Neglecting the 21 in the denominator since it is small compared to 2×10^5 and rearranging yields

$$21 \times 2^{10} \le 2 \times 10^5 - 10^{(5-CMR/20)}$$

Solving for *CMR* gives

$$CMR \ge -5 \text{ dB}$$

Since *CMR* is by definition positive, any amplifier with a gain of 2×10^5 will work, and we learn from this example that the grounded positive input amplifier configuration is least sensitive to signal-generated common mode voltages.

Case II: $V_1 = 0$ (same resistor values as for Case I). The circuit is shown in Fig. 2-40. For this case the ideal gain is 21. Using Eq. (2-94) with $V_1 = 0$ we proceed as follows:

$$\left| \frac{R_1(2 \times 10^5 \times 10^{-CMR/20} - 1) + 20R_1(2 \times 10^5 \times 10^{-CMR/20} - 2 - 20)}{R_1 + 20R_1 + 2 \times 10^5 R_1 - 10^5 R_1 10^{-CMR/20}} V_2 \right| \le \frac{10}{2^{10}}$$

or,

$$\left| \frac{4.2 \times 10^6 \times 10^{-CMR/20} - 441}{21 + 2 \times 10^5 - 10^5 \times 10^{-CMR/20}} V_2 \right| \le \frac{10}{2^{10}}$$

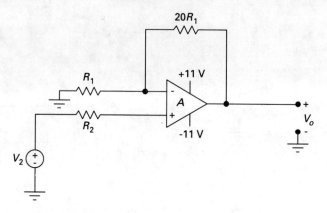

Figure 2-40 Noninverting amplifier for Example 2-18 (case (b)).

As before, neglect the 21 in the denominator and also use the largest V_2 value, which is

$$V_2 \max = \pm \frac{10}{21} = \pm 0.476 \text{ V}$$

Before we can remove the absolute value bars, we observe that the denominator is always positive ($CMR > 0$), but the numerator can be negative. In case the numerator is negative we take V_2min, which is negative, so that their product is positive. Thus we can write

$$\left| \frac{\left(4.2 \times 10^6 \times 10^{-CMR/20} - 441\right)(\pm 0.476)}{10^5 \left(2 - 10^{-CMR/20}\right)} \right| \leq \frac{10}{2^{10}}$$

The critical CMR value for which the numerator becomes negative is at

$$4.2 \times 10^6 \times 10^{-CMR/20} - 441 = 0$$

which yields

$$CMR_{crit} = 79.6 \text{ dB}$$

The inequality above may be rewritten as

$$(\pm 0.476)\left(4.2 \times 10^6 \times 10^{-CMR/20} - 441\right) \leq \frac{1}{2^{10}} \times 10^6 \left(2 - 10^{-CMR/20}\right)$$

where the bottom sign is used when $CMR > 79.6$ dB. Then the preceding yields two equations:

$$10^{-CMR/20} \leq 0.0010842, \text{ when } CMR < 79.6 \text{ dB}$$

and

Chap. 2 Amplifiers and Their Integration into Signal Processing Systems

$$10^{-CMR/20} \leq -0.000872, \text{ when } CMR > 79.6 \text{ dB}$$

The second of these is not possible, because the left side is always greater than zero, so we know $CMR < 79.6$ dB. Then for $CMR < 79.6$ dB, solving for CMR we obtain

$$CMR \geq 59 \text{ dB}$$

which agrees with our assumption that the lower limit is less than 79.6 dB. Note that the positive gain configuration places the severest restriction on CMR.

Exercises

E2.10a For the operational amplifier configuration shown in Fig. E2.10a determine the relationship between E_1, A, CMR, and N that will allow one to compute the CMR. (Hint: The equation developed in the text can be specialized to this case.) Assume amplifier terminal currents are zero. For the case $V_{CC} = |V_{EE}| = 10$ V, $A = 10^5$, and $R_f = 40R_1$ compute the minimum CMR for a 12-bit system.

E2.10b Shown in Fig. E2.10b is a difference amplifier configuration that has a common mode signal between the two grounds, as indicated. Using the output ground as reference, derive the equation relating v_o to v_1, v_2, and E_{cm}. Assume $A = \infty$ and input currents to the op amp are zero. What condition eliminates the common mode output?

E2.10c For the unity gain buffer amplifier determine the relationship between the input e_1, A, N, and CMR. For $A = \infty$, and V_2 for worst case (drive output to limit) make a table showing N and CMR for $4 \leq N \leq 14$.

E2.10d Determine the common mode rejection of the overall amplifier assuming that A is large (see Fig. E2.10d). Denote the CMR of the op amp alone by CMR_0.

E2.10e For the circuit shown in Fig. E2.10e, what are allowable values of R_f if $A = 2 \times 10^5$, $CMR = 40$ dB, $V_o^{span} = 20$ V, and an 8-bit data word is used? If $E_1 = \pm 0.5$ V, can this amplifier be used? If not, what would you suggest?

E2.10f For practical amplifiers $A \gg A_{cm}$ and $A \gg 1$. Also, since CMR is usually greater than 40 dB, we would have

$$A10^{-CMR/20} \ll A$$

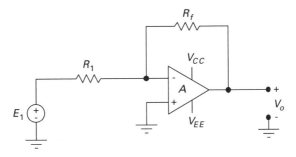

Figure E2.10a

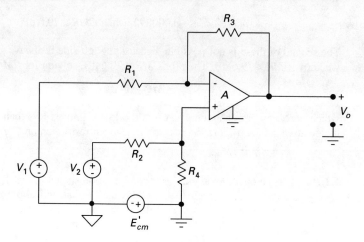

Figure E2.10b

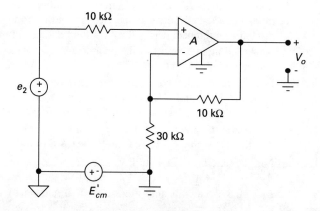

Figure E2.10d

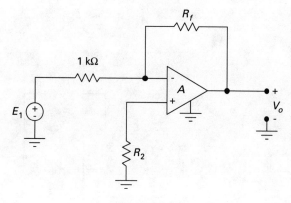

Figure E2.10e

Chap. 2   Amplifiers and Their Integration into Signal Processing Systems   195

For these approximations obtain a reduced expression for the error of Example 2-17, Eq. (2-94).

2.11 OPERATIONAL AMPLIFIER CURRENT OFFSET AND VOLTAGE OFFSET

Operational amplifiers typically have input impedances (resistances) ranging from 100 kΩ (bipolar) to $10^{12}\,\Omega$ (1 TΩ) (*FET*). Many amplifiers also require some DC path to ground through both the positive and negative input terminals to provide proper biasing and referencing. Thus even if the inputs are grounded, an output may result due to currents flowing in the input lines. Additionally, the potential levels at each of the terminals may effectively be different and produce an output voltage.

In order to discuss these effects we consider the inverting amplifier configuration. The implication here is that each configuration must be examined to determine the effects of current and voltage offsets. The basic structure is shown in Fig. 2-41 and is selected because of its practical value.

Since the offset current sources are shown explicitly, there will be zero currents into the minus and plus terminals of the internal amplifier of gain A. The value of A is the same as that for the actual amplifier and is on the order of 10^5.

Now for the internal amplifier, $V_0 = A\,(V^+ - V^-)$ as usual. From the figure we have:

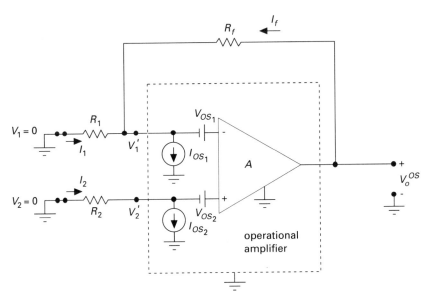

Figure 2-41 Current offsets and voltage offsets in an operational amplifier. Basic structure.

$$V^- = V_{os_1} + V_1'$$
$$V^+ = V_{os_2} + V_2'$$
$$I_1 + I_f = I_{os_1} \quad (2\text{-}95)$$
$$I_2 = I_{os_2}$$

Next, using Ohm's law, we have

$$I_1 = \frac{-V_1'}{R_1}$$

$$I_f = \frac{V_o - V_1'}{R_f}$$

$$I_1 = \frac{-V_2'}{R_2}$$

Substituting these three results into the current equations of Eq. (2-95) we have that set reducing to

$$V^- = V_{os_1} + V_1'$$
$$V^+ = V_{os_2} + V_2'$$
$$-\frac{V_1'}{R_1} + \frac{V_o - V_1'}{R_f} = I_{os_1}$$
$$-\frac{V_2'}{R_2} = I_{os_2}$$

We next write this last set of equations in standard form and include the amplifier basic equation $V_o = A\,(V^+ - V^-)$ as a first equation:

$$V_o + AV^- - AV^+ + 0 \cdot V_1' + 0 \cdot V_2' = 0$$
$$0 \cdot V_o + V^- + 0 \cdot V^+ - V_1' + 0 \cdot V_2' = V_{os_1}$$
$$0 \cdot V_o + 0 \cdot V^- + V^+ + 0 \cdot V_1' - V_2' = V_{os_2}$$
$$\frac{1}{R_f}V_o + 0 \cdot V^- + 0 \cdot V^+ - \left(\frac{1}{R_f} + \frac{1}{R_1}\right)V_1' + 0 \cdot V_2' = I_{os_1}$$
$$0 \cdot V_o + 0 \cdot V^- + 0 \cdot V^+ + 0 \cdot V_1' - \frac{1}{R_2}V_2' = I_{os_2} \quad (2\text{-}96)$$

Next solve this set for V_0, which is not too much fun, but at least there are lots of nice zeros. The solution is

Chap. 2 Amplifiers and Their Integration into Signal Processing Systems 197

$$V_0 = \frac{R_1 + R_f(V_{os_2} - V_{os_1})}{1 + \frac{R_1 + R_f}{AR_1}} + \frac{I_{os_1}R_f - \frac{R_2(R_1 + R_f)}{R_1}I_{os_2}}{1 + \frac{R_1 + R_f}{AR_1}} \qquad (2\text{-}97)$$

As a practical matter, it is usual to define the difference terms $V_{os_2} - V_{os_1}$ and $I_{os_2} - I_{os_1}$ as the input offset parameters (see definitions in Table 2-2). Thus in the numerator of the second term of the preceding equation we add and subtract $R_f I_{os_2}$ to obtain

$$V_0 = \frac{\frac{R_1 + R_f}{R_1}(V_{os_2} - V_{os_1})}{1 + \frac{R_1 + R_f}{AR_1}}$$

$$+ \frac{-R_f(I_{os_2} - I_{os_1}) + \left[R_f I_{os_2} - \frac{R_2(R_1 + R_f)}{R_1}I_{os_2}\right]}{1 + \frac{R_1 + R_f}{AR_1}}$$

This can be written as three terms rather than two and also simplified by defining $V_{os} = V_{os_2} - V_{os_1}$ and $I_{os} = I_{os_2} - I_{os_1}$, which are the usual manufacturer's specifications. Note that these quantities can be positive or negative. Then the output offset effect is, for this configuration,

$$V_o^{os} = V_o = \frac{\frac{R_1 + R_f}{R_1}V_{os}}{1 + \frac{R_1 + R_f}{AR_1}} - \frac{R_f}{1 + \frac{R_1 + R_f}{AR_1}}I_{os} + \frac{\left[R_f - R_2\left(1 + \frac{R_f}{R_1}\right)\right]I_{os_2}}{1 + \frac{R_1 + R_f}{AR_1}} \qquad (2\text{-}98)$$

What we obviously want to do is minimize this output since it is not a desired one. However, R_2 is the only real independent variable. Since I_{os_2} is not normally known, we observe that the last term can be made zero by setting the numerator to zero. This yields

$$R_2 = \frac{R_1 R_f}{R_1 + R_f} \qquad (2\text{-}99)$$

With this result V_o^{os} becomes

$$V_o^{os} = \frac{\frac{R_1 + R_f}{R_1}V_{os}}{1 + \frac{R_1 + R_f}{AR_1}} - \frac{R_f}{1 + \frac{R_1 + R_f}{AR_1}}I_{os} \qquad (2\text{-}100)$$

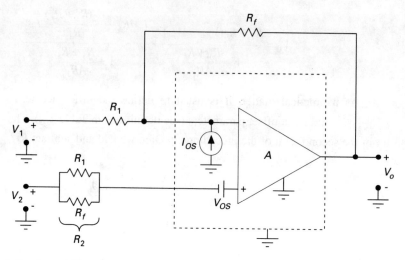

Figure 2-42 Input offset parameter equivalent circuit that shows manufacturer specification values.

The worst case is when V_{os} and I_{os} have opposite signs.

From this last equation we obtain two important results. First, we can redraw the circuit of Fig. 2-41 to get the form that gives circuit values showing the input offset parameters as specified by manufacturers. This is shown in Fig. 2-42. Note particularly that the equivalent input offset current direction is toward the negative terminal now. The current I_{os} flows out into R_1 and R_f (internal amplifier input current is zero) which makes the negative terminal positive with respect to ground, thereby producing a negative output voltage contribution as required by the right term in Eq. (2-100). These offset quantities are often called offset values referred to the input (RTI).

The second result we get is a criterion for selecting an amplifier in terms of V_{os} and I_{os}. Although it is possible to make V_o^{os} zero at a given temperature using external compensation as will be shown next, we can determine at least whether it is even necessary to provide this initial zero set or if we can select an amplifier that has values which make V_o^{os} zero. For the worst-case output error the two terms in Eq. (2-100) add, and this error must be less than 1/2 LSB to be negligible. Thus we require

$$\frac{\left(1+\dfrac{R_f}{R_1}\right)}{1+\dfrac{1}{A}\left(1+\dfrac{R_f}{R_1}\right)}|V_{os}| + \frac{R_f}{1+\dfrac{1}{A}\left(1+\dfrac{R_f}{R_1}\right)}|I_{os}| \leq \frac{V_o^{span}}{2^{N+1}} \qquad (2\text{-}101)$$

Since A is usually much greater than the design gain $R_f/R_1 + 1$, this may be simplified to

$$\left(1+\frac{R_f}{R_1}\right)|V_{os}| + R_f|I_{os}| < \frac{V_o^{span}}{2^{N+1}} \qquad (2\text{-}102)$$

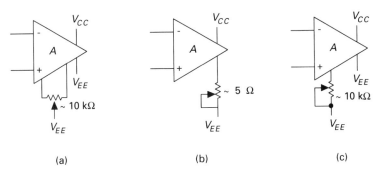

Figure 2-43 Offset nulling using op amp terminals other than input. (a) Offset null terminal pair. (b) Power supply offset nulling. (c) Offset null single terminal.

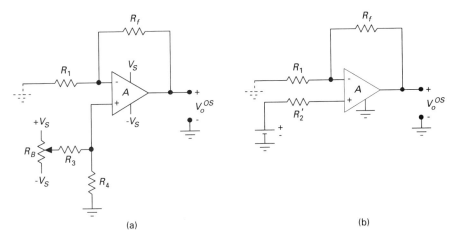

Figure 2-44 Bias method for offset trim. (a) Basic offset configuration. (b) Thevenin equivalent of bias offset (a).

In most operational amplifiers there are two external pin connections that can be used to set V_o^{os} equal to zero at a given temperature. These are called offset null terminals, which are identified in Fig. 2-18. However, when several op amps are placed on a single chip there are not enough pins to provide offset null for each amplifier, so Eq. (2-102) provides the criterion. There are other ways of providing offset nulling using the power supply and input terminals. Fig. 2-43 shows the connections using the offset null terminals and the power supply terminals. It is also important to know that for some op amps the single terminal nulling may have a separate pin from the power supply, as shown in Fig. 2-44. In some respects this technique is better than using the null terminals, since it is possible to provide the required R_2 value shown in Fig. 2-42 at the same time. That this is so can be explained by considering the Thevenin equivalent circuit of the bias network consisting of $\pm V_s$, R_B, R_3, and R_4. As the bias potentiometer setting changes, the equivalent resistance R_2' changes. Thus one can not only cancel V_o^{os} but also set an optimum R_2' for reducing the right term in Eq. (2-98). Recall that the other two schemes shown in Fig. 2-43 still require an R_2 value given by Eq. (2-99), although these two features could be adjusted for some "best" performance.

As a final note one should be aware that the static or offset performances of bipolar and FET op amps are quite different. The bipolar types generally have higher offset and bias current values and lower offset voltages which tend to be more temperature variable. (This is discussed in the next section.) The bias currents of most FET amplifiers are nearly of the same magnitude as the current offset, so the inclusion of resistor R_2 may not always be necessary. For the FET amplifiers one normally finds that offset voltages rather than currents require most of the attention. The temperature variation of voltages is most important for these type amplifiers.

Exercises

E2.11a Design an amplifier that has a signal applied only at V_1 (Fig. 2-42, with $A = \infty$) that is to have a gain of -100 and whose maximum input signal is 100 mV (unipolar-positive). The operational amplifier has an offset current of 50 nA and an offset voltage of 50 µV RTI. Answer the following:

a. What is the DC level of V_o^{os} due to the offset sources only for the worst case?

b. Refer your output offset voltage to the input and compare it with the maximum applied signal.

c. Decrease each of your resistor values by a factor of 10 and determine whether any improvement is obtained. Repeat for an increase by a factor of 10.

d. For your original circuit, what is the smallest allowable input signal that can be used if an 8-bit data digitization is to be used (offset not nulled out)?

e. If the impedance the circuit drives is to be 330 kΩ (resistive), what is the output required current drive for your amplifier?

E2.11b For time-varying voltages it is proposed to use capacitive coupling, as shown in Fig. E2.11b. Comment on the effect of C on the output offset. (Where does the DC offset I_{os} flow?) Would this configuration work for op amps that required DC continuity external? Explain.

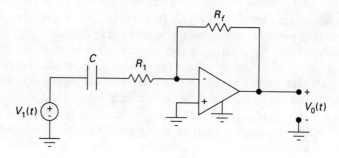

Figure E2.11b

Chap. 2 Amplifiers and Their Integration into Signal Processing Systems

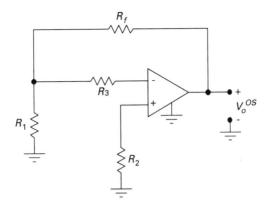

Figure E2.11c

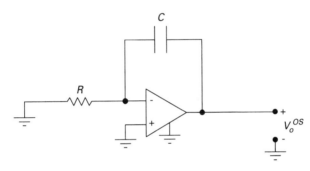

Figure E2.11d

E2.11c For the inverting amplifier shown in Fig. E2.11c, what is the output offset caused by V_{os} and I_{os} in terms of these and resistor values? What should R_3 be for minimum output offset? Use the op amp model of Fig. 2-41. Here is a partial answer:

$$R_3 = R_2 - \frac{R_1 R_f}{R_1 + R_f}$$

E2.11d Derive an equation for the output voltage of the integrator in Fig. E2.11d due to the input offset voltage V_{os}. For an N-bit data representation, what is the longest allowable integration time in terms of V_{os}, N, R, C, and the output span V_o^{span}?

E2.11e Show that the bias offset configuration in Fig. 2-44 can be adjusted for optimum value of R_2' and can cancel input offset at the same time.

E2.11f For the summer-amplifier shown in Fig. E2.11f, determine the value of R_3 that results in an output offset voltage that is a function of V_{os} and I_{os} variables only. Use Fig. 2-41 as the amplifier model.

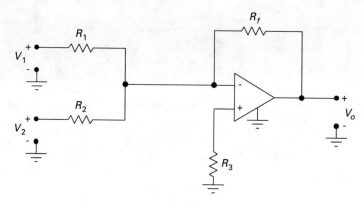

Figure E2.11f

2.12 OPERATIONAL AMPLIFIER TEMPERATURE CHARACTERISTICS

Since the operational amplifier is a solid-state device, its parameters vary with temperature. The most important ones are input offset current temperature dependence and input offset voltage temperature dependence. These are denoted respectively by

$$D_T^{I_{os}} = \frac{\Delta I_{os}}{\Delta T}$$

and

$$D_T^{V_{os}} = \frac{\Delta V_{os}}{\Delta T}$$

Since V_{os} and I_{os} are at the amplifier input, these are called the drift values referred to the input (RTI). It is important to know, however, how the manufacturer determines the Δs and any references used. One precise method for specifying the temperature variation is the so-called "butterfly" definition used by some manufacturers. The amplifier is first zeroed at some reference (usually room) temperature T_R. The temperature is then raised to a high limit of T_H and then lowered to T_L, which span the desired temperature limits. The offset values are measured at these temperatures and a curve, as shown in Fig. 2-45, might be observed, where offset voltage is used as an example. Similar definitions apply to offset current drift. From the measurements two computations are made:

1. High temperature: $\dfrac{\Delta V_{os}^H}{T_H - T_R}$

2. Low temperature: $\dfrac{\Delta V_{os}^L}{T_R - T_L}$

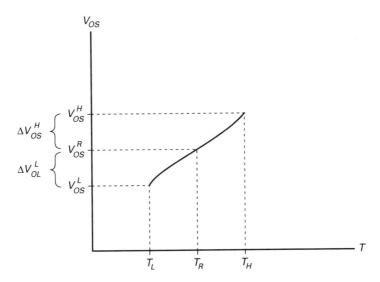

Figure 2-45 Temperature offset voltage variation and definitions.

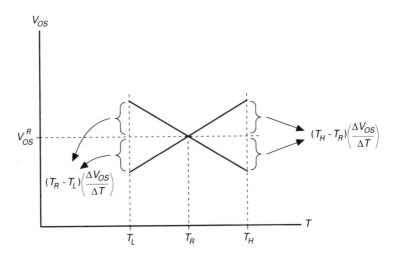

Figure 2-46 Temperature drift butterfly characteristic for V_{os}.

The honest amplifier rating would then be the larger of the two values, and this would be the reported value of $\Delta V_{os}/\Delta T$. Suppose, for example, that the high temperature reading was the largest. Then if one used linear interpolation one would be reasonably confident of being within the rating. This is shown in Fig. 2-46 for the extremes of possible variation, whether the parameter increased or decreased with temperature. This is the so-called "butterfly" characteristic definition (RTI).

Example 2-19

An inverting amplifier has a gain of –50, and the output operating range is 0 to +10 volts. The feedback resistor is 50 kΩ. The op amp has a voltage offset drift (RTI) of 3 µV per °C, and the original amplifier has been offset nulled at 27° C. For 12-bit data words, what is the maximum allowable ambient temperature? Neglect offset current drift for simplicity. The gain to V_{os} is that for a noninverting amplifier, $1 + R_f/R_1$.

For a temperature drift ΔT the output offset voltage is

$$V_o^{os} = \left| (3 \times 10^{-6} \Delta T)(50 + 1) \right| = 1.5 \times 10^{-4} \Delta T$$

This error must be less than 1/2 LSB so,

$$1.53 \times 10^{-4} \Delta T \leq \frac{10 - 0}{2^{12+1}} = 1.22 \times 10^{-3}$$

Therefore,

$$\Delta T_{max} = 8° \text{ C}$$

Thus the maximum operating temperature (assuming the amplifier has been nulled at 27° C) is

$$T_{max} = 35° \text{ C}$$

Note that if the amplifier had not been nulled, the 27° C output offset voltage would need to be added to the left side of the inequality.

For amplifier modules such as instrumentation amplifiers and isolation amplifiers, the value of V_{os} (RTI) is found to be a function of overall amplifier gain. This will be discussed in the section on instrumentation amplifiers.

Exercises

E2.12a Suppose that the offset output voltage of an amplifier has been set to 1/10 LSB for an N-bit system. The amplifier has the configuration shown in Fig. 2-42 (inverting configuration), and we assume that I_{os} is zero. Develop the equation for the maximum allowable input offset drift. Compute the value for $\Delta T = 20°$ C, a 10-bit system, an amplifier overall gain of 10, and an output voltage range of –10 V to +10 V. Compare your result with available operational amplifiers.

E2.12b For the amplifier shown in Fig. E2.12b, compute the output offset voltage using the network shown directly and assuming the internal amplifier has $A = \infty$ (principle of the virtual short applies internally). $V_{os} = 2$ mV and $I_{os} = 200$ nA. Which of the offset parameters dominates at the output? Compute the currents in all resistors. (See Fig. 2-41.) What number of bits would be justified if no offset nulling were used? If that number of bits were used and the amplifier were nulled, what is the required voltage offset drift for operation from 10° C to 45° C?

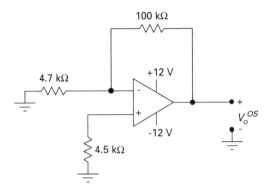

Figure E2.12b

E2.12c In Example 2-19 suppose the amplifier were operating bipolar −10 V to +10 V output. Compute the maximum operating temperature. Repeat for −5 V to +5 V operation.

2.13 OPERATIONAL AMPLIFIER INPUT IMPEDANCES

An operational amplifier has two input impedances which must be considered in amplifier design, especially if any sources that drive the amplifier have Thevenin equivalent impedances, since the resulting voltage divider reduces the voltage seen by the amplifier. These impedances are the differential impedance and common mode impedance. The definitions of these impedances in terms of the input terminals are shown in Fig. 2-47. For bipolar input op amps the resistances R_D are on the order of $10^6 \,\Omega$ to $10^9 \,\Omega$, with capacitances on the order of 1 to 50 pF. The corresponding common mode impedances range from $10^8 \,\Omega$ to $10^{10} \,\Omega$, with similar capacitances. For frequencies in the tens of kilohertz and below the capacitances are neglected (see exercise E2.13a). Operational amplifiers having FET input stages in their design have resistor values on the order of 10^9 to $10^{12} \,\Omega$ for R_D and 10^{10} to $10^{12} \,\Omega$ for R_{cm}.

Since the computations with the model of Fig. 2-47 involve the usual circuit theory techniques, the details of the effects of these impedances are deferred to the exercises of this section.

Exercises

E2.13a From a manufacturer's catalog obtain values for R_{cm}, C_{cm}, R_D, and C_D and compute the 3-dB frequencies of the parallel networks Z_{cm} alone and Z_D alone when driven by a sinusoidal current source.

E2.13b The operational amplifier in Fig. E2.13b is modeled the same as that in Fig. 2-47. Neglecting the capacitances, what is the resistance seen by the source for A finite and for A infinite?

E2.13c Repeat exercise E2.13b for Fig. E2.13c.

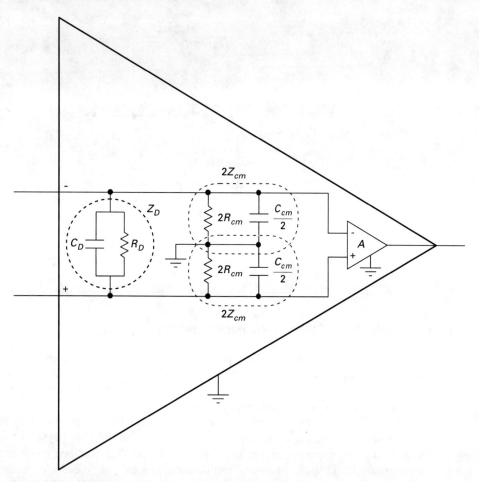

Figure 2-47 Definitions of amplifier common mode impedance, Z_{cm}, and differential impedance, Z_D.

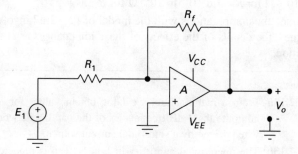

Figure E2.13b

E2.13d For exercise E2.13b, what is the maximum internal source impedance that E_1 may have in order that the voltage at the left of R_1 be within 1/2 LSB of the true value?

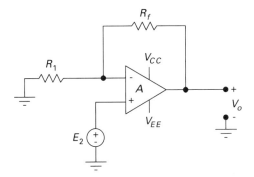

Figure E2.13c

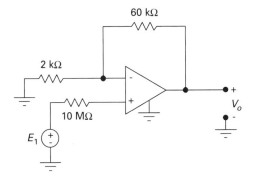

Figure E2.13f

E2.13e Repeat exercise E2.13d for the amplifier of exercise E2.13c.

E2.13f The common mode impedance of an op amp is listed as 150 MΩ, and the differential impedance is 2 MΩ. The op amp low-frequency gain is 2×10^5. Calculate the overall midband gain of the amplifier of Fig. E2.13f. Compare with ideal gain.

2.14 OTHER AMPLIFIER PARAMETERS

The lengths of the lists of amplifier parameters given in Section 2-5 and the important definitions given in Table 2-2 indicate the impossibility of giving a detailed presentation of each parameter and its effect on input signals and the number of bits and sample rates used to represent the analog signals. In the preceding sections we have simply picked a few of the important ones and shown the basic ideas for analysis of their effects. To avoid spending disproportionate text space on amplifiers, we leave the analyses of the other parameters as exercises. This approach puts the parameter in a practical context, so it should be readily accessible to anyone who has followed the preceding sections.

Some of the exercises also have situations where the interaction of the amplifiers with other elements external to the amplifiers is considered.

It should also be pointed out that the parameters that have been discussed are not necessarily all of the most important ones. They are perhaps those which involve a little more physical explanation and interpretation in mathematical terms. The authors hope that those selected have given the reader the capability of extending the analyses to other parameters.

For a good discussion of definitions and how the parameters are measured, see [4], particularly Appendices A and B.

Exercises

E2.14a A transducer has a first-order frequency response with a 3-dB frequency of f_{3dB}. An amplifier is to be used to extend the operating bandwidth by compensating for the transducer falloff. The amplifier has a second-order response of the form

$$A(\omega) = \frac{\omega_p^2 A_o}{\sqrt{(\omega_p^2 - \omega^2)^2 + (2\zeta\omega_p\omega)^2}}$$

What value of ζ is allowed, and at what resonant frequency ω_p should the amplifier be designed to operate? At what point must the two curves (for transducer and amplifier) cross for correct performance of a following digital processor? Use an 8-bit digital processor for all answers. What is the percentage of bandwidth increase? No closed expressions. Give equations which would be solved.

E2.14b The output impedance of an operational amplifier is Z_o ohms (assumed resistive). The equivalent circuit is shown in Fig. E2.14b(a). For finite A, determine the Thevenin equivalent (output) impedances of the two feedback amplifiers shown in part (b) of the figure. For a 741 op amp with $Z_o = 75\ \Omega$ and an open loop gain of 10^5, compute the output impedances of the circuits for gains of 20. Do your results depend upon the resistances individually; that is, what happens if all resistors are increased by a factor of 10?

E2.14c In exercise E2.14b, an equation was derived for the output impedances of feedback amplifier configurations. It showed that the manufacturer-specified amplifier output impedance can be reduced using feedback. In this exercise, we investigate the effect of this resistance on the number of bits N used for data representation. The problem is that the circuit output impedance and the input impedance of the following unit form a voltage divider that reduces the actual signal V_o. If amplitude is important, how many bits N would be justified for the system as shown in Fig. E2.14c? Next, derive an equation for the required amplifier gain to compensate for this signal reduction. Let the desired gain, neglecting the output voltage divider, be G_o and the required gain be G_R. G_R will be a function of G_o, R_o, and R_{in}.

E2.14d Suppose that a unity gain amplifier is designed using an op amp that has an input offset voltage of V_{os}. The amplifier is then offset trimmed for zero output offset, V_o^{os}. Using the definition of power supply rejection ratio (PSRR) given in Table 2-2, determine the required value of PSRR for an N-bit system.

Chap. 2   Amplifiers and Their Integration into Signal Processing Systems

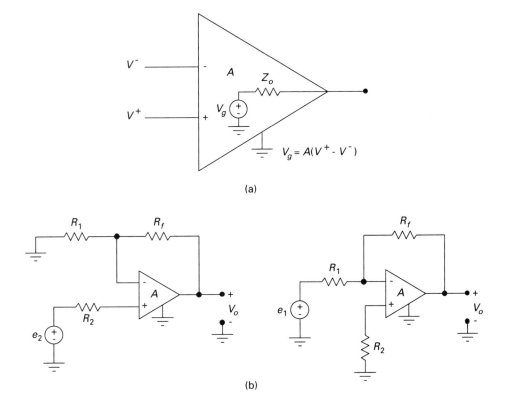

Figure E2.14b

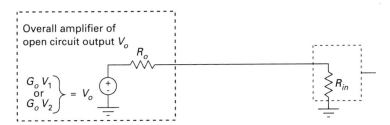

Figure E2.14c

2.15 INSTRUMENTATION AMPLIFIERS

When the input signals from transducers are very small, say on the order of millivolts or less as may come from an EEG or EKG, it is often difficult to design an appropriate amplifier to tight specifications beginning from "scratch" with basic op amps and components. The instrumentation amplifier with its integrated amplifiers and components on a chip provides a way to avoid design problems. Generally speaking, these will be more expensive, but the cost frequently offsets the required design and construction problems using the basic operational amplifier.

The net result of today's solid-state technology is to make obsolete the building of one's own instrumentation amplifier from op amps and components, just as the introduction of the op amp made building an amplifier from transistors and RC components unnecessary except in special applications.

The particular properties of instrumentation amplifiers that distinguish them from operational amplifiers include:

1. Generally a small gain range (1 to 1000), frequently fixed gain, committed to one specific task.
2. Usually have higher *CMR* values than op amps, often 100 dB or more.
3. Lower noise and drift parameters. Thus these work well on mV and µV signals such as one gets from thermocouples, strain gauges, and high Z probes.
4. Smaller gains (1 to 1000) and moderate bandwidths. They often have several op amps within their design.
5. High-precision components possible on IC chip, < 1% tolerance, and low temperature coefficients. Linearity often 0.002% max.
6. Conductive isolation can be obtained to allow separate grounds for input and output circuits. This class of amplifier is called the isolation amplifier and is discussed in a following section.

In addition to the above, instrumentation amplifiers may have external control terminals. The usual ones are the Sense terminal, and the Reference terminal. The sense terminal is often used to modify the amplifier gain using external resistors (hence the use of the variable resistor symbol on the instrumentation amplifier). The reference terminal is often used to shift the voltage reference of the output. The work that follows will show the electrical configuration of the terminals.

Symbolically, the instrumentation amplifier is often represented by the circuit shown in Fig. 2-48. The sense and reference terminals, if present, are never left open and are frequently high-impedance lines, so they draw small currents. The basic gain relationship is

$$E_S - E_R = G\,(V_2 - V_1) \qquad (2\text{-}103)$$

where values of G range from 1 to 1000.

At first glance it may seem a bit paradoxical that the gain relationship does not even yield the output voltage. If, however, we connect the sense terminal to V_o and the reference terminal to ground, we would have $E_S = V_o$ and $E_R = 0$. The gain relation would then become

$$V_o = G\,(V_2 - V_1) \qquad (2\text{-}104)$$

(E_S tied to V_o, E_R grounded).

Some instrumentation amplifiers have no sense and/or reference lines. If there is no sense line then we set $E_S = V_o$ in Eq. (2-103) and use

$$V_o - E_R = G\,(V_2 - V_1) \qquad (2\text{-}105)$$

(amplifier with no sense line).

If the instrumentation amplifier has no reference line or sense line we set $E_R = 0$. This yields the same result as Eq. (2-104) above if there is no sense terminal:

Chap. 2 Amplifiers and Their Integration into Signal Processing Systems

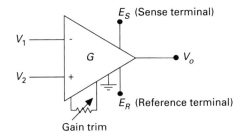

Figure 2-48 Symbol for the instrumentation amplifier.

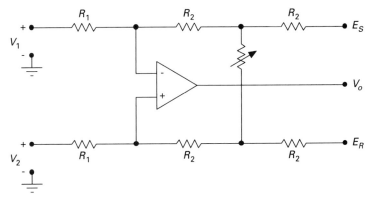

Figure 2-49 Basic instrumentation amplifier. [22]

$$V_o = G(V_2 - V_1) \tag{2-106}$$

(amplifier with no sense line or reference line).

The fundamental structure of the instrumentation amplifier from which other forms are obtained is given in Fig. 2-49. We first compute the gain function for the particular case $E_S = V_o$ and $E_R = 0$, as shown in Fig. 2-50, where the various currents and voltages required to write KCLs and KVLs are also indicated. All voltages are measured with respect to ground.

Writing KCL at the nodes and then using Ohm's law to find currents we have the following:

At E_a: $\quad I_1 = I_- + I_1' = I_1'(I_- = 0) \rightarrow \dfrac{V_1 - E_a}{R_1} = \dfrac{E_a - E_a'}{R_2} \quad$ (2 unknowns)

At E_a': $\quad I_1' = I_c - I_o \rightarrow \dfrac{E_a - E_a'}{R_2} = \dfrac{E_a' - E_b'}{KR_2} - \dfrac{V_o - E_a'}{R_2} \quad$ (2 new unknowns)

At E_b: $\quad I_2 = I_+ + I_2' = I_2'(I_+ = 0) \rightarrow \dfrac{V_2 - E_b}{R_1} = \dfrac{E_b - E_b'}{R_2} \quad$ (1 new unknown)

At E_b': $\quad I_2' = I - I_c \rightarrow \dfrac{E_b - E_b'}{R_2} = \dfrac{E_b'}{R_2} - \dfrac{E_a' - E_b'}{KR_2} \quad$ (0 new unknowns)

Amplifier gain ($E_S = V_o$, $E_R = 0$): $\quad V_o = AE_b - AE_a \quad$ (0 new unknowns)

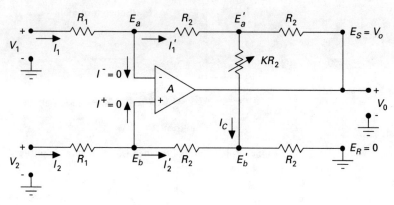

Figure 2-50 Basic instrumentation amplifier connected for the gain calculation. [22]

We consider V_1 and V_2 to be input sources and solve for V_o. Putting the preceding equations in standard form we have

$$\left(\frac{R_1}{R_2}+1\right)E_a - \frac{R_1}{R_2}E_a' + 0 \cdot E_b + 0 \cdot E_b' + 0 \cdot V_o = V_1$$

$$1 \cdot E_a - \left(\frac{1}{K}+2\right)E_a' + 0 \cdot E_b + \frac{1}{K}E_b' + 1 \cdot V_o = 0$$

$$0 \cdot E_a + 0 \cdot E_a' + \left(\frac{R_1}{R_2}+1\right)E_b - \frac{R_1}{R_2}E_b' + 0 \cdot V_o = V_2$$

$$0 \cdot E_a + \frac{1}{K}E_a' + 1 \cdot E_b - \left(\frac{1}{K}+2\right)E_b' + 0 \cdot V_o = 0$$

$$AE_a + 0 \cdot E_a' - AE_b + 0 \cdot E_b' + 1 \cdot V_o = 0$$

After some lengthy algebra (or matrix) work, the solution for V_o is

$$V_o = \frac{Num}{Denom}$$

where

$$Num = A\,(V_2 - V_1)\left[\left(\frac{1}{K}+2\right)^2\left(\frac{R_1}{R_2}+1\right) - \frac{1}{K^2}\left(\frac{R_1}{R_2}+1\right) - \frac{R_1}{R_2}\left(\frac{1}{K}+2\right)\right]$$

$$- A\,(V_2 - V_1)\frac{R_1}{R_1+R_2}$$

and

Chap. 2   Amplifiers and Their Integration into Signal Processing Systems

$$Denom = A\left[\left(\frac{R_1^2}{R_2^2}+\frac{R_1}{R_2}\right)\left(\frac{1}{K}+2\right)-\frac{R_1^2}{R_2^2}-\frac{R_1^2}{KR_2^2}-\frac{R_1}{KR_2}\right]$$

$$+\left(\frac{R_1}{R_2}+1\right)\left[\left(\frac{1}{K}+2\right)^2\left(\frac{R_1}{R_2}+1\right)-\frac{R_1}{K^2R_2}-\frac{1}{K^2}-\frac{R_1}{KR_2}-\frac{2R_1}{R_2}\right]$$

$$-\left(\frac{R_1^2}{R_2^2}+\frac{R_1}{R_2}\right)\left(\frac{1}{K}+2\right)+\frac{R_1^2}{R_2^2} \qquad (2\text{-}107)$$

From this the gain is obtained as

$$G=\frac{V_o}{V_2-V_1}$$

$$=\frac{2\left(\frac{1}{K}+1\right)\left(\frac{R_1}{R_2}+2\right)}{\frac{R_1}{R_2}\left(\frac{R_1}{R_2}+2\right)+\frac{1}{A}\left[\frac{R_1}{R_2}\left(\frac{2R_1}{KR_2}+\frac{R_1}{R_2}+\frac{6}{K}+4\right)+4\left(\frac{1}{K}+1\right)\right]} \qquad (2\text{-}108)$$

For large A (that is, an ideal operational amplifier with $A \to \infty$),

$$G=\frac{V_o}{V_2-V_1}=2\frac{R_2}{R_1}\left(1+\frac{1}{K}\right) \qquad (2\text{-}109)$$

Eq. (2-109) shows the advantage of integrated circuit technology, since the resistance ratio is the key quantity. Since R_1 and R_2 are made simultaneously they will most likely vary in a similar manner. Notice also that the gain can be trimmed easily either with an internally laser-trimmed resistance or an external variable resistance.

The basic circuit, as exercise E2.15a reveals, is critically dependent upon the two R_1 values being equal and the R_2 values being equal. Also, the input resistance is fairly low, being on the order of R_1 in value at each terminal (see exercise E2.15b). One might think that a very large R_1 would solve the input resistance problem, but if the amplifier A were a practical bipolar op amp, the signal V_1 would be reduced by an input divider effect. If large R_1 values are used with an FET input op amp, the noise becomes excessive and the operating bandwidth decreases significantly. The sense and reference lines also draw current, which is undesirable.

One solution to the input impedance problem is to buffer V_1 and V_2 with unity gain isolators, as shown in Fig. 2-51. This circuit still requires good component matching as well as good input op amp *CMR* in order to maintain an overall high *CMR*. The gain expression is still given by Eq. (2-108) or (2-109).

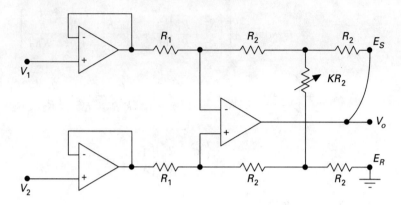

Figure 2-51 Basic instrumentation amplifier with buffered inputs. [22]

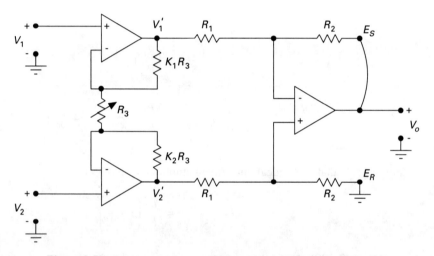

Figure 2-52 Instrumentation amplifier having isolation stage gain. [22]

Component sensitivity reduction and an improved *CMR* for the instrumentation amplifier can be obtained using a configuration that produces differential gain on the input stages. This next modification is shown in Fig. 2-52. The gain is given by

$$G = \frac{R_2}{R_1}(1 + K_1 + K_2) \quad (2\text{-}110)$$

To show the improvement in the instrumentation amplifier *CMR*, first consider V_1 and V_2 tied together and connected to a voltage E_{cm} with respect to ground. Now if we suppose that the gains of the two input op amps are very large, then the principle of the virtual short requires that all four input op amp terminals are at voltage E_{cm}. Thus both the top and bottom of resistor R_3 are at the same potential, E_{cm}, so there is no current in R_3. Since there is no current in R_3 then there are no currents in either K_1R_3 or K_2R_3 (we assume zero op amp input current). Then there are no voltage drops across these resistors, so that $V_1' = V_2' = E_{cm}$; that is, the common mode gain is unity. However, there is a

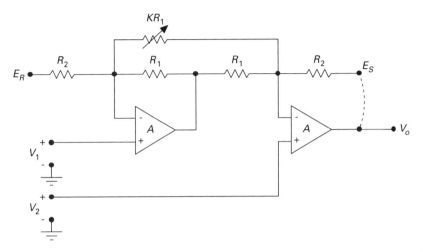

Figure 2-53 Modified buffered instrumentation amplifier for signals with small common mode component. [22]

gain to the voltages V_1 and V_2 so that the output amplifier would see increased desired signal and the same common mode signal as at the input.

Next suppose that we have an application where we expect low common mode voltages. We can preserve the high input impedance and save an op amp by using the instrumentation amplifier shown in Fig. 2-53. This has the disadvantage, as already mentioned, of requiring good resistor match for high *CMR*, but for applications where low *CMR* can be anticipated this is a perfectly good alternative to obtaining high input impedance. Note where the reference terminal is here. For the sense line tied to the output V_o and the reference line tied to ground (assuming $A = \infty$)

$$G = 1 + \frac{R_2}{R_1}\left(1 + \frac{2}{K}\right) \tag{2-111}$$

It was pointed out earlier that for many instrumentation amplifiers E_R and E_S are high-impedance lines. An examination of the preceding circuits shows that the E_R and E_S lines are low-impedance lines. One method for increasing the impedances is to use current feedback as implemented by one manufacturer [22] and shown in Fig. 2-54. The gain can be directly varied by changing one of two resistors, making the gain set a very easy process. Actually, both resistors could be trimmed to bring the amplifier gain to a precise value.

Qualitatively the effect of R_{gain} on the amplifier operation is explained as follows. Let us suppose that the value of R_{gain} increases. Since R_{gain} is in the emitters of the two transistors, there is increased negative voltage feedback to the emitter-base junctions. Thus the gains of the two input transistors are reduced. This means that for a given set of inputs V_1 and V_2 the voltages on the input operational amplifiers are smaller so that the output is reduced. Then, the gain must decrease with increasing R_{gain} value. This is in agreement with the equation for gain given in Fig. 2-54. (See exercise 2.15f for a thought exercise on the effect of the value of R_{scale} on gain.) Although the figure shows bipolar

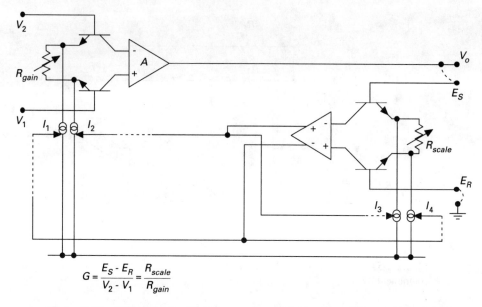

Figure 2-54 Current feedback used to produce high Z at E_R and E_S terminals. [22]

transistors, an FET configuration could also be used to obtain the high-impedance sense and reference lines.

Since the instrumentation amplifier is a two-input differential amplifier, its performance is subject to the same parameters as the operational amplifier. One noteworthy difference is that the RTI offset voltage drift is a function of the gain for which the amplifier is set. In most cases the input offset voltage drift is specified at gains of 1 and 1000. For other gains a simple linear interpolation is adequate (see exercise E2.15a).

Exercises

E2.15a For the simple basic fixed-gain instrumentation amplifier, (Fig. 2-50) determine the maximum tolerance in percent that may exist between the two resistors labeled R_1 in order to preserve a common mode rejection of -100 dB. Assume the amplifier A to be ideal; meaning that $A = \infty$, input currents $= 0$, and output impedance $= 0$. Also assume all R_2 to be identical. (Note the fact that we have used a negative dB value implies we have used the second definition of *CMR* given in the text). Your solution will be in terms of circuit parameters, R_1, K, and others. This exercise requires considerable algebra.

E2.15b For the basic instrumentation amplifier of Fig. 2-50, determine the input impedance seen by source V_1 under two conditions:

  a. source $V_2 = 0$
  b. source $V_2 \neq 0$

Assume that the op amp is ideal.

E2.15c Analyze the operation of the two input stages of the buffered instrumentation amplifier with gain and show:

  a. Common mode voltage gain $= 1$.
  b. V_1 and V_2 have gains >1. (Derive expressions for outputs V_1' and V_2'.)
  From these results derive Eq. (2-110).

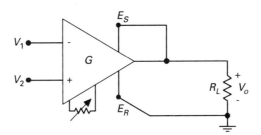

Figure E2.15f

E2.15d For the current feedback amplifier shown in Fig. 2-54, qualitatively describe how an increase in the value of R_{scale} produces an increase in gain of the amplifier.

E2.15e Determine the equation for the overall system gain defined by

$$H = \frac{V_o}{V_2 - V_1}$$

for the instrumentation amplifier configuration shown in Fig. E2.15f. Assume sense and reference line currents are zero. What does this tell you about using an instrumentation amplifier to feed a variable load? What do you think would limit the load values?

E2.15f List the basic reasons why one would want to have an instrumentation amplifier immediately following the transducer in a general signal processing system.

2.16 APPLICATIONS OF INSTRUMENTATION AMPLIFIERS

We first consider the application of the instrumentation amplifier to a "balanced" system such as might be found in a strain gauge transducer. This is shown in Fig. 2-55. Note that the total common mode voltage (E_{cm}) is E_{cm2} plus the raise in potential caused by V_1 and V_2, both being above transducer ground. For example, if $V_1 = V_2$, the common mode level above transducer ground is V_1. In general, the common mode generated component is (see Section 2.9)

$$1/2(V_1 + V_2)$$

Thus

$$E_{cm} = E_{cm2} + 1/2(V_1 + V_2)$$

and the corresponding total output voltage is given by

$$V_o = E_S = G(V_2 - V_1) + A_{cm}E_{cm} \qquad (2\text{-}112)$$

The relationship of these results, as well as other amplifier properties, to the number of bits N used to represent the digitized data follows exactly the results discussed for

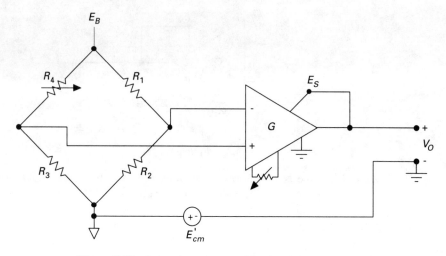

Figure 2-55 Instrumentation amplifier in a bridge transducer.

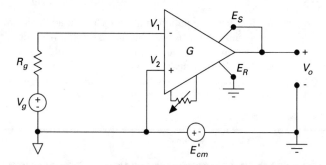

Figure 2-56 Circuit for illustrating the computation of common mode input voltage for single-ended input.

the operational amplifier having finite gain A. Now, however, the value of G is small (usually ≤ 1000), so finite gain must be used in all cases.

Next let us examine the unbalanced or single-ended input configuration shown in Fig. 2-56. For these applications the prime use of instrumentation amplifiers is to eliminate common mode effects, which are often outside the capability of the op amp.

For this configuration $V_2 \equiv 0$ so that the common mode voltage is simply

$$E_{cm} = E_{cm2} + \frac{V_1}{2} = E_{cm2} + \frac{V_g}{2}$$

The last equality has assumed that the amplifier input impedance is infinite. This, of course, will need to be checked to see if it is significant. The output voltage will be (again assuming $V_1 = E_g$)

$$V_o = G\,(V_2 - V_1) + A_{cm}E_{cm} = G\,(V_2 - V_1) + \left(E_{cm2} + \frac{V_g}{2}\right)A_{cm} \qquad (2\text{-}113)$$

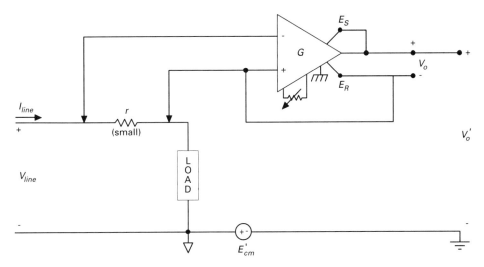

Figure 2-57 Using an instrumentation amplifier to monitor high side currents.

Another important application of the instrumentation amplifier is for cases where high-voltage isolation is required in addition to a common mode rejection. As an example consider a situation where one may want to monitor the current in the high side of a line, where perhaps the load ground terminal cannot be disturbed. This is shown in Fig. 2-57. In these applications the use of the reference line is very important. If the output voltage were to be taken with respect to the output ground, then it (V_o') would not be the voltage across the sampling resistor r, since the load voltage would be included. Also, if the reference E_R were tied to ground, then the line voltage, V_{line}, would be a large CMV to the amplifier in addition to any E_{cm2} present. Also, if V_{line} were of the order of the amplifier supply voltage, then the amplifier might be driven completely out of its linear range, since any inputs, even common mode, can affect the quiescent operating levels of the amplifier internal components. To see this, one must examine the internal operational amplifier configuration, as shown in Fig. 2-58. Notice that the power supply ground reference is also important. If E_{cm} is large (plus or minus), the operating balance of the amplifier is completely destroyed. If E_{cm} were large enough, the amplifier could be burned out even though the differential input is zero. Thus we see that both power supply (grounds) and instrumentation amplifier referencing are critical in some applications. In such applications a separate supply is required for the amplifier whose ground is not connected to the system ground.

An obvious choice for the reference connection is to one side of the sampling resistor r. This avoids having V_L affect the output V_o. If this choice is made then the power supply grounds must also be somewhere near that value. In these cases battery operation is often the only possible approach or else an isolating DC-DC converter may be required.

It should also be observed that if the voltage V_o is to be used elsewhere (which it must if the amplifier is to perform any useful function) then an isolation amplifier is necessary so that the output ground can again be used.

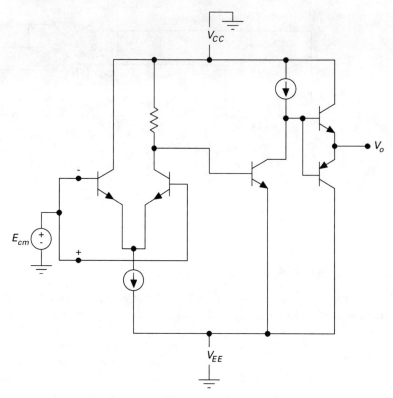

Figure 2-58 Relationship of common mode voltage to the operational amplifier operating point.

For the reference terminal connected to the resistance r, suppose the right-hand side, the output V_o, would be given by (with respect to the reference terminal)

$$V_o = -GI_{line}\, r \qquad (2\text{-}114)$$

The reference terminal E_R can be used to provide output offset to compensate for metal-metal contact potentials, provide offset for plotter pens, and provide a bias for a relay or comparator switch point. One example is shown in Fig. 2-59. This example also shows a case when it would be desirable to have a high-impedance input to a reference terminal. This avoids having to provide a reference voltage that supplies a large current.

As an example of the use of the sense terminal, suppose that the instrumentation amplifier is not capable of delivering the current required by a load. The approach would be to let the instrumentation amplifier feed a current driver. The sense line, being in the negative feedback path within the instrumentation amplifier, can be used to compensate for booster amplifier offset drifts, gain errors, and so forth. The circuit configuration is shown in Fig. 2-60. The output of the unit would be given by

$$V_o = E_S = G_o(V_2 - V_1)$$

since $E_R = 0$.

To show how the sense line can be used to compensate for booster amplifier drift, suppose that the output voltage from the booster amplifier increased in value so that V_o

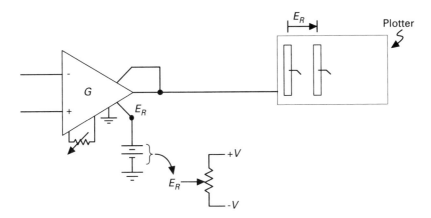

Figure 2-59 Sense line used to provide plotter pen offset bias.

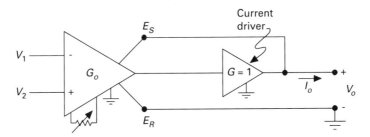

Figure 2-60 Sense line used to provide current-boosting output.

increased. This increase in V_o is felt (sensed) by the sense line, which applies an increased voltage on the negative terminal of the op amp within the instrumentation amplifier (see Fig. 2-49 for example). This increase in voltage causes the output of the instrumentation amplifier to decrease, which in turn reduces the voltage output of the unity gain booster toward its original value.

As our next application, suppose we want to deliver to a floating impedance a current that is independent of the impedance value, as in Fig. 2-61. For this application we need high-impedance inputs into the sense and reference lines, so we assume $I_S = I_R = 0$. This makes the current in the floating load the same as the current in the current set resistor R. The analysis for the ideal case (see also exercise E2.16c) is straightforward, as follows.

Using Ohm's law:

$$I = \frac{E_S - E_R}{R} = \frac{E_S}{R}$$

Since the sense current is zero,

$$I_L = I = \frac{E_S}{R}$$

The amplifier operation requires (for $E_R = 0$)

222 Amplifiers and Their Integration into Signal Processing Systems Chap. 2

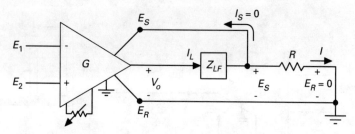

Figure 2-61 Voltage-controlled current source to a floating load impedance for which load current is independent of load, Z_{LF}.

$$E_S = G(V_2 - V_1)$$

The equation for the load current becomes

$$I_L = \frac{G(V_2 - V_1)}{R} \qquad (2\text{-}115)$$

which shows that the load current depends only upon the input signals V_1 and V_2 and not on load value.

Exercises

E2.16a For the current booster system of Fig. 2-60, suppose that the current driver were to start delivering excess current. Show how the instrumentation amplifier will compensate for this. Does it matter whether the impedance into the sense line is high or low? Why or why not?

E2.16b Suppose that the current output of the current driver (Fig. 2-61) is to be used to drive a circuit that requires current input (a logarithmic amplifier, for example). For the amplifier operating parameters given below, what is the minimum ratio

$$\frac{R}{|Z_{LF}|}?$$

(Hint: What limits R and what limits $|Z_{LF}|$?) Maximum output current the amplifier will deliver is denoted by I_{om}, and the supply voltages are $\pm V_S$. Evaluate this ratio for the case $G = 100$, $I_{om} = 10$ mA, $V_S = 10$ V. Your answer may be in terms of $V_2 - V_1$. Is there a worst-case value for $V_2 - V_1$?

E2.16c For the current driver of Fig. 2-61, what is the equation for the maximum allowable value of the sense terminal current (assume $Z_{REF} = \infty$) for N-bit data representation of load current? Assume I_{om} as the maximum output current.

E2.16d For the case where we wish to supply a current to a grounded load impedance that is independent of the impedance value (shown in Fig. E2.16d), derive the expression for load current as a function of G and $(V_2 - V_1)$. Is there a relationship in terms of size between R and Z_{LG}? Assume that the sense and reference lines have infinite input impedances.

E2.16e A pressure transducer, instrumentation amplifier, and A/D converter are used as parts of the solid-state decompression and depth computer shown in Fig. E2.16e. The pressure transducer has a DC offset voltage of +2.5 V at 0 Pa and

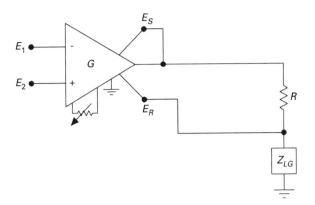

Figure E2.16d

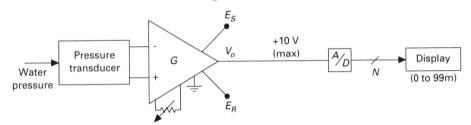

Figure E2.16e

a sensitivity of 1.5 µV/Pa. The amplifier supply voltages are ±15 V. Determine the amplifier connections necessary to correct the offset voltage. Also determine the gain G necessary to display depths to 99 meters in sea water. (Density: 1025 Kg/m³.) What is the allowable input pressure range?

E2.16f An instrumentation amplifier (output span = V_o^{span}) is to be used to amplify the signal from an unbalanced bridge. The bridge is fed by a V_B volt DC source, and the bridge is capable of detecting a signal difference (min) of 1 mV. Suppose also that $R_2 = R_3$ and $R_1 = R_{4max}$ and that $R_{4mm} = 1/2R_1$. If we assume that the CMV is the main source of error and is not to exceed 0.1% of the output signal at minimum signal level, what is the required CMR of the amplifier? There are two possible answers (both in terms of circuit parameters) depending upon which definition of CMR is used. Work using both definitions. What is the number of bits N allowed for data?

E2.16g An unbalanced system utilizes an instrumentation amplifier that has a CMR of -60 dB. What is ground line if the amplifier must detect a 1-mV signal? You are free to specify some rule-of-thumb value or level for which the common mode output should be below the signal output. Amplifier gain is G.

E2.16h The output voltage of a transducer is 20 mV (maximum, unipolar) and is to be amplified by an amplifier having a gain of 100. There is a 1/2 volt amplitude, 60-Hz, common mode signal at the amplifier input. What is the required CMR of the amplifier so that at the amplifier output the common mode signal is 1% of the total output? How many bits in the data word would be required to reject the 1% output noise? Next, suppose the digital processor uses 12-bit data words. What is the required CMR?

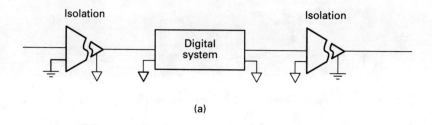

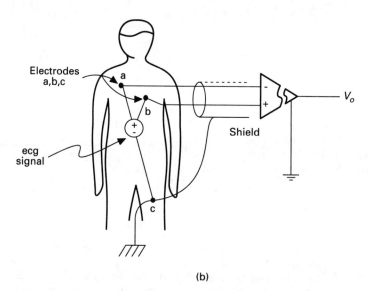

Figure 2-62 Typical isolation amplifier applications. (a) Digital system isolation. (b) Patient isolation.

2.17 ISOLATION AMPLIFIERS

As mentioned in Section 2.15 there is frequently a need for an amplifier that can isolate one circuit from another by breaking any metallic connection (normally a ground) existing between them. This is called galvanic or conductive isolation. Also, one circuit may be at a very high common potential reference which cannot be removed by simple common mode rejection of the amplifier. Amplifiers that perform these functions form a special class of instrumentation amplifiers called isolation amplifiers. The power supplies required by the input and output sections must also be conductively isolated.

Two very common applications of isolation amplifiers are shown in Fig. 2-62. In many data gathering systems coupled to a digital processing system there tend to be large ground interferences and common mode problems. It is then desirable to protect the digital system from these and vice versa to keep pulse transients out of analog elements.

In medical electronics it is a matter of patient safety to guard against any possibility of power line or recording equipment voltages inadvertently reaching the patient. As little as 100 mA of current can cause death.

The basic characteristics of isolation amplifiers, whether they be of the transformer coupled or optically isolated type, are the following:

Chap. 2 Amplifiers and Their Integration into Signal Processing Systems

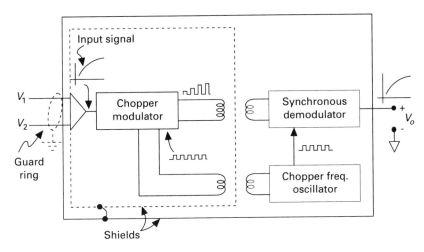

Figure 2-63 Basic system for chopper modulator isolation amplifier.

1. They have committed gains like instrumentation amplifiers, usually gains of 1 to 1000.
2. They will tolerate very high common mode reference voltages between input and output sections: as high as 5 kV. These large voltages have very little effect on the amplifier output. For example, a 120-V DC value placed between the two grounds could not be tolerated by an unisolated amplifier, since that would raise the input base (or gate) terminals far above the usual ±10 or ±15-V supply voltages.
3. They generally have only two leads with no bias current capability.
4. They can amplify millivolt and microvolt level signals with very low noise injection. The low noise property, specifically immunity to RFI or EMI, is obtained by a built-in electromagnetic shield to protect the low-level input circuitry. Guard rings are also used on PC boards.
5. Their frequency responses are often not the best, at most covering the audio range, at present, without special external circuits.

The detailed operation of optically coupled isolators is outside the realm of this text, but there are several good descriptions of these in most of the large product guides of the manufacturers and electronics textbooks.

A short description of the idea behind the transformer-coupled isolator is of interest because the concept has other applications. A block diagram of some common elements in a transformer-coupled isolator is given in Fig. 2-63. The problem, of course, is that the transformer cannot respond to low-frequency signals which are often encountered in data acquisition systems. What is frequently done is to convert the input signal into a higher-frequency signal, couple it across the transformer, and then reconstruct it at the output.

Initially, the incoming signal (or difference signal) is amplified by the FET amplifier and then applied to a modulator device called a chopper. The output waveform from the chopper oscillator is transformer coupled to the modulator, where it produces an amplitude-modulated wave which can more effectively couple across the the transformer. Then, the synchronous demodulator removes the chopper frequency (usually followed by a filter to remove any transients that may be introduced). There are many different ways of implementing this process. Optical isolation of ground can also be used. This technique is used with either analog or digital signals.

Exercises

E2.17a Design a circuit that will operate as a chopper modulator in the system shown in Fig. 2-63. Is there any relationship between the oscillator frequency and the input signal? Describe any relationship you propose. The input to the chopper modulator (FET output) is ±5 V.

E2.17b Looking at Fig. 2-62(b), explain why the required CMR_1 is only 60 dB, whereas the output CMR_2 is required to be 120 dB. (Hint: The amplitudes of CMV_1 and CMV_2 are involved.) For a 10-bit data representation of output data from the ECG unit (maximum 2 V), what is the largest permissible value of CMV_2?

2.18 INSTRUMENTATION AMPLIFIER OFFSET DRIFT

Any change in instrumentation amplifier internal component values due to temperature, time (aging), or signal level will produce an amplifier output offset. As described briefly in Section 2.15, the offset voltage of these amplifiers is a function of gain also. It is common practice to give drift specifications in terms of an equivalent input voltage that would produce the measured value of output drift. Thus the specification depends upon amplifier gain, since we would divide the output drift voltage by the gain to obtain the effective input. This is called a referred to input (RTI) specification. Formally one would write the drift RTI as

$$\text{voltage (current) } drift_{RTI} \equiv \frac{\text{output drift}}{\text{amplifier gain}} \tag{2-116}$$

Not all manufacturers use the RTI scheme, so one must examine each specification sheet carefully to determine the convention used.

The specification of drift is in the basic units, which depends upon the variable producing the drift. For the temperature drift this would be

$$drift_{RTI} = \frac{\frac{\Delta V_{out}}{G}}{\Delta T} \quad \frac{\text{volts}}{°C} \tag{2-117}$$

whereas for time drift we would use

$$drift_{RTI} = \frac{\frac{\Delta V_{out}}{G}}{\Delta t} \quad \frac{\text{volts}}{s} \tag{2-118}$$

In many cases the drift is specified at the extreme gains of the amplifier (normally 1 and 1000, for example). To obtain the drift at another gain value, one makes a linear interpolation, as indicated in Fig. 2-64. Using the linear interpolation scheme we would obtain the output drift at gain G as

Chap. 2   Amplifiers and Their Integration into Signal Processing Systems   227

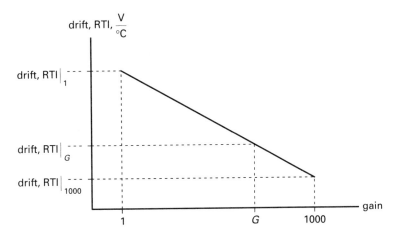

Figure 2-64   Linear interpolation on temperature drift parameter.

$$\text{output drift}\Big|_G = \frac{Drift_{RTI}\Big|_{1000} \times 1000 - Drift_{RTI}\Big|_1 \times 1}{1000 - 1} \times (G-1)$$

$$+ Drift_{RTI}\Big|_1 \times 1 \frac{V \text{ (or } A\text{)}}{°C \text{ (or sec)}}$$

(2-119)

The actual output voltage (or current) change at gain G caused by a temperature change ΔT would then be

$$\Delta V_o = \text{output drift}\Big|_G \times \Delta T$$

An example of the use of these results is given in Section 2.19, where we discuss an amplifier error analysis.

As a closing note, one should be aware of the fact that many models have an external pair of terminals that can be used to set the gain at values other than 1 or 1000. Usually a resistance is connected between the terminals and set to give the desired gain. Most often the specification sheets contain formulas for computing the resistance value. The temperature variation of the resistor may yield an additional source of error.

Exercise

E2.18a   The drift RTI of an amplifier is 150 μV/°C at a gain of 1 and 2 μV/°C at a gain of 1000. What is the actual output voltage change at a gain of 100 caused by a temperature change of 5° C?

E2.18b   An amplifier specification sheet lists the offset drift referred to the input as ±200 μV/°C at a gain of 1 and ±3 μV/°C at a gain of 1000. If the instrumentation amplifier were set up for a gain of 50, what is the permissible temperature drift if the amplitude of the signal is important and 8 bits are being used for the data words?

2.19 SELECTION OF INSTRUMENTATION AMPLIFIERS OR OPERATIONAL AMPLIFIERS

Due to the myriad of parameters available in operational amplifiers and instrumentation amplifiers, one may well ask, How do I know whether to select an instrumentation amplifier or to design my own amplifier using operational amplifiers? The question is a very important one, since instrumentation amplifiers often are more expensive. The apparent trade-offs, then, involve a consideration of cost versus the design time and the ease with which an amplifier can be designed to meet the given system requirements. Obviously, the more stringent the amplifier requirements the more one is forced to the dedicated instrumentation amplifier.

It is not possible to set down a simple checklist of all the variables that might influence the decision. In some cases one particular parameter requirement may clearly point to the instrumentation amplifier, such as *CMR* of 120 dB at low noise.

Exercise

E2.19a From manufacturers' specification sheets for operational amplifiers and instrumentation amplifiers, make a list of three or four important parameters, such as *CMR*, offset drift, and slew rates. Compare the values and their ranges.

2.20 ERROR ANALYSIS AND ERROR BUDGETS FOR AMPLIFIERS [22]

We conclude this chapter by taking a specific case design and analyzing all the error sources [22]. The resulting list of error sources is called the error budget. We consider here a bridge circuit and its following amplifier. Let the subscripts d and D refer to difference quantities and *CM* refer to common mode quantities. For simplicity we assume that the bridge is balanced and determine the Thevenin equivalent circuits for V_1 and V_2. Having done this, the equivalent circuit may be drawn as in Fig. 2-65.

From basic definitions:

$$V_d = V_2 - V_1$$

$$E_{cm} = \frac{V_1 + V_2}{2}$$

(for the case where external common mode is small) and

$$CMRR = \frac{G}{A_{cm}}$$

For small bridge imbalances, as is often the case,

$$V_1 \approx V_2 \approx \frac{V_B}{2}$$

(assumes all bridge legs are of equal value). For $R_L = \infty$,

$$V_o = GV_d + A_{cm}E_{cm} = G(V_2 - V_1) + A_{cm}\frac{V_1 + V_2}{2} \approx G(V_2 - V_1) + \frac{G}{CMRR}\frac{V_B}{2}$$

Chap. 2 Amplifiers and Their Integration into Signal Processing Systems

We next list the operating conditions for the desired system and then list the amplifier specifications for easy reference.

Operating Conditions	Amplifier Specs
$E_{in} = V_d = \pm 5$ mV (bridge imbalance)	$Drift_{RTI}\big|_{G=1000} = 11\ \dfrac{\mu V}{°C}$ (offset)
$V_o = \pm 5$ V, $V_B = 10$ V ($G = 1000$)	Gain drift $= \dfrac{\pm 0.01\%}{°C}$ ($G = 1000$)
	$I_B = 20$ nA
$E_{cm} = \dfrac{V_B}{2} = +5$ V	
$\Delta T = \pm 20°$ C about the reference temperature 25° C	$Z_d = 300$ MΩ
$T_A = 25°$ C	$Z_{CM} = 1$ GΩ
$R_L = 10$ kΩ	$CMRR = 10^5$, $CMR = 100$ dB, $G = 1000$
	Nonlinearity $= \pm 0.01\%$ at 10 V out
$R_S' = R_S = 500$ Ω	$Z_o = 10$ Ω

We next compute the error contribution by each amplifier parameter and construct the error budget.

Error Source	Computation	Value	%F.S.
Gain drift (+5 to +45° C)	$\left(\dfrac{0.01\%/°C}{100}\right) \Delta T V_o$		
	$= (.0001)(\pm 20)(5)$	± 10 mV	0.2%
Offset drift (+5 to +45° C)	$(11\ \mu V/°C)\, G \Delta T$		
	$= (11\ \mu V/°C)(1000)(\pm 20)$	± 220 mV	4.4%
Linearity error (worst case)	$\left(\dfrac{\pm 0.01\%}{100}\right) E_{o_{max}}$		
	$= (\pm 0.0001)(\pm 5)$	$\pm .5$ mV	0.01%
Common mode error	$E_{cm} A_{cm}$		
	$= E_{cm} \dfrac{G}{CMRR} = (5)\dfrac{(1000)}{10^5}$	± 50 mV	1%
Offset error	adjust to zero	0	0%
Input loading error (referred to output)	$\dfrac{R_s}{R_s + R_d} V_d$		
	$= \dfrac{500}{3 \times 10^8 + 500}(\pm 5\ \text{mV})(1000)$	± 8 mV	0.16%
Output loading error	$\dfrac{R_o}{R_o + R_L} V_{o_{max}}$	± 5 mV	0.1%
	$= \dfrac{10}{10 + 10^4}(\pm 5\ \text{mV})$		

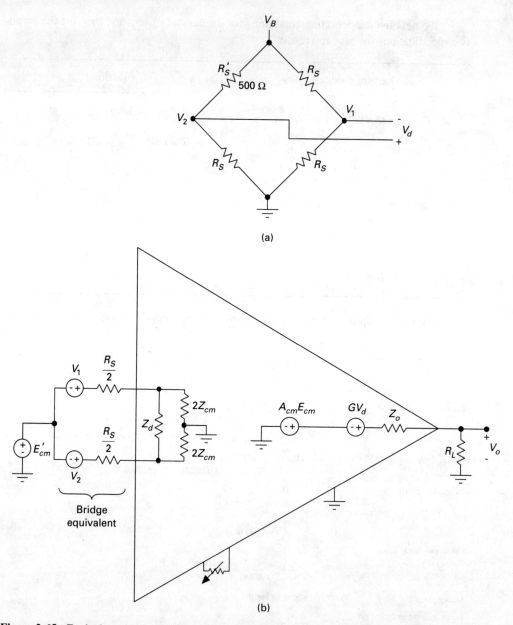

Figure 2-65 Equivalent circuit of instrumentation amplifier with bridge transducer [22]. (a) Strain gauge bridge. (b) Bridge connected to amplifier. Reprinted by permission of Analog Devices, Inc.

Now, if one is pessimistic and assumes all the errors combine by simply adding (Murphy's law case; that is, the worst will happen) then we have the following results:

Error at T = 25 °C: (T constant)   ±63.5 mV   or   ±1.27%
(Sum of temperature-independent errors)
Error over temperature: range +5° C to +45° C   ± 293.5 mV   or   ±5.87%

Chap. 2 Amplifiers and Their Integration into Signal Processing Systems

Note that temperature is what dominates in this particular system. Suppose we originally plan to use 8-bit data conversion on the output of the amplifier. The error could be, even at constant temperature of 25° C, as much as 63.5 mV. For this to be below 1/2 LSB would require

$$63.5 \text{ mV} \leq 1/2 \frac{10}{2^N}$$

$$2^N \leq \frac{5}{63.5 \text{ mV}} = 78.7$$

$$N \leq \frac{\log_{10} 78.7}{\log_{10} 2} = 6.3$$

Thus N would need to be 6 or less. This means that even at constant temperature the last two bits of an 8-bit system are uncertain.

Now suppose we let temperature vary over the extremes of ±20° C. Then we would have

$$N \leq \frac{\log_{10} \frac{5}{293.5} \text{ mV}}{\log_{10} 2} = 4$$

The temperature parameter has in effect destroyed the meaning of the lower 4 bits of an 8-bit word. For this particular system we would then need to select another amplifier whose drift is smaller. Instrumentation amplifiers having drifts of 0.5 μV/°C (RTI) are available, fortunately.

Using root sum squared error yields

$$error = \sqrt{10^2 + 220^2 + .5^2 + 50^2 + 8^2 + 5^2} = \pm 226 \text{ mV}$$

This still results in a value of 4 for N.

Exercises

E2.20a For the example of this section, determine the allowable drift (RTI) at $G = 1000$ in order to use 8-bit data conversion. Repeat for $N = 10$. Select an amplifier from a manufacturer's guide of products (for 8 bits) that will be sufficient. Do not overlook the other specs as well. Construct an error budget for your new selection.

E2.20b For the example of this section, suppose that the power supply V_B is a 10-V battery. How much can the battery voltage drop before the output voltage is reduced by 1/2 LSB for an 8-bit word? This tells you when you have to replace the battery. We assume here that only R_S' changes value. Suppose we define the output sensitivity as dV_d/dR'_S. What is the smallest $\Delta R'_S$ that has any meaning for this system, and what is the corresponding ΔV_d?

GENERAL EXERCISES

2.1 Prove that for n identical stages of amplification which each have a frequency characteristic

$$G(f) = \frac{A_o}{1 + j\frac{f}{f_2}}$$

the overall 3-dB bandwidth is

$$f_{2(n)} = f_2 \left[2^{1/n} - 1 \right]^{1/2}$$

Plot $f_{2(n)}/f_2$ as a function of n, and plot $|G(f)|^n$ as a function of f/f_2 for the cases $n = 1, 2, 3, 4$, and mark the 3-dB frequencies f_{2o}^n. Comment on the physical significance of these results.

2.2 Three amplifiers are placed in cascade that have gains of

$$A_1 = \frac{2}{1 + j\frac{f}{2 \times 10^5}}$$

$$A_2 = \frac{4}{1 + j\frac{f}{4 \times 10^5}}$$

$$A_3 = \frac{6}{1 + j\frac{f}{6 \times 10^5}}$$

Compute the overall 3-dB bandwidth and the DC gain.

2.3 For a 741 op amp, design R_f and R_1 to obtain a noninverting gain of 100. What is the bandwidth of your amplifier? ($A_o = 10^5$, $f_2 = 10$ Hz) What is the input impedance of your amplifier?

2.4 Design an inverting amplifier with a gain of -50 using an op amp with a 4-MHz GBW (gain-bandwidth product). What are the amplifier bandwidth and input impedance?

2.5 If six identical stages having GBW $= 10^7$ Hz are cascaded, what is the overall midband (DC) gain and the operating bandwidth?

2.6 Design an amplifier using op amps that have a gain bandwidth product of 5×10^6 Hz so that the overall gain, DC, is 500 and the operating bandwidth is 500 kHz. How many stages are required, and what is the individual stage gain?

2.7 Design an amplifier that can be used in a 10-bit data word signal processing system to provide amplitude detection of frequencies up to 5 kHz with a gain of +20. The maximum signal size is 50 mV at 5 kHz. The op amp has $A_o = 2 \times 10^5$ and $f_2 = 5$ Hz. Note that the supply voltages are the unknowns here and assume $V_{CC} = |V_{EE}|$. Is this a practical amplifier system?

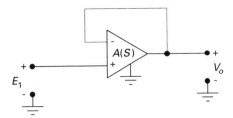

Figure PR2.10

2.8 Design a noninverting amplifier that can be used to detect a 20-kHz signal of 100 µV minimum amplitude for a system using an op amp with a gain bandwidth product of 2×10^6 and a DC gain of 50,000. The signal processing system uses 8-bit data words and has a 5-V signal maximum limit for a 60-mV input. The solution to this problem is not unique, as with many design problems. What is the 3-dB frequency of your amplifier?

2.9 To determine the extent to which a pole factor $s + (K + \Delta K)$ may be replaced by $S + K$ for N-bit data words, determine the relationship between N, K, ΔK such that the difference between the step responses of $\alpha/s + K$ and $\alpha/s + (K + \alpha K)$, with α a constant, is less than 1/2 LSB at any point.

2.10 Determine the transfer function (including the effect of nonideal op amp gain $A(s)$) of the voltage follower circuit shown in Fig. PR2.10. Input currents = 0.

2.11 An analog signal is band-limited and has the following frequency spectrum:

$$F(\omega) = \frac{SM\left(\dfrac{\omega}{f_h}\right)}{\left(\dfrac{\omega}{f_h}\right)} \left[U(\omega + 2\pi f_h) - U(\omega - 2\pi f_h) \right]$$

Will an op amp connected for unity positive gain and a slew rate of $2\pi F_h$ V/S be slew rate limited? Are there any particular amplitudes that will ensure no limiting? If so, give them.

2.12 Many A/D converters work on a 10-V full-scale basis (±10 or ±5 for most bipolar systems). Suppose an engineer is working in the audio frequency range to 10 kHz. What must the engineer select for the slew rate and gain-bandwidth product of an op amp for an input signal maximum of 100 mV?

2.13 You want to track an audio signal that is frequency band-limited to 6 kHz, and the A/D converter following works over the voltage range 0 to ±10 V. An amplifier is needed to boost the signal so that the 8 bits will cover the total A/D span. What is the minimum slew rate of the op amp? Can the slope of a complex audio signal ever exceed that of a signal 6-kHz sine wave? If so, explain and give an example. If not, discuss whether or not the minimum SR will guarantee proper performance.

2.14 The summer amplifier shown in Fig. PR2.14 has two different frequency inputs. The amplifier has a first-order low-pass characteristic $A(s)$.
   a. What value of φ yields the largest output amplitude?
   b. For part (a) what is the largest allowable V_m?

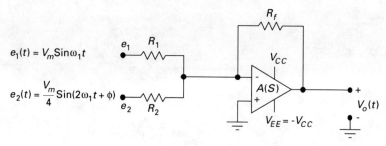

Figure PR 2.14

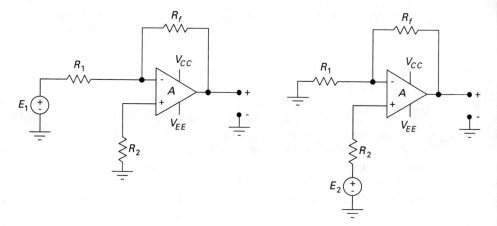

Figure PR 2.17

    c. For the original signals, what value of φ gives the largest output signal slope? Find the largest slope value.

    d. What is the frequency response limiting slope of the amplifier? (Use two step inputs whose combined effects just produce voltage saturation.)

2.15 Use the superposition theorem for linear circuits (apply E'_{cm} and then apply V_2 and V_1 together) and show that the effective common mode voltage E_{cm} can also be expressed as

$$E_{cm}^{eff} = \frac{A}{A_{cm}}(V_2 - V_1) + E'_{cm} + \frac{V_1 + V_2}{2}$$

(Hint: Determine total output voltage and divide by A_{cm} to get the effective input voltage.)

2.16 Suppose that a manufacturer specification sheet listed values for V_{os}, I_{os}, and A. An engineer desires an ideal gain of -20 for an inverter-amplifier. Using R_2 as given by Eq. (2-97), what values of R_1 and R_f minimize the output offset V_o^{os}? Is this result valid for all normal operating temperatures? Explain why or why not. Assume V_{os} and I_{os} are both positive.

Chap. 2 Amplifiers and Their Integration into Signal Processing Systems

2.17 For the amplifiers in Fig. PR2.17, what are the equations for the resistor tolerance(s) as a function of desired gain G_o ($A = \infty$) and the number of bits N used to represent the data? See Fig. PR2.17. Neglect amplifier input and output impedances and offset current and voltage.

2.18 Derive the exact gain expression for Fig. 2-53 assuming A is finite. Show that for $A \to \infty$ Eq. (2-111) results. Determine the minimum value of A for which N-bit data words can be used if the output voltage span is V_o^{span}.

2.19 For the configuration of Fig. 2-49, determine the sensitivity of gain with respect to the factor K. Compute the sensitivity for typical component values.

2.20 Derive Eq. (2-16) beginning with the equations for $\overline{V}(d)$ and $\overline{I}(d)$. What does $\overline{Z}(d)$ reduce to for the lossless case ($\alpha = 0$)?

REFERENCES

1. Franco, S., *Design with Operational Amplifiers and Analog Integrated Circuits*, McGraw-Hill, New York, 1988.
2. Kennedy, E. J., *Operational Amplifier Circuits: Theory and Applications*, Holt, Rinehart and Winston, New York, 1988.
3. Wait, J. V., L. P. Huelsman, and G. A. Korn, *Introduction to Operational Amplifiers: Theory and Applications*, McGraw-Hill, New York, 1975.
4. Graem, J., G. Tolbey, and L. P. Huelsman, *Operational Amplifiers: Design and Applications*, McGraw-Hill, New York, 1971.
5. Garrett, P., *Analog Systems for Microprocessors and Minicomputers*, Reston Publishing Company (Prentice Hall), Reston, VA, 1978.
6. Sheingold, D. H. (ed.), *Analog-Digital Conversion Notes*, Analog Devices, Norwood, MA, 1977.
7. Sheingold, D. H. (ed.), *Transducer Interfacing Handbook*, Analog Devices, Norwood, MA, 1980.
8. Anderson, E. M., *Electric Transmission Line Fundamentals*, Reston Publishing Company (Prentice Hall), Reston, VA, 1985.
9. Edwards, T. C., *Foundations for Microstrip Circuit Design*, John Wiley and Sons, New York, 1981.
10. Howe, H., Jr., *Stripline Circuit Design*, Artech House, Dedham, MA, 1974.
11. Davidson, C. W., *Transmission Lines for Communications*, Halsted Press (John Wiley and Sons), New York, 1978.
12. Dworsky, L. N., *Modern Transmission Line Theory and Applications*, Wiley-Interscience, John Wiley and Sons, New York, 1979.
13. Blood, B., "Improve Fast-Logic Designs. Terminated Lines Reduce Reflections to Overcome Line-Length and Fan-out Limitations," *Electronic Design*, v.10, May 10, 1973, pp. 90–93.
14. Davidson, E. E., and R. D. Lane, "Diodes Damp Line Reflections Without Overloading Logic," *Electronics*, February 19, 1976, pp. 123–127.
15. Goodenough, F., "Op Amp or Instrumentation Amp: When Do You Use Which and Why?" *Electronic Design News*, May 20, 1977, pp. 109–113.
16. Bailey, D. C., "An Instrumentation Amplifier is Not An Op Amp," *Electronic Products*, September 18, 1972.
17. *MECL System Design Handbook*, Motorola, 1971.

18. *MECL Integrated Circuits Data Book*, Motorola, 1972.
19. Pease, R. A., "Low Noise Composite Op Amp Beats Monolithics," *EDN (Electronic Design News), May 5, 1980*, pp. 179ff.
20. Taub, H., and D. Schilling, *Digital Integrated Electronics*, McGraw-Hill, New York, 1977 (particularly Appendix A).
21. Jordan, E. C. (ed.), *Reference Data For Engineers: Radio, Electronics, Computer, and Communications,* 7th ed., Howard W. Sams & Co. (Macmillan), Indianapolis, IN, 1985.
22. Sheingold, D. H. (ed.), *Analog-Digital Conversion Handbook,* Analog Devices, Norwood, MA, 1972.

APPENDIX 2.1 OPERATIONAL AMPLIFIERS

SPECIFICATIONS (typical @ +25°C and ±15V unless otherwise noted)

MODEL	234J	234K	234L	235J	235K	235L
OPEN LOOP GAIN						
DC, 2k ohm load	10^7 V/V min	*	*	5×10^7 V/V min	**	**
RATED OUTPUT						
Voltage	±10V min	*	*	*	*	*
Current	±5mA min	*	*	*	*	*
Load Capacitance Range	0-1000pF min	*	*	0.01µF	**	**
FREQUENCY[1]						
Unity Gain, Small Signal	2.5MHz	*	*	1MHz	**	**
Full Power Response	500kHz min	*	*	5kHz min	**	**
Slew Rate	30V/µs	*	*	0.3V/µs min	**	**
SETTLING TIME to 0.01%						
20kΩ load, 10V step	4µs	*	*	N/A	N/A	N/A
INPUT OFFSET VOLTAGE						
Initial Offset[2]	±50µV max	±20µV max	±20µV max	±25µV max	**	±15µV max
vs. Temp, 0 to +70°C	±1.0µV/°C max	±0.3µV/°C max	±0.1µV/°C max	±0.5µV/°C max	±0.25µV/°C max	±0.1µV/°C max
vs. Supply Voltage	±0.2µV/%	*	*	±0.1µV/%	**	**
vs. Time	±2µV/month	*	*	±5µV/year	**	**
vs. Turn On, 10 sec to 10 min	±3µV	*	*	*	*	*
INPUT BIAS CURRENT						
Initial, @ +25°C	±100pA max	*	*	*	±50pA max	±50pA max
vs. Temp, 0 to +70°C	±4pA/°C max	±2pA/°C max	±1pA/°C max	1pA/°C max	0.5pA/°C max	0.5pA/°C max
vs. Supply Voltage	±0.5pA/%	*	*	0.2pA/%	**	**
INPUT IMPEDANCE						
Inverting Input to Signal Ground	300k ohms	*	*	*	*	*
INPUT NOISE						
Voltage, 0.01 to 1Hz	0.7µV p-p	*	*	0.5µV p-p	2µV p-p max	2µV p-p max
0.1 to 10Hz	1.5µV p-p	*	*	3.5µV p-p	**	**
10Hz to 10kHz	2µV rms	*	*	5µV rms	**	**
Current, 0.01 to 1Hz	2pA p-p	*	*	10pA p-p	**	**
0.1 to 10Hz	4pA p-p	*	*	30pA p-p	**	**
INPUT VOLTAGE RANGE						
(−) Input to Signal Ground	±15V max	*	*	*	*	*
POWER SUPPLY (V dc)[3]						
Rated Performance	±15V @ 5mA	*	*	*	*	*
Operating	±(12 to 18)V	*	*	*	*	*
TEMPERATURE RANGE						
Rated Specifications	0 to +70°C	*	*	*	*	*
Operating	−25°C to +85°C	*	*	*	*	*
Storage	−25°C to +100°C	*	*	−55°C to +125°C	**	**

NOTES
*Specifications same as model 234J.
**Specifications same as model 235J.
[1] Model 235 overload recovery, 10 sec typ.
[2] Externally adjustable to zero.
[3] Recommended power supply: Analog Devices model 904, ±15V dc @ 50mA.

OUTLINE DIMENSIONS
Dimensions shown in inches and (mm).

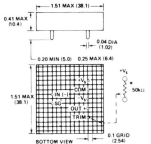

NOTES:
*Connect Trim Terminal to Common if Trim Pot is not used.
1. SG Tied to Common.
2. Mating Socket AC1010.
3. Weight: 27 grams.

Source: Reprinted with permission of Analog Devices, Inc.

SPECIFICATIONS (typical @ +25°C and $V_S = \pm 15V$ dc unless otherwise specified)

Model	AD381JH / AD382JH	AD381KH / AD382KH	AD381LH / AD382LH	AD381SH / AD382SH
OPEN LOOP GAIN				
$V_{OUT} = \pm 10V, R_L \geq 2k\Omega$ (AD381)	60,000 min	100,000 min	**	**
$V_{OUT} = \pm 10V, R_L = 200\Omega$ (AD382)	25,000 min	35,000 min	**	**
$R_L = 10k\Omega$ (AD382)	100,000	150,000	**	**
OUTPUT CHARACTERISTICS (AD382)				
Voltage @ $R_L = 200\Omega$	$\pm 12V (\pm 10V$ min$)$	*	*	Note 1
Voltage @ $R_L = 10k\Omega$	$\pm 13V (\pm 12V$ min$)$	*	*	*
Short Circuit Current, Continuous	80mA	*	*	*
OUTPUT CHARACTERISTICS (AD381)				
Voltage @ $R_L = 1k\Omega$, T_A = min to max	$\pm 12V (\pm 10V$ min$)$	*	*	Note 2
Voltage @ $R_L = 2k\Omega$, T_A = min to max	$\pm 12V (\pm 10V$ min$)$	*	*	*
Voltage @ $R_L = 10k\Omega$, T_A = min to max	$\pm 13V (\pm 12V$ min$)$	*	*	*
Short Circuit Current, Continuous	20mA	*	*	*
DYNAMIC RESPONSE				
Unity Gain, Small Signal	5MHz	*	*	*
Full Power Response	500kHz	*	*	*
Slew Rate, Unity Gain	$30V/\mu s$ ($20V/\mu s$ min)	*	*	*
Settling Time: 10V Step to 0.1%	700ns	*	*	*
10V Step to 0.01%	$1.2\mu s$	$1.2\mu s$ ($2.0\mu s$ max)	**	**
INPUT OFFSET VOLTAGE	1.0mV max	0.5mV max	0.25mV max	*
vs. Temperature, T_A = min to max[3]	$15\mu V/°C$ max	$10\mu V/°C$ max	$5\mu V/°C$ max	$10\mu V/°C$ max
vs. Supply	$200\mu V/V$ max	$100\mu V/V$ max	**	**
INPUT BIAS CURRENT[4]				
Either Input	20pA (100pA max)	10pA (50pA max)(* for AD381)	** (* for AD381)	** (* for AD381)
Input Offset Current	5pA	*	*	*
INPUT IMPEDANCE				
Differential	$10^{12}\Omega \| 7pF$	*	*	*
Common Mode	$10^{12}\Omega \| 7pF$	*	*	*
INPUT VOLTAGE RANGE				
Differential[5]	$\pm 20V$	*	*	*
Common Mode	$\pm 12V (\pm 10V$ min$)$	*	*	*
Common-Mode Rejection, $V_{IN} = \pm 10V$	70dB min	80dB min	**	**
POWER SUPPLY				
Rated Performance	$\pm 15V$	*	*	*
Operating	$\pm (5$ to $18)V$	*	*	*
Quiescent Current AD382	3.4mA (6mA max)	*	*	*
AD381	3.2mA (5mA max)	*	*	*
VOLTAGE NOISE				
0.1Hz–10Hz	$2\mu V$ p-p	*	*	*
10Hz	$35nV/\sqrt{Hz}$	*	*	*
100Hz	$22nV/\sqrt{Hz}$	*	*	*
1kHz	$18nV/\sqrt{Hz}$	*	*	*
10kHz	$16nV/\sqrt{Hz}$	*	*	*
TEMPERATURE RANGE				
Operating, Rated Performance	0 to +70°C	*	*	−55°C to +125°C
Storage	−65°C to +150°C	*	*	*
Thermal Resistance – θ_{JA} (AD382)	100°C/W	*	*	*
Thermal Resistance – θ_{JC} (AD382)	70°C/W	*	*	*

NOTES

[1] The AD381SH has an output voltage of $\pm 12V (\pm 10V$ min) for a $1k\Omega$ load from T_{min} to +70°C. From +70°C to +125°C the output current is 7mA.

[2] The AD382SH has an output voltage of $\pm 12V (\pm 10V$ min) for a 200Ω load from T_{min} to +100°C. To +125°C the output current is 35mA.

[3] Input Offset Voltage Drift is specified with the offset voltage unnulled. Nulling will induce an additional $3\mu V/°C$ for every mV of offset nulled.

[4] Bias Current specifications are guaranteed maximum at either input after 5 minutes of operation at T_A = +25°C. For higher temperatures, the current doubles every 10°C.

[5] Defined as the maximum safe voltage between inputs, such that neither exceeds $\pm 10V$ from ground.

*Specifications same as J grade.
**Specifications same as K grade.

SPECIFICATIONS (typical @ +25°C and ±15V dc unless otherwise noted)

MODEL	AD510JH	AD510KH	AD510LH	AD510SH
OPEN LOOP GAIN				
$V_{OS} = \pm 10V$, $R_L \geq 2k\Omega$	250,000 min	10^6 min	**	**
T_{min} to T_{max}	125,000 min	500,000 min	**	250,000
OUTPUT CHARACTERISTICS				
Voltage @ $R_L \geq 2k\Omega$, T_{min} to T_{max}	±10V min	*	*	*
Load Capacitance	1000pF	*	*	*
Output Current	10mA min	*	*	*
Short Circuit Current	25mA	*	*	*
FREQUENCY RESPONSE				
Unity Gain, Small Signal	300kHz	*	*	*
Full Power Response	1.5kHz	*	*	*
Slew Rate, Unity Gain	0.10V/μs	*	*	*
INPUT OFFSET VOLTAGE				
Initial Offset, $R_S \leq 10k\Omega$	100μV max	50μV max	25μV max	**
vs. Temp., T_{min} to T_{max}	3.0μV/°C max	1.0μV/°C max	0.5μV/°C max	**
vs. Supply	25μV/V max	10μV/V max	**	**
T_{min} to T_{max}	40μV/V max	15μV/V max	**	20μV/V max
INPUT OFFSET CURRENT				
Initial	5nA max	4nA max	2.5nA max	**
T_{min} to T_{max}	8nA max	6nA max	4nA max	10nA max
INPUT BIAS CURRENT				
Initial	25nA max	13nA max	10nA max	**
T_{min} to T_{max}	40nA max	20nA max	15nA max	30nA max
vs. Temp, T_{min} to T_{max}	±100pA/°C	±50pA/°C	±40pA/°C	**
INPUT IMPEDANCE				
Differential	4MΩ	6MΩ	**	**
Common Mode	100MΩ‖4pF	*	*	*
INPUT NOISE				
Voltage, 0.1Hz to 10Hz	1μV p-p	*	*	*
f = 10Hz	18nV/√Hz	*	*	*
f = 100Hz	13nV/√Hz	*	*	*
f = 1kHz	10nV/√Hz	*	*	*
Current, f = 10Hz	0.5pA/√Hz	*	*	*
f = 100Hz	0.3pA/√Hz	*	*	*
f = 1kHz	0.3pA/√Hz	*	*	*
INPUT VOLTAGE RANGE				
Differential or Common Mode max safe	±V_S	*	*	*
Common Mode Rejection, V_{in} = ±10V	94dB min	110dB min	**	**
Common Mode Rejection, T_{min} to T_{max}	94dB	100dB min	**	**
POWER SUPPLY				
Rated Performance	±15V	*	*	*
Operating	±(5 to 18)V	*	*	±(5 to 22)V
Current, Quiescent	4mA max	3mA max	**	**
TEMPERATURE RANGE				
Operating Rated Performance	0 to +70°C	*	*	−55°C to +125°C
Storage	−65°C to +150°C	*	*	*
PACKAGE OPTIONS:[1] TO-99 Style (H08B)	AD510JH	AD510KH	AD510LH	AD510SH

NOTES
*Specifications same as AD510JH.
**Specifications same as AD510KH.

SPECIFICATIONS (@ +25°C and $V_S = \pm 15V$ dc)

Model	AD547J Min	AD547J Typ	AD547J Max	AD547K Min	AD547K Typ	AD547K Max	AD547L Min	AD547L Typ	AD547L Max	AD547S Min	AD547S Typ	AD547S Max	Units
OPEN LOOP GAIN[1]													
$V_O = \pm 10V, R_L \geq 2k\Omega$	100,000			250,000			250,000			250,000			V/V
T_{min} to $T_{max}, R_L = 2k\Omega$	100,000			250,000			250,000			100,000			V/V
OUTPUT CHARACTERISTICS													
Voltage @ $R_L = 2k\Omega$, T_{min} to T_{max}	±10	±12		±10	±12		±10	±12		±10	±12		V
Voltage @ $R_L = 10k\Omega$, T_{min} to T_{max}	±12	±13		±12	±13		±12	±13		±12	±13		V
Short Circuit Current		25			25			25			25		mA
FREQUENCY RESPONSE													
Unity Gain Small Signal		1.0			1.0			1.0			1.0		MHz
Full Power Response		50			50			50			50		kHz
Slew Rate, Unity Gain		3.0			3.0			3.0			3.0		V/μs
INPUT OFFSET VOLTAGE[2]													
Initial Offset			1.0			0.5			0.25			0.5	mV
Input Offset Voltage vs. Temp.[3]		5			2			1.0			5.0		μV/°C
Input Offset Voltage vs. Supply, T_{min} to T_{max}			200			100			100			100	μV/V
INPUT BIAS CURRENT													
Either Input[4]		10	50		10	25		10	25		10	25	pA
Offset Current		5			2			2			2		pA
INPUT IMPEDANCE													
Differential[5]		$10^{12}\|6$			$10^{12}\|6$			$10^{12}\|6$			$10^{12}\|6$		MΩ‖pF
Common Mode		$10^{12}\|6$			$10^{12}\|6$			$10^{12}\|6$			$10^{12}\|6$		MΩ‖pF
INPUT VOLTAGE RANGE													
Differential		±20			±20			±20			±20		V
Common Mode	±10	±12		±10	±12		±10	±12		±10	±12		V
Common Mode Rejection	76			80			80			80			dB
INPUT NOISE													
Voltage 0.1Hz to 10Hz		2			4			4			4		μV p-p
f = 10Hz		70			70			70			70		nV/√Hz
f = 100Hz		45			45			45			45		nV/√Hz
f = 1kHz		30			30			30			30		nV/√Hz
f = 10kHz		25			25			25			25		nV/√Hz
POWER SUPPLY													
Rated Performance		±15			±15			±15			±15		V
Operating	±5		±18	±5		±18	±5		±18	±5		±18	V
Quiescent Current		1.1	1.5		1.1	1.5		1.1	1.5		1.1	1.5	mA
TEMPERATURE RANGE													
Operating, Rated Performance	0		+70	0		+70	0		+70	−55		+125	°C
Storage	−65		+150	−65		+150	−65		+150	−65		+150	°C
PACKAGE[6]													
TO-99 Style (H08B)	AD547JH			AD547KH			AD547LH			AD547SH			

NOTES

[1] Open Loop Gain is specified with V_{OS} both nulled and unnulled.
[2] Input Offset Voltage specifications are guaranteed after 5 minutes of operation at $T_A = +25°C$.
[3] Input Offset Voltage Drift is specified with the offset voltage unnulled. Nulling will induce an additional 3μV/°C/mV of nulled offset.
[4] Bias Current specifications are guaranteed at maximum at either input after 5 minutes of operation at $T_A = +25°C$. For higher temperatures, the current doubles every 10°C.
[5] Defined as the maximum safe voltage between inputs, such that neither exceed ±10V from ground.
[6] See Section 19 for package outline information.

SPECIFICATIONS (@ +25°C and $V_S = \pm 15V$ dc)

Model	AD544J Min	AD544J Typ	AD544J Max	AD544K Min	AD544K Typ	AD544K Max	AD544L Min	AD544L Typ	AD544L Max	AD544S Min	AD544S Typ	AD544S Max	Units
OPEN LOOP GAIN[1]													
$V_O = \pm 10V$, $R_L \geq 2k\Omega$	30,000			50,000			50,000			50,000			V/V
T_{min} to T_{max}, $R_L = 2k\Omega$	20,000			40,000			40,000			20,000			V/V
OUTPUT CHARACTERISTICS													
Voltage @ $R_L = 2k\Omega$, T_{min} to T_{max}	±10	±12		±10	±12		±10	±12		±10	±12		V
Voltage @ $R_L = 10k\Omega$	±12	±13		±12	±13		±12	±13		±12	±13		V
Short Circuit Current		25			25			25			25		mA
FREQUENCY RESPONSE													
Unity Gain Small Signal		2.0			2.0			2.0			2.0		MHz
Full Power Response		200			200			200			200		kHz
Slew Rate, Unity Gain	8.0	13.0		8.0	13.0		8.0	13.0		8.0	13.0		V/μs
Settling Time to 0.01%		3.0			3.0			3.0			3.0		μs
Total Harmonic Distortion		0.0025			0.0025			0.0025			0.0025		%
INPUT OFFSET VOLTAGE[2]													
Initial Offset			2.0			1.0			0.5			1.0	mV
Input Offset Voltage vs. Temp. or T_{min} to T_{max}			20			10			5			15	μV/°C
Input Offset Voltage vs. Supply, T_{min} to T_{max}			200			100			100			100	μV/V
INPUT BIAS CURRENT[3]													
Either Input		10	50		10	25		10	25		10	25	pA
Offset Current		5			2			2			2		pA
INPUT IMPEDANCE													
Differential[4]		$10^{12}\|6$			$10^{12}\|6$			$10^{12}\|6$			$10^{12}\|6$		MΩ|pF
Common Mode		$10^{12}\|3$			$10^{12}\|3$			$10^{12}\|3$			$10^{12}\|3$		MΩ|pF
INPUT VOLTAGE RANGE													
Differential		±20			±20			±20			±20			V
Common Mode	±10	±12		±10	±12		±10	±12		±10	±12		V
Common Mode Rejection	74			80			80			80			dB
INPUT NOISE													
Voltage 0.1Hz to 10Hz		2			2			2			2		μV p-p
$f = 10Hz$		35			35			35			35		nV/$\sqrt{Hz}$
$f = 100Hz$		22			22			22			22		nV/$\sqrt{Hz}$
$f = 1kHz$		18			18			18			18		nV/$\sqrt{Hz}$
$f = 10kHz$		16			16			16			16		nV/$\sqrt{Hz}$
POWER SUPPLY													
Rated Performance		±15			±15			±15			±15			V
Operating	±5		±18	±5		±18	±5		±18	±5		±18	V
Quiescent Current		1.8	2.5		1.8	2.5		1.8	2.5		1.8	2.5	mA
TEMPERATURE RANGE													
Operating, Rated Performance	0		+70	0		+70	0		+70	−55		+125	°C
Storage	−65		+150	−65		+150	−65		+150	−65		+150	°C
PACKAGE[5]													
TO-99 Style (H08B)		AD544JH			AD544KH			AD544LH			AD544SH		

NOTES
[1] Open Loop Gain is specified with V_{OS} both nulled and unnulled.
[2] Input Offset Voltage specifications are guaranteed after 5 minutes of operation at $T_A = +25°C$.
[3] Bias Current specifications are guaranteed at maximum at either input after 5 minutes of operation at $T_A = +25°C$. For higher temperatures, the current doubles every 10°C.
[4] Defined as voltage between inputs, such that neither exceeds ±10V from ground.
[5] See Section 19 for package outline information.

Specifications subject to change without notice.

Specifications shown in boldface are tested on all production units at final electrical test. Results from those tests are used to calculate outgoing quality levels. All min and max specifications are guaranteed, although only those shown in boldface are tested on all production units.

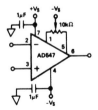

Standard Null Circuit

SPECIFICATIONS (@ +25°C and $V_S = \pm 15V$ dc)

Model	AD642J Min	AD642J Typ	AD642J Max	AD642K Min	AD642K Typ	AD642K Max	AD642L Min	AD642L Typ	AD642L Max	AD642S Min	AD642S Typ	AD642S Max	Units
OPEN LOOP GAIN													
$V_O = \pm 10V$, $R_L \geq 2k\Omega$	100,000			250,000			250,000			250,000			V/V
T_{min} to T_{max}, $R_L = 2k\Omega$	100,000			250,000			250,000			100,000			V/V
OUTPUT CHARACTERISTICS													
Voltage @ $R_L = 2k\Omega$, T_{min} to T_{max}	±10	±12		±10	±12		±10	±12		±10	±12		V
Voltage @ $R_L = 10k\Omega$, T_{min} to T_{max}	±12	±13		±12	±13		±12	±13		±12	±13		V
Short Circuit Current		25			25			25			25		mA
FREQUENCY RESPONSE													
Unity Gain Small Signal		1.0			1.0			1.0			1.0		MHz
Full Power Response		50			50			50			50		kHz
Slew Rate, Unity Gain		3.0			3.0			3.0			3.0		V/µs
INPUT OFFSET VOLTAGE[1]													
Initial Offset			2.0			1.0			0.5			1.0	mV
Input Offset Voltage T_{min} to T_{max}			3.5			2.0			1.0			3.5	mV
Input Offset Voltage vs. Supply, T_{min} to T_{max}			200			100			100			100	µV/V
INPUT BIAS CURRENT[2]													
Either Input		10	75		10	35		10	35		10	35	pA
Offset Current		5			2			2			2		pA
MATCHING CHARACTERISTICS[3]													
Input Offset Voltage			1.0			0.5			0.25			0.5	mV
Input Offset Voltage T_{min} to T_{max}			3.5			2.0			1.0			3.5	mV
Input Bias Current			35			25			25			35	pA
Crosstalk		−124			−124			−124			−124		dB
INPUT IMPEDANCE													
Differential		10^{12}‖6			10^{12}‖6			10^{12}‖6			10^{12}‖6		MΩ‖pF
Common Mode		10^{12}‖6			10^{12}‖6			10^{12}‖6			10^{12}‖6		MΩ‖pF
INPUT VOLTAGE RANGE													
Differential[4]		±20			±20			±20			±20		V
Common Mode	±10	±12		±10	±12		±10	±12		±10	±12		V
Common Mode Rejection	76			80			80			80			dB
INPUT NOISE													
Voltage 0.1Hz to 10Hz		2			2			2			2		µV p-p
f = 10Hz		70			70			70			70		nV/√Hz
f = 100Hz		45			45			45			45		nV/√Hz
f = 1kHz		30			30			30			30		nV/√Hz
f = 10kHz		25			25			25			25		nV/√Hz
POWER SUPPLY													
Rated Performance		±15			±15			±15			±15		V
Operating	±5		±18	±5		±15	±5		±15	±5		±15	V
Quiescent Current			2.8			2.8			2.8			2.8	mA
TEMPERATURE RANGE													
Operating, Rated Performance	0		+70	0		+70	0		+70	−55		+125	°C
Storage	−65		+150	−65		+150	−65		+150	−65		+150	°C
PACKAGE[5]													
TO-99 Style (H08B)	AD642JH			AD642KH			AD642LH			AD642SH			

NOTES
[1] Input Offset Voltage specifications are guaranteed after 5 minutes of operation at $T_A = +25°C$.
[2] Bias Current specifications are guaranteed at maximum at either input after 5 minutes of operation at $T_A = +25°C$. For higher temperatures, the current doubles every 10°C.
[3] Matching is defined as the difference between parameters of the two amplifiers.
[4] Defined as the maximum safe voltage between inputs, such that neither exceeds ±10V from ground.
[5] See Section 19 for package outline information.

SPECIFICATIONS[1] (typical @ +V_S = ±15V, R_L = 2kΩ & T_A = +25°C unless otherwise specified)

MODEL	AD522AD	AD522BD	AD522SD
GAIN			
Gain Equation	$1 + \frac{2(10^5)}{R_g}$	*	*
Gain Range	1 to 1000	*	*
Equation Error			
G = 1	0.2% max	0.05% max	**
G = 1000	1.0% max	0.2% max	**
Nonlinearity, max (see Fig. 4)			
G = 1	0.005%	0.001%	**
G = 1000	0.01%	0.005%	**
vs. Temp, max			
G = 1	2ppm/°C (1ppm/°C typ)	*	*
G = 1000	50ppm/°C (25ppm/°C typ)	*	*
OUTPUT CHARACTERISTICS			
Output Rating	±10V @ 5mA	*	*
DYNAMIC RESPONSE (see Fig. 6)			
Small Signal (−3dB)			
G = 1	300kHz	*	*
G = 100	3kHz	*	*
Full Power GBW	1.5kHz	*	*
Slew Rate	0.1V/μs	*	*
Settling Time to 0.1%, G = 100	0.5ms	*	*
to 0.01%, G = 100	5ms	*	*
to 0.01%, G = 10	2ms	*	*
to 0.01%, G = 1	0.5ms	*	*
VOLTAGE OFFSET			
Offsets Referred to Input			
Initial Offset Voltage			
(adjustable to zero)			
G = 1	±400μV max (±200μV typ)	±200μV max (±100μV typ)	±200μV max (±100μV typ)
vs. Temperature, max (see Fig. 3)			
G = 1	±50μV/°C (±10μV/°C typ)	±25μV/°C (±5μV/°C typ)	±10μV/°C (±10μV/°C typ)
G = 1000	±6μV/°C	±2μV/°C	±6μV/°C
1 < G < 1000	$\pm(\frac{50}{G} + 6)$μV/°C	$\pm(\frac{25}{G} + 2)$μV/°C	$\pm(\frac{100}{G} + 6)$μV/°C
vs. Supply, max			
G = 1	±20μV/%	*	*
G = 1000	±0.2μV/%	*	*
INPUT CURRENTS			
Input Bias Current			
Initial max, +25°C	±25nA	*	*
vs. Temperature	±100pA/°C	*	*
Input Offset Current			
Initial max, +25°C	±20nA	*	*
vs. Temperature	±100pA/°C	*	*
INPUT			
Input Impedance			
Differential	10^9 Ω	*	*
Common Mode	10^9 Ω	*	*
Input Voltage Range			
Maximum Differential Input, Linear	±10V	*	*
Maximum Differential Input, Safe	±20V	*	*
Maximum Common Mode, Linear	±10V	*	*
Maximum Common Mode Input, Safe	±15V	*	*
Common Mode Rjection Ratio,			
Min @ ±10V, 1kΩ Source			
Imbalance (see Fig. 5)			
G = 1 (dc to 30Hz)	75dB (90dB typ)	80dB (100dB typ)	75dB (90dB typ)
G = 10 (dc to 10Hz)	90dB (100dB typ)	95dB (110dB typ)	90dB (110dB typ)
G = 100 (dc to 3Hz)	100dB (110dB typ)	100dB (120dB typ)	100dB (120dB typ)
G = 1000 (dc to 1Hz)	100dB (120dB typ)	110dB (>120dB typ)	100dB (>120dB typ)
G = 1 to 1000 (dc to 60Hz)	75dB (88dB typ)	80dB (88dB typ)	*
NOISE			
Voltage Noise, RTI (see Fig. 4)			
0.1Hz to 100Hz (p-p)			
G = 1	15μV	*	*
G = 1000	1.5μV	*	*
10Hz to 10kHz (rms)			
G = 1	15μV	*	*
TEMPERATURE RANGE			
Specified Performance	−25°C to +85°C	*	−55°C to +125°C
Operating	−55°C to +125°C	*	*
Storage	−65°C to +150°C	*	*
POWER SUPPLY			
Power Supply Range	±(5 to 18)V	*	*
Quiescent Current, max @ ±15V	±10mA	±8mA	**

243

SPECIFICATIONS (typical @ + 25°C, & V_S = 15V unless otherwise noted)

MODEL	AD293A	AD293B	AD294A
GAIN			
Range	1 to 1000V/V	*	*
Formula (Input)	$G_{IN} = \left(1 + \frac{100k}{R_G}\right)$; $R_G \geq 1k\Omega$; G_{IN} max = 100		
(Output)	$G_{OUT} = \left(1 + \frac{R_A}{R_B}\right)$; $1 \leq G_{OUT} \leq 10$; G_{OUT} max = 10		
Deviation from Formula			
G = 1	± 1.0%	*	*
G > 1	± 3.0%	*	*
vs. Temperature (−25°C to + 85°C)[1,2] (Gain = 1)	± 60ppm/°C max	*	*
(Gain > 1)	± 120ppm/°C max	*	*
Nonlinearity (± 5V swing)[2]	± 0.1% max	± 0.05% max	*
INPUT VOLTAGE RATINGS			
Linear Differential Range	± 10V min	*	± 5V min
Max Safe Differential Input			
Continuous	120V rms max	*	*
1 Minute	240V rms max	*	*
Max CMV (Inputs to Outputs)			
Continuous (ac or dc)	± 2500V peak	*	± 3500V peak
ac, 60Hz, 1 minute Duration	2500V rms	*	3500V rms
Pulse, 10ms Duration, 1 pulse/10 sec	—	—	± 8000V peak
CMR (60Hz), G = 10V/V			
$R_S \leq 1k\Omega$ Balanced Source Impedance	108dB	*	*
$R_S \leq 1k$ Source Impedance Imbalance	100dB min	*	*
$R_S \leq 5k$ Balanced Source Impedance	—	—	100dB
$R_S \leq 5k$ Source Impedance Imbalance	—	—	95dB min
Leakage Current, Input to Output			
@ 115V ac, 60Hz	2μA rms max	*	*
Input Impedance, G = 1			
Differential	150pF‖$10^8\Omega$	*	*
Overload	100kΩ	*	*
Common Mode	30pF‖($5 \times 10^{10}\Omega$)	*	*
Input Bias Current			
Initial @ + 25°C	2nA (5nA max)	*	*
vs. Temperature	20pA/°C	*	*
Input Noise			
Voltage			
0.05Hz to 100Hz	10μV p-p	*	*
10Hz to 1kHz	5μV rms	*	*
Current			
0.05Hz to 100Hz	50pA p-p	*	*
FREQUENCY RESPONSE			
Small Signal (−3dB) G = 1V/V to 100V/V	2.5kHz	*	*
Full Power, 20V p-p Output (10V p-p AD294)			
G = 1V/V (G_{IN} = 1V/V, G_{OUT} = 1V/V)	200Hz	*	*
G = 100V/V (G_{IN} = 100V/V, G_{OUT} = 1V/V)	100Hz	*	*
G = 10V/V (G_{IN} = 1V/V, G_{OUT} = 10V/V)	1.5kHz	*	*
Slew Rate	9.1V/ms	*	*
OFFSET VOLTAGE, REFERRED TO INPUT			
Initial, @ + 25°C, max	$\left(\pm 3 \pm \frac{22}{G_{IN}}\right)$mV	*	*
vs. Temperature			
(0 to + 70°C)	$\left(\pm 3 \pm \frac{150}{G_{IN}}\right)$μV/°C	*	$\left(\pm 10 \pm \frac{300}{G_{IN}}\right)$μV/°C max
(−25°C to + 85°C) max	$\left(\pm 10 \pm \frac{500}{G_{IN}}\right)$μV/°C max	$\left(\pm 5 \pm \frac{250}{G_{IN}}\right)$μV/°C	$\left(\pm 10 \pm \frac{1000}{G_{IN}}\right)$μV/°C
vs. Supply Voltage	$\left(\pm 0.01 \pm \frac{3}{G_{IN}}\right)$mV/V	*	*
RATED OUTPUT			
Voltage, 2kΩ Load	± 10V min	*	*
Output Impedance	< 1Ω	*	*
Output Ripple, (dc to 100kHz) Bandwidth	4mV p-p	*	*
POWER SUPPLY[3]			
Voltage, Rated Performance	± 15V dc ± 3%	*	*
Voltage, Operating[4]	± 12V dc ± 18V dc	*	*
Current, Quiescent ($V_S = \pm 15V$)	+ 1mA, − 1mA	*	*
(+ V_{OSC} = + 15V)	+ 11mA		
ISOLATED POWER	− 13V dc @ 200μA	*	*
TEMPERATURE RANGE			
Rated Performance	− 25°C to + 85°C	*	*
Operating	− 40°C to + 100°C	*	*

3
Filters

Filters are designed to eliminate unwanted frequency components from a signal. These undesired components are generally referred to as noise. Some noises might be present at the input, others might be introduced by circuit elements, others might consist of interference signals picked up at various points of the system, and still others are introduced by the sampling process. All but the latter noise source have been considered in earlier chapters. It is necessary to use an analog filter preceding the digital sampling process, thus a knowledge of filter basics is appropriate at this time.

The topic of filters is very broad, and there are many textbooks available that provide detailed coverage of filter design. In this chapter we introduce some of the fundamentals of filter theory and basic concepts. There are, however, several very different types of filters. The traditional filter is the passive analog circuit containing resistors, capacitors, and perhaps inductors. The active filter adds amplifying elements such as transistors or op amps to eliminate the need for inductors and reduce the size of the capacitors. Active filters can be implemented with either discrete circuits or with integrated circuits. The switched capacitor filter is a popular form of active circuit that is implemented on an IC chip.

After a continuous signal is converted to digital samples, it is possible to use digital methods to eliminate unwanted components. This is referred to as digital filtering and is treated in most digital signal processing textbooks.

The intent of this chapter is to consider the origin of unwanted components arising from the sampling process (aliasing) and the elimination of this and other noise in the signal by filtering. This chapter will concentrate on the specification of various types of popular filter responses. An understanding of filter responses will allow a system designer to purchase off-the-shelf IC filters or to design discrete circuit filters as outlined

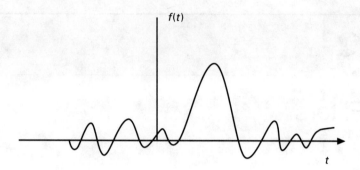

Figure 3-1 A continuous function of time.

in the following chapter. No attempt is made to consider the design of IC filters, since the acquisition system designer will never be involved in this process. Nor will digital filter methods be considered in this text, which is directed toward the acquisition rather than the processing of digital signals.

3.1 ALIASING

A continuous signal, $f(t)$, is shown in Fig. 3-1. If this signal is aperiodic, the Fourier transform or integral can be used to identify the important frequency range of the signal [1]. The Fourier integral transforms the time domain to the frequency domain through the equation

$$F(j\omega) = \int_{-\infty}^{\infty} f(t) e^{-j\omega t} \, dt \qquad (3\text{-}1)$$

The variable ω takes on positive and negative values. If $f(t)$ is the known or observable signal that is to be converted to a sequence of digital codes, the frequency content of $f(t)$ is found by applying Eq. (3-1). It is instructive to compare the frequency content of the actual signal to the frequency content of the sampled signal.

A frequency spectrum for the signal $f(t)$ might appear as shown in Fig. 3-2. Theoretically, the frequency components extend to infinity, however, for most practical waveforms, the energy contained in the higher-frequency components becomes negligible. A typical application will be concerned only with frequencies ranging from zero to some upper limit ω_c, ignoring all components above this value. For a telephone, for example, the upper frequency to be transmitted is approximately 3300 Hz, consequently all higher-energy components are eliminated.

Let us assume that the continuous signal $f(t)$ is sampled at intervals of T, to create the discrete signal $f^*(t)$. This signal is shown in Fig. 3-3. The sampling process can be considered as the amplitude modulation of $p(t)$, a train of unit pulses of width b, as indicated in Fig. 3-4 [2]. The discrete waveform is given by $f^*(t) = f(t)p(t)$.

The train of pulses $p(t)$ can be expressed in terms of unit step functions as

$$p(t) = \sum_{k=-\infty}^{\infty} u(t - kT) - u(t - kT - b) \qquad (3\text{-}2)$$

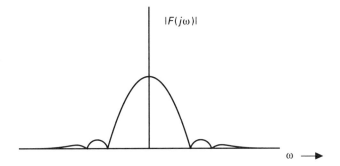

Figure 3-2 Frequency spectrum.

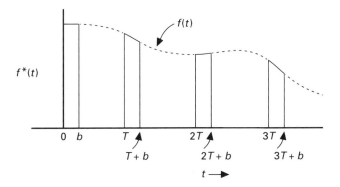

Figure 3-3 A sample waveform.

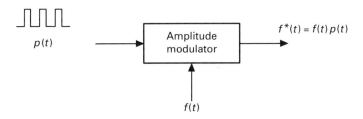

Figure 3-4 One representation of the sampling process.

where $u(t)$ is the unit step function. Since this is a periodic waveform, a Fourier series can be found allowing $p(t)$ to be expressed in exponential form as

$$p(t) = \sum_{n=-\infty}^{\infty} C_n e^{jn\omega_s t} \qquad (3\text{-}3)$$

where $\omega_s = 2\pi/T$ and is the sampling frequency in radians per second. The coefficients C_n are obtained by using the complex exponential form of the Fourier series. The result is [2]

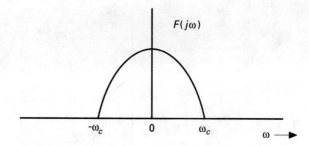

Figure 3-5 Spectrum of continous signal $f(t)$.

$$C_n = \frac{b}{T} \frac{\sin n\omega_s b/2}{(n\omega_s b/2)} e^{-jn\omega_s b/2} \tag{3-4}$$

The effect of aliasing can be found by comparing the Fourier transform of $f^*(t)$ to that of $f(t)$. For a given function $f(t)$, the important components of $F(j\omega)$ might appear as shown in Fig. 3-5, which represents a signal whose frequency content is limited to $\pm\omega_c$. This is a band-limited signal, as represented by the spectrum in Fig. 3-5.

The Fourier transform of the discrete signal $f^*(t)$ is found from

$$F^*(j\omega) = \int_{-\infty}^{\infty} f*(t) e^{-j\omega t}\, dt = \int_{-\infty}^{\infty} f(t) p(t) e^{-j\omega t}\, dt \tag{3-5}$$

Substituting for $p(t)$ from Eq. (3-3) gives

$$F^*(j\omega) = \int_{-\infty}^{\infty} f(t) \left(\sum_{n=-\infty}^{\infty} C_n e^{jn\omega_s t} \right) e^{-j\omega t}\, dt \tag{3-6}$$

This equation can be rearranged by changing the order of integration and summation and noting that C_n is a constant. The expression then becomes

$$F^*(j\omega) = \sum_{n=-\infty}^{\infty} C_n \int_{-\infty}^{\infty} f(t) e^{jn\omega_s t} e^{-j\omega t}\, dt \tag{3-7}$$

The shifting theorem [1] tells us that the Fourier transform of $f(t)e^{at}$ is

$$F(j\omega - a)$$

This can be compared to Eq. (3-1), which indicates that the Fourier transform of $f(t)$ is $F(j\omega)$.

Letting $jn\omega_s = a$ allows this theorem to be applied to Eq. (3-7) and results in

$$F^*(j\omega) = \sum_{n=-\infty}^{\infty} C_n F(j\omega - jn\omega_s) \tag{3-8}$$

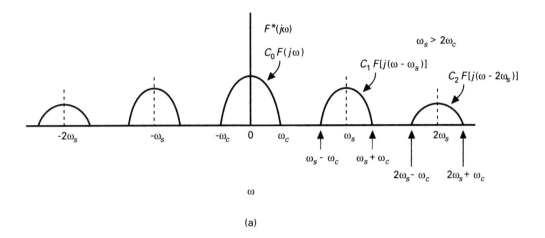

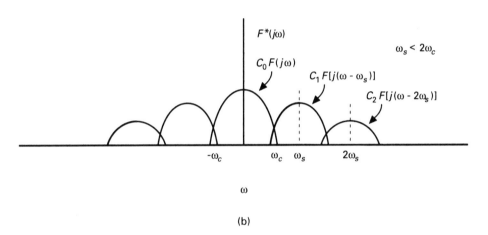

Figure 3-6 Frequency spectra of $f^*(t)$ for: (a) $\omega_s > 2\omega_c$; (b) $\omega_s < 2\omega_c$.

Equation (3-8) indicates that the $n = 0$ term of $F^*(j\omega)$ contains the exact same spectrum as does $F(j\omega)$. This spectrum is multiplied by C_0, which is found from Eq. (3-4) to be b/T. $F^*(j\omega)$ also contains an infinite number of additional spectra, similar to that of $F(j\omega)$, but centered about frequencies of $n\omega_s$, where $n = \pm 1, \pm 2, \pm 3, \ldots$. These spectra are multiplied by the coefficients $C_1, C_2, C_3, \ldots$, respectively. Figure 3-6 demonstrates these results for two values of sampling rate where $F(j\omega)$ is that shown in Fig. 3-5.

We note in Fig. 3-6, as indicated by Eq. (3-4), that the spectra centered about $n\omega_s$ fall off in magnitude with n, that is, C_1 is smaller than C_0, C_2 is smaller than C_1, and so on. If the pulse train $p(t)$ approaches a train of ideal impulse functions, with infinite amplitude, zero width, and unit area, C_n is a constant independent of n and so also are the maximum spectra magnitudes [2]. This is an important special case that will be discussed later in this section.

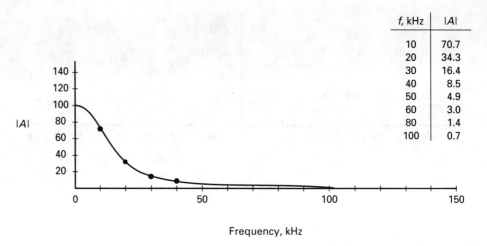

Figure 3-7 Amplifier frequency response.

We see that the difference between the continuous waveform $f(t)$ and the sampled waveform $f*(t)$ is the additional components contained in the sampled waveform. These additional components that do not exist in the continuous waveform are called *aliases*.

In Fig. 3-6(a) the sampling frequency ω_s is chosen to be more than two times the highest significant frequency of the continuous waveform ω_c. For this choice of sampling frequency, no aliased frequencies are folded into the frequency range of actual signal frequencies. If ω_s is chosen to be smaller than $2\omega_c$, the spectra of aliased frequencies fold into the zero to ω_c range.

The frequency $2\omega_c$ represents the lowest possible sampling frequency without adding aliased frequency components to the actual signal spectrum. The frequency $2\omega_c$ is called the Nyquist sampling frequency [2]. We must remember that the spectrum of $f(t)$ will not abruptly stop at ω_c. Most engineering waveforms show a reasonably rapid falloff of magnitude with frequency, but there will be components above ω_c with finite amounts of energy. To account for these smaller components above the frequency ω_c, most accurate systems sample at a rate that is 2.5 to 5 times the highest desired component of ω_c. If a sharp falloff (and expensive) filter is used, the sampling rate may approach a value of 2.5 times the highest frequency, while a less sharp falloff filter will require a higher sampling rate.

In applications that do not allow these higher sampling rates, very sharp filters are used to attenuate the high-frequency components to a negligible level. Rarely, however, does a practical sampling rate drop below a frequency of $3\omega_c$.

While the plots of Fig. 3-6 represent signal spectra, it is possible to use the results of the sampling theorem (Eqs. (3-4) and (3-8)) to compare frequency response curves before and after sampling. For example, the curve of Fig. 3-7 may represent the frequency response of an amplifier voltage gain. From this curve we can calculate the output voltage for a given input if we know the signal frequency. A sinusoidal input voltage of 0.1-V peak value applied at a 20-kHz frequency will result in a 3.43-V peak output, since the magnitude of the voltage gain is 34.3 at 20 kHz.

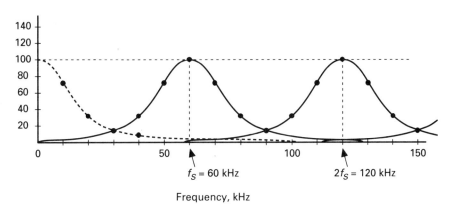

Figure 3-8 Aliased components.

If the output of the amplifier is to be converted to digital by sampling at a 60-kHz rate, the curves of Fig. 3-8 can be used to evaluate the magnitudes of the aliased components. These curves assume ideal impulse sampling (pulse width $b = 0$). The input signal of 0.1 V at 20 kHz would now result in an output component with a magnitude 3.43 V due to the amplifier gain, plus a second component of magnitude 3.43 V at a frequency of 40 kHz due to aliasing. There are also an infinite number of 3.43-V components at frequencies of 80 kHz, 100 kHz, 140 kHz, 160 kHz, 200 kHz, and so on. These components appear 20 kHz below and above each multiple of the sampling frequency.

An input signal of 0.1 V at 40 kHz would result in an output due to amplifier gain of 0.85 V. All aliased components would have this same magnitude and would appear at frequencies of 20 kHz, 80 kHz, 100 kHz, 140 kHz, 160 kHz, and so on.

At this point of the discussion, one might properly question the validity of the assumption of ideal impulse sampling. After all, an ADC samples the analog value and holds the converted digital value until the next conversion is completed. This conversion scheme can be depicted as shown in Fig. 3-9. Of course, the converted value would be a digital code, but it has a finite constant value between samples rather than the zero value assumed in impulse sampling. The computer then reads the ADC output values after every sample. However, all popular digital processing methods are based on the assumption that the values were obtained by ideal impulse sampling. Since the computer treats the data as if it were impulse sampled, the converted digital sequence contains the aliased components.

Example 3-1

An amplifier with the frequency response of Fig. 3-7 is used to amplify signals having a frequency range of 80 kHz. Assuming no component within the frequency range will be larger than 0.1 V peak, what sampling frequency should be used to eliminate all aliased components at the output above a magnitude of 0.3 V in the 0–20 kHz band?

Solution: We note from the frequency response curve that a gain of 3 occurs at a frequency of 60 kHz. If the sampling frequency is set at 80 kHz, the aliased component at 20 kHz will be less than 0.3 V for a 0.1-V input.

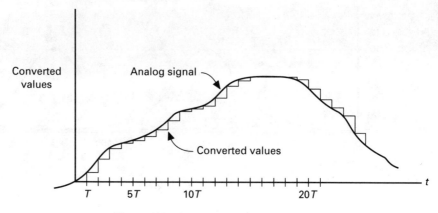

Figure 3-9 Conversion of an analog signal.

Exercises

- **E3.1a** Construct a curve similar to that of Fig. 3-8 for a sampling frequency of $f = 50$ kHz.
- **E3.1b** Using the curve constructed in E3.1a, calculate the magnitudes and frequencies contained in the output for an input to the amplifier of 0.1 V at 30 kHz.
- **E3.1c** Repeat E3.1b for an input of 0.1 V at 10 kHz.

3.2 INTRODUCTION TO FILTERS

3.2.1 The Need for Filters

The previous section indicated that the sampling process introduces new frequency components into the sampled signal that do not exist in the continuous waveform of interest. The conversion of a time-varying analog signal to a series of digital values represents a process that approximates ideal sampling. The output digital values from the converter can be used in various ways, but they represent the discrete values of the waveform at the instants of sampling.

The measurand, or variable being measured by a transducer, will have some upper limit of frequency response that is significant. For example, a high-fidelity microphone may be required to respond to frequencies up to 20 kHz, while the mouthpiece of a telephone may need only to produce output frequencies up to 3300 Hz. A scale that weighs large loads may have a frequency response that does not exceed 2–3 Hz. As the analog signal is converted to voltage or current, amplified, and converted to digital values, various noise sources contribute to system inaccuracy. Some of the noise sources generate frequency components that do not overlap into the desired range of frequencies of the measurand. The effects of these noise sources can be eliminated while measurand frequencies are retained by using electronic filters.

Filters are designed to greatly attenuate unwanted frequency components while passing desired components with little or no attenuation. We will consider the specification and design of filters in this and the next chapter. For our purposes, we will divide the noise sources into two classifications: aliased components and nonaliased components. The frequencies of the aliased components can be controlled by the sampling rate, while no such control can be exercised over the other noise sources.

Nonaliased noise sources have been treated earlier. Some examples are resistive noise, amplifier noise, electromagnetic interference, and the inevitable 60-Hz hum that seems to exist in most electronic systems. If the frequency components of these sources are largely out of the range of desired measurand frequencies, a filter can be used to decrease the overall noise of the system. If the highest-frequency component of the measurand is 5 Hz, a filter cutoff frequency of 10 Hz might be used. This filter would pass all measurand frequencies but would attenuate all higher-frequency components that arise in the system ahead of the filter.

A certain amount of filtering is inherent in many systems without adding a separate filter. A transducer may only respond to frequencies below some limit. An amplifier may have an upper 3-dB frequency, above which the gain falls rapidly with frequency. While some systems operate effectively without a separate filter, some highly accurate or noisy systems would be rendered useless if a filter were not utilized.

If unwanted frequency components fall in the range of desired measurand frequencies, a filter cannot be used to rid the signal of these unwanted components. Attenuation of these components by a filter would be accompanied by an attenuation of the signal frequencies. No net gain in signal-to-noise ratio would result.

Aliased noise can be eliminated or greatly reduced by using appropriate filtering and choosing a proper value of sampling frequency. A filter can be used in two ways to decrease the effects of aliasing. When frequencies up to some frequency ω_c are of interest, but higher frequencies are present, a filter can be used to attenuate the higher frequencies before sampling takes place. After sampling, a filter can again be used to reduce the aliased components.

The first filter is often called a presampling or antialiasing filter. This filter can be realized only by an electronic circuit. The second filter, used to reduce the aliased components, can be an electronic circuit, in some cases, or a digital filter. If the signal is being converted for further processing by a computer, the computer can implement the digital filter. Since the electronic filter can be used for both purposes, but the digital filter can only be used for one, the remainder of this chapter will consider the electronic filter.

The discussion on aliasing in connection with Fig. 3-6 indicated that as long as the sampling frequency ω_s is greater than $2\omega_c$, no unwanted components are aliased into the actual signal frequency range of 0 to ω_c. This is based on the assumption that the analog signal contains zero energy above ω_c. A component at a frequency of $2\omega_c$ in the original signal would result in an aliased component at $0.5\omega_c$, even if the sampling frequency is chosen to be $2.5\omega_c$. This aliased component could be eliminated or reduced by filtering the original analog signal. The small component at $2\omega_c$ can be attenuated drastically, while the components from 0 to ω_c are unattenuated by a properly designed filter. The filter output signal now closely approximates the assumption that the signal contains zero energy above ω_c.

The frequency ω_c may be controlled by the transducer or by a filter inserted somewhere between the transducer and the analog-to-digital converter. A microphone might respond to all frequencies up to 10 kHz before the output begins to fall with frequency.

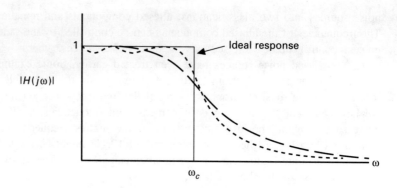

Figure 3-10 Ideal and actual responses.

To transmit conversational-quality speech requires only that frequencies up to 5 kHz be reproduced. If this signal is to be converted to a digital code and then transmitted, the microphone output is filtered to attenuate or eliminate components above 5 kHz. The frequency ω_c is established by the filter to be $2\pi \times 5$ kHz $= 31.4$ krad/s. A ratio of ω_s to ω_c of 4:1 would now lead to a sampling frequency of 20 kHz rather than the 40 kHz that would be selected if ω_c were 10 kHz.

3.2.2 Filter Specifications

Figure 3-10 shows a plot of the transfer function magnitude of an ideal filter and that of two actual filters. The ideal response could be easily specified as no attenuation in the passband below ω_c and 100% attenuation in the stopband (above ω_c). Such a sharp transition between passband and stopband is impossible to achieve with actual filters. Instead a response such as the other two curves of Fig. 3-10 must be used in practice. In the actual filter there is not such an obvious delineation between the passband and the stopband. There is a transition region between the region of low attenuation and that of high attenuation. Generally, the cutoff frequency is taken as the frequency at which the magnitude of the transfer function has dropped to some specified percentage of the low-frequency value. A common percentage used is 70.7% (3-dB point), but this is not the only possible value.

Some filters specify the passband as those frequencies below ω_c and the stopband as those frequencies above ω_c. Other filters define the passband in the same way but use a second specification for the stopband. The stopband is defined as those frequencies for which $|H(j\omega)|$ is below some specified value, such as 10% or 1% of the value at $\omega = 0$. The region between passband and stopband is the transition region.

Magnitude is not the only quantity of interest in every filter application. There are occasions when the phase response is either the major interest or of comparable interest to the magnitude response. In cases where the phase distortion must be controlled, phase response becomes significant.

There are also several types of response other than the low-pass response of Fig. 3-10. Figure 3-11 shows the high-pass, band-pass, and band-reject filter responses that are each important in particular applications. The following sections will consider the synthesis of filters that realize several of these responses.

Chap. 3 Filters

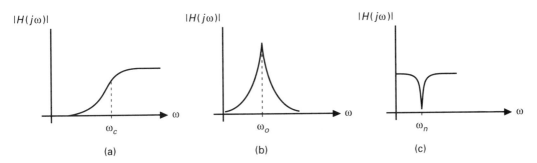

Figure 3-11 (a) High-pass response. (b) Band-pass response. (c) Band-reject or notch filter.

3.2.3 Filter Functions

There are three popular responses that will be considered in this chapter. These are the Butterworth or maximally flat magnitude response, the Chebyshev or equiripple response, and the linear phase or maximally flat time delay response.

Each of these responses has particular characteristics that make it appropriate for specific applications. The Butterworth low-pass response is quite constant near $\omega = 0$ and falls rather sharply in the transition region. The response is monotonic. The Chebyshev response is not monotonic in the passband. The response ripples in the passband but falls more sharply in the transition region than does a comparable Butterworth filter. The linear phase filter is monotonic but does not fall as sharply as does the Butterworth filter; however, the phase distortion is less than other types of response.

Exercise

E3.2a Explain why a presampling filter is needed to follow an amplifier if the maximum input signal frequency is 2 kHz and the sampling rate is 8 kHz. Assume that the amplifier output is noisy, with the noise bandwidth extending from 0 to 10 kHz.

3.3 SYNTHESIS FUNDAMENTALS

3.3.1 Normalization/Denormalization

A particular filter frequency response, such as a 3-pole Butterworth, may have an infinite number of transfer functions, depending on the cutoff frequency. In order to consolidate this high number of functions into a single function, frequency normalization is used. In this method it is assumed that a particular response is normalized to have a cutoff frequency of $\omega_c = 1$ rad/s. Furthermore, for similar reasons, in filters that are terminated in a resistance it is assumed that this termination has been normalized to a value of 1 Ω.

If a low-pass filter is to be realized with a cutoff frequency of 12 kHz and a 610-Ω termination, the filter is first designed with a cutoff frequency of 1 rad/s and a termination of 1 Ω. The element values calculated are normalized values. These values are then denormalized to produce a cutoff frequency of 12 kHz with a 610-Ω terminating resistance.

The key to the process of impedance normalization or denormalization is that transfer functions are based on ratios of impedances. The transfer function of a network remains constant as impedances change, if all impedances change by the same ratio. In order to denormalize a resistance R_n to a new value R, all impedances are scaled up by the factor k_i where $k_i = R/R_n$.

This means that inductors are increased by a factor of k_i, since inductive impedance varies directly with L. Capacitor values are decreased by the factor k_i, since capacitive impedance varies inversely with C. Thus the impedance scaling equations are

$$R = k_i R_n \tag{3-9a}$$

$$L = k_i L_n \tag{3-9b}$$

$$C = \frac{C_n}{k_i}. \tag{3-9c}$$

These equations allow element values to be changed while keeping the transfer function constant.

Frequency normalization or denormalization is based on the fact that corner frequencies depend on the relative values of reactance and resistance. If a cutoff frequency is related to a capacitor and resistance by

$$\omega_{cn} C_n = R_n$$

then a new cutoff frequency that is k_ω times higher than ω_{cn} can be produced by using a capacitance that is a factor of k_ω times smaller than C_n. The capacitive reactance of the denormalized circuit at the new cutoff frequency of $k_\omega \omega_{cn}$ is made to equal the capacitive reactance of the normalized circuit at ω_{cn}. The frequency denormalizing equations are

$$C = \frac{C_n}{k_\omega} \tag{3-10a}$$

$$L = \frac{L_n}{k_\omega} \tag{3-10b}$$

Both impedance denormalization and frequency denormalization can be accomplished in one step by using the following equations:

$$R = k_i R_n \tag{3-11a}$$

$$L = \frac{k_i}{k_\omega} L_n \tag{3-11b}$$

$$C = \frac{C_n}{k_i k_\omega} \tag{3-11c}$$

Example 3.2

The circuit of Fig. 3-12 has a corner frequency of 1 rad/s. Find the element values required to denormalize this frequency to 12 kHz with a terminating resistance of 610 Ω.

Chap. 3 Filters

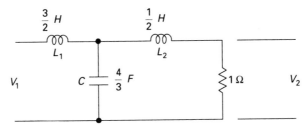

Figure 3-12 A low-pass filter.

Solution: The two normalizing factors in this problem are

$$k_i = \frac{610}{1} = 610$$

and

$$k_\omega = \frac{12{,}000 \times 2\pi}{1} = 7.54 \times 10^4$$

The terminating resistance becomes 610 Ω and the denormalized values of L_1, L_2, and C are found from Eq. (3-11) to be

$$L_1 = \frac{610}{7.54 \times 10^4} 1.5 = 12.1 \text{ mH}$$

$$L_2 = \frac{610}{7.54 \times 10^4} 0.5 = 4.05 \text{ mH}$$

$$C = \frac{1.333}{610 \times 7.54 \times 10^4} = 0.0290 \text{ μF}$$

These values result in a filter with a cutoff frequency of 12 kHz.

Filter design is based on transfer functions that result in normalized filters. Handbooks publish transfer functions that are normalized for the types of responses of interest. When the appropriate response has been selected, the element values are denormalized to achieve the desired cutoff frequency and impedance level.

3.3.2 Finding Poles from the Magnitude Response

When we are analyzing a given transfer function to find the magnitude (amplitude) of the frequency response, the procedure is rather simple [3]. For example, the transfer function

$$H(j\omega) = \frac{1}{1 + j\omega/\omega_2}$$

has a magnitude given by

$$|H(j\omega)| = \frac{1}{\left[1 + (\omega/\omega_2)^2\right]^{1/2}}$$

The frequency response can be plotted by substituting various values of ω into this equation and calculating corresponding values of $|H(j\omega)|$.

The synthesis of filters involves the reverse process. In this case, a desired *magnitude* response is given and the complete transfer function is to be found. Once the transfer function is known, it can be realized with actual circuit elements. This collection of elements comprise the filter.

There are various problems that may arise in the synthesis process. We may obtain a transfer function that is expressed as a function of $j\omega$ or s that cannot be realized with practical element values. For example, we might find that a negative value of inductance or resistance is required in the filter. Another problem is that there are often several different available methods to use in realizing the filter, resulting in different configurations or element values.

The procedures to be suggested in this and following sections have been developed over a period of several years. They will result in a physically realizable filter and will allow the choice of a desired configuration of elements. Designs are generally based on the Laplace transform representation of the transform function, $H(s)$. The concept of analytic continuation allows a function $H(s)$ to be determined from a given magnitude function $|H(j\omega)|$. We will demonstrate the method by assuming that we are given a transfer function magnitude of

$$|H(j\omega)| = \frac{1}{\left[1 + \omega^4/\omega_2^4\right]^{1/2}}$$

where ω_2 is a known constant. This function might be derived from a knowledge of the necessary frequency response of the desired filter. The problem is to find a transfer function $H(s)$ that can then be used to design or synthesize a physically realizable circuit.

We will use the expression $H(s)$ or $H(j\omega)$ to represent the transfer function of a filter. For most of the filters to be considered in this text, this transfer function will represent the ratio of the output voltage to the input voltage, or

$$H(s) = \frac{E_{out}(s)}{E_{in}(s)}$$

From complex variable theory we recognize that

$$|H(j\omega)|^2 = H(j\omega)H(-j\omega) \tag{3-12}$$

For sinusoidal inputs, $s = j\omega$, and Eq. (3-12) can be written as

$$|H(j\omega)|^2 = H(s)H(-s)|_{s=j\omega} \tag{3-13}$$

Chap. 3   Filters

The square of the transfer function magnitude for this example is

$$|H(j\omega)|^2 = \frac{1}{1+\dfrac{\omega^4}{\omega_2^4}}$$

$$= \left.\frac{1}{1+\dfrac{(s/j)^4}{\omega_2^4}}\right|_{s=j\omega} = \left.\frac{1}{1+\dfrac{s^4}{\omega_2^4}}\right|_{s=j\omega} = H(s)H(-s)\bigg|_{s=j\omega}$$

We now want to find the transfer function $H(s)$ that results in this magnitude function. We can construct $H(s)$ by locating the poles of $H(s)$. In order to do this we first find the poles of $H(s)H(-s)$ by finding the roots of the denominator. Equating the denominator to zero gives

$$\frac{s^4}{\omega_2^4} = -1 = 1\angle 180° = \left(\frac{s}{\omega_2}\right)^4$$

and allows us to find four roots. We find these roots by recalling from complex number theory that the nth root of a complex number expressed in polar form is

$$(A\angle\Theta)^{1/n} = A^{1/n}\angle\Phi \tag{3-14}$$

The angle Φ is given by

$$\Phi = \left[\frac{\Theta}{n} + \frac{360k}{n}\right] \text{ (in degrees)}$$

where $k = 0,1,2,\ldots, n-1$. For this case $A = 1$, $n = 4$, and $\Theta = 180°$. The four roots for s/ω_2 are then

$$1\angle 45°, 1\angle 135°, 1\angle 225°, \text{ and } 1\angle 315°,$$

or

$$0.707 + j0.707,\ -0.707 + j0.707,\ -0.707 - j0.707,\text{ and } 0.707 - j0.707$$

The poles of $H(s)H(-s)$ are then equal to these values multiplied by ω_2. We will assume that ω is normalized by ω_2. The locations of these normalized poles are shown in Fig. 3-13. The function can be written in terms of factors as

$$H(s)H(-s) = \frac{1}{(s-0.707-j0.707)(s+0.707-j0.707)(s+0.707+j0.707)(s-0.707+j0.707)}$$

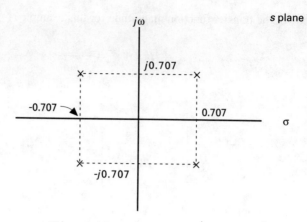

Figure 3-13 Pole locations of $|H(s)H(-s)|$.

The four factors in the preceding equation must be associated with $H(s)$ and $H(-s)$.

In order that $H(s)$ be a physically realizable transfer function, its poles must be in the left-hand plane (LHP) of the Laplace plane. Poles in the RHP lead to an impractical rising exponential in the output of the transfer function when excited by a finite impulse. Thus the LHP poles are associated with $H(s)$ and the RHP poles are associated with $H(-s)$. This gives a transfer function of

$$H(s) = \frac{1}{(s + 0.707 - j0.707)(s + 0.707 + j0.707)}$$
$$= \frac{1}{s^2 + \sqrt{2}s + 1}$$

This function can now be synthesized with passive elements such as resistors, inductors, and capacitors. The topic of synthesis is discussed in Chapter 5. The principle of analytic continuation thus allows us to find the roots of $H(s)H(-s)$ and choose the appropriate ones to determine $H(s)$.

The next two sections are concerned with the problems of selecting a proper magnitude response for a given application and synthesizing a circuit after obtaining $H(s)$.

Exercises

E3.3a Rework Example 3.2 if a corner frequency of 20 kHz and a terminating resistance of 1 kΩ are required.

E3.3b Find a transfer function that results in a magnitude response of

$$|H(j\omega)| = \frac{1}{\left[1 + \omega^6\right]^{1/2}}$$

3.4 TRANSFER FUNCTIONS FOR FILTERS

The process of transfer function selection begins with the attempt to duplicate the ideal filter characteristics of Fig. 3-10 with a practical transfer function. One of the requirements a function must satisfy is that it be expressible as the ratio of two polynomials in s. All actual circuits possess a transfer function that can be expressed in this manner; thus the function that approximates the ideal characteristics must also be a ratio of polynomials in s.

It is well known that actual circuit transfer functions cannot yield discontinuities in the frequency response [3]. Consequently, the actual transfer function can only approximate the ideal function. The sense in which the actual function best matches the ideal function allows latitude for several choices.

One popular choice attempts to approximate the ideal function with one that is as flat as possible in the passband. While this function results in a flat function near $\omega = 0$, it does not have a sharp transition to the stopband and is not very flat as the cutoff frequency is approached. This approximation is called the maximally flat magnitude response, or Butterworth response.

A second approximation allows the response to ripple in the passband in order to achieve a sharper (steeper) transition region. This is called the equiripple response, or Chebyshev response, and maximizes the rate of falloff in the transition region.

The third type of response does not approximate the ideal magnitude response as well as the other two, but it exhibits a linear phase characteristic in the passband. This is called a Bessel filter and is selected when phase information rather than amplitude is the prime consideration.

All of these approximations become more accurate as higher-order polynomials are used. However, more circuit elements are required to realize these higher-order functions. Thus some compromise between accuracy and filter size is made.

3.4.1 Butterworth (Maximally Flat) Response

The major condition applied to the low-pass Butterworth approximation is that as many of the derivatives of the function as possible are set to zero at the $\omega = 0$ point. This condition makes the function as flat as possible near the origin.

The Butterworth approximation also applies the two constraints

$$|H(j0)| = 1$$

and

$$|H(j\infty)| = 0$$

A low-pass response with zeros only at $\omega = \infty$ can be represented by

$$H(s) = \frac{1}{1 + a_1 s + a_2 s^2 + \ldots + a_n s^n} \qquad (3\text{-}15)$$

The square of the magnitude is $H(s)H(-s)|_{s=j\omega}$ and is expressible as

$$|H(j\omega)|^2 = \frac{1}{1 + d_2\omega^2 + d_4\omega^4 + \ldots + d_{n-2}\omega^{2n-2} + d_{2n}\omega^{2n}} \tag{3-16}$$

We note from this equation that at $\omega = 0$ the square of the magnitude (and the magnitude) is unity. As $\omega \to \infty$ the magnitude rapidly approaches zero, behaving asymptotically as

$$\frac{1}{d_{2n}\omega^{2n}}$$

if d_{2n} is nonzero.

We cannot make the slope of the response equal zero throughout the passband, but we can set the slope equal to zero at $\omega = 0$. We can also make the slope depart from zero more slowly as ω increases by setting higher-order derivatives of $|H(j\omega)|^2$ with respect to ω equal to zero. This can be done most easily by dividing the denominator of Eq. (3-16) into unity. The result of this operation is

$$|H(j\omega)|^2 = 1 - d_2\omega^2 + (d_2^2 - d_4)\omega^4 - (d_2^3 - 2d_2d_4 + d_6)\omega^6 + \ldots$$

The first derivative of $|H(j\omega)|^2$ is

$$-2d_2\omega + 4(d_2^2 - d_4)\omega^3 - 6(d_2^3 - 2d_2d_4 + d_6)\omega^5 + \ldots$$

The slope at $\omega = 0$ is thus always zero.

The second derivative is

$$-2d_2 + 12(d_2^2 - d_4)\omega^2 - 30(d_2^3 - 2d_2d_4 + d_6)\omega^4 + \ldots$$

This derivative can be set equal to zero if d_2 is set equal to zero.

The third and all odd derivatives are always zero at $\omega = 0$ for any values of coefficients. The higher-order even derivatives can be zeroed by setting

$$d_4 = d_6 = \ldots = 0$$

If all coefficients were set equal to zero, however, the magnitude could not approach zero at $\omega = \infty$. In fact, it would always be unity. Consequently, d_{2n} must be nonzero, but all others are chosen to be zero; that is

$$d_2 = d_4 = \ldots = d_{2n-2} = 0$$

The choice of d_{2n} is somewhat arbitrary, but if chosen to be unity, the response becomes a Butterworth response. This simplifies the final expression for the square of the magnitude to

Chap. 3 Filters

TABLE 3-1 BUTTERWORTH (MAXIMALLY FLAT) FUNCTIONS

n	$\|H(j\omega)\|$	$H(s)$
1	$\dfrac{1}{\sqrt{1+\omega^2}}$	$\dfrac{1}{s+1}$
2	$\dfrac{1}{\sqrt{1+\omega^4}}$	$\dfrac{1}{s^2 + 1.41421s + 1.00000}$
3	$\dfrac{1}{\sqrt{1+\omega^6}}$	$\dfrac{1}{s^3 + 2.00000s^2 + 2.00000s + 1.00000}$
4	$\dfrac{1}{\sqrt{1+\omega^8}}$	$\dfrac{1}{s^4 + 2.61313s^3 + 3.41421s^2 + 2.61313s + 1.00000}$
5	$\dfrac{1}{\sqrt{1+\omega^{10}}}$	$\dfrac{1}{s^5 + 3.23607s^4 + 5.23607s^3 + 5.23607s^2 + 3.23607s + 1.00000}$

$$|H(j\omega)|^2 = \frac{1}{1+\omega^{2n}} \qquad (3\text{-}17)$$

The 3-dB point of the Butterworth transfer function occurs at $\omega_c = 1$.

The concept of analytic continuation can now be applied to the Butterworth response to find a physically realizable transfer function. In fact, the previous section demonstrated this procedure for a second-order Butterworth response. The same procedure has been applied to the various orders up to at least-tenth order, with the results tabulated [4]. This eliminates the need to recalculate the transfer function each time a filter is desired. These values up to a fifth-order response are tabulated in Table 3-1.

The coefficients of $H(s)$ are tabulated to five decimals in Table 3-1 because of high accuracy requirements in the synthesis process. Figure 3-14 shows the plots of Butterworth frequency responses of $|H(j\omega)|$ for different values of n.

Example 3-3

Derive the expression for $H(s)$ for a third-order Butterworth response.

Solution: The square of the magnitude for a third-order function is, from Eq. (3-17),

$$|H(j\omega)|^2 = \frac{1}{1+\omega^6}$$

Substituting s/j for ω gives the analytic continuation form

$$H(s)H(-s) = \frac{1}{1-s^6}$$

The poles for this function are shown in Fig. 3-15. We associate the LHP poles with $H(s)$ and write

$$H(s) = \frac{1}{\left(s + \dfrac{1}{2} - j\dfrac{\sqrt{3}}{2}\right)(s+1)\left(s + \dfrac{1}{2} + j\dfrac{\sqrt{3}}{2}\right)}$$

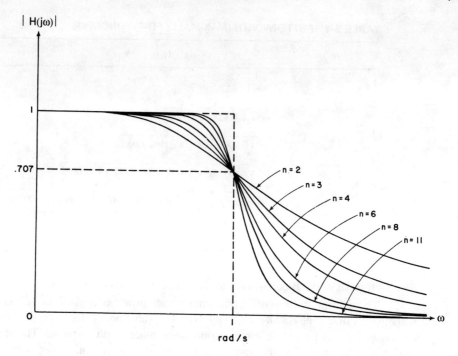

Figure 3-14 Butterworth responses. From David E. Johnson, *Introduction to Filter Theory*, (c) 1976, p. 44. Reprinted by permission of Prentice Hall, Englewood Cliffs, New Jersey.

Multiplying the factors together gives

$$H(s) = \frac{1}{s^3 + 2s^2 + 2s + 1}$$

which agrees with the information in Table 3-1.

3.4.2 Chebyshev (Equiripple) Response

A response that has equal amplitude ripples in the passband but falls sharply above $\omega = 1$ is given by

$$|H(j\omega)| = \frac{1}{\left[1 + \varepsilon^2 \, C_n^2(\omega)\right]^{1/2}} \qquad (3\text{-}18)$$

To cause equal ripples, ε is chosen to be a constant, and $C_n(\omega)$ is a polynomial that varies between 0 and 1 as ω increases from 0 to 1 and becomes much larger than 1 as ω exceeds 1. Depending on the order of the polynomial, $C_n(\omega)$ can vary between 0 and 1 several times in the passband.

While the Butterworth filter achieves maximal flatness about the $\omega = 0$ point, the Chebyshev response that we now consider approximates the ideal filter in a different way. This response allows some prescribed deviation from the ideal curve in the passband and the fastest possible rate of cutoff in the transition region. The Chebyshev

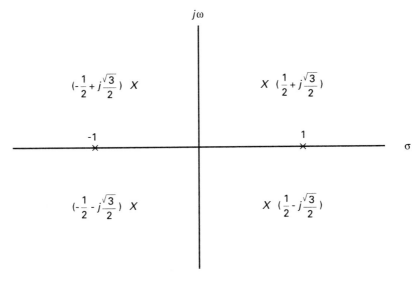

Figure 3-15 Poles for $H(s) H(-s)$.

TABLE 3-2 CHEBYSHEV POLYNOMIALS

n	$C_n(\omega)$	ω for $C_n = 0$	ω for $C_n = \pm 1$
1	ω	0	1
2	$2\omega^2 - 1$	0.707	0, 1
3	$4\omega^3 - 3\omega$	0, 0.866	0.500, 1
4	$8\omega^4 - 8\omega^2 + 1$	0.383, 0.924	0, 0.707, 1
5	$16\omega^5 - 20\omega^3 + 5\omega$	0, 0.588, 0.951	0.314, 0.810, 1

polynomials have been known for years and were originally named after an individual named Tschebycheff, or Chebyshev. The subscript n indicates the order of the polynomial. Table 3-2 shows the Chebyshev polynomials from first to fifth order and indicates the values of ω for which C_n equals 0 or ± 1.

As C_n varies between 0 and ± 1 in the passband, the magnitude of the transfer function varies between 1 when $C_n = 0$ and

$$\frac{1}{\sqrt{1+\varepsilon^2}}$$

when $C_n = \pm 1$. The ripple width (maximum value minus minimum value) is given by

$$RW = 1 - \frac{1}{\sqrt{1+\varepsilon^2}} \tag{3-19a}$$

It is common to express this in dB by

$$RW(\text{dB}) = 20 \log_{10} \sqrt{1+\varepsilon^2} \tag{3-19b}$$

TABLE 3-3 RIPPLE WIDTH AS A FUNCTION OF ε

ε	0.34931	0.50885	0.76478	0.99763
RW (dB)	0.5	1.0	2.0	3.0
Passband min.	0.944	0.891	0.794	0.708

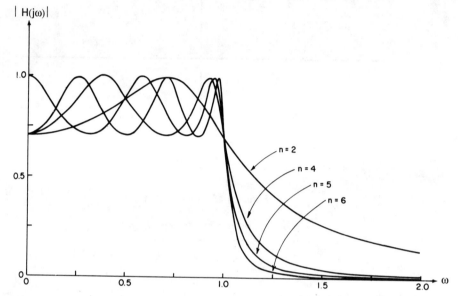

Figure 3-16 Chebyshev responses with $\varepsilon = 1$ (3-dB ripple width). From David E. Johnson, *Introduction to Filter Theory*, (c) 1976, p. 55. Reprinted by permission of Prentice Hall, Englewood Cliffs, New Jersey.

The value of *RW* depends on ε, and a few typical values are given in Table 3-3 along with the minimum values of $|H(j\omega)|$ in the passband.

Figure 3-16 shows Chebyshev responses for a ripple width of 3 dB and various values of n. The ripple width is typically smaller than the 3 dB shown in Fig. 3-16. We note that the number of peaks in the response equals $n/2$.

At frequencies above 1, the term $\varepsilon^2 C_n^2$ becomes large compared to 1 and hence is the dominant term. For example, at $\omega = 1.5$, the fourth-order polynomial is $C_4(1.5) = 23.5$. The leading term of C_n dominates the polynomial value as ω increases. For higher values of ω the magnitude of the transfer function can be approximated by

$$|H(j\omega)| \approx \frac{1}{\varepsilon 2^{n-1} \omega^n}$$

We will designate the decibel magnitude of the Chebyshev transfer function by MH_C and the decibel magnitude of the Butterworth function by MH_B. Thus

$$MH_C = 20 \log \frac{1}{\varepsilon 2^{n-1} \omega^n}$$
$$= -20 n \log \omega - 6(n-1) - 20 \log \varepsilon$$

This can be compared to a Butterworth magnitude of

$$MH_B = -20\, n \log \omega$$

The last two terms in MH_C indicate a higher falloff than exhibited by the Butterworth response. The third term in MH_C will always be negative when ε is smaller than unity, thus $-20 \log \varepsilon$ will be positive. This term detracts from a sharp falloff, but the second term typically contributes more to the falloff than the third term loses. As an example, for a third-order Chebyshev response with $\varepsilon = 0.5$, the magnitude of the transfer function is

$$MH_C = MH_B - 12 + 6 = MH_B - 6$$

The third-order Chebyshev response is always 6 dB lower than the third-order Butterworth response for large values of ω. This additional attenuation takes place primarily in the transition region.

We note from Fig. 3-16 that the cutoff frequency for the Chebyshev filter is the point where the function drops below a value of

$$\frac{1}{\sqrt{1+\varepsilon^2}}$$

For values of ε less than 1, the magnitude at $\omega = 1$ is less than 3 dB down from the magnitude at $\omega = 0$. The Butterworth response is always 3 dB down at $\omega = 1$. Thus when the Butterworth and the Chebyshev responses are compared, one must not compare on the basis of the magnitude at the normalized cutoff frequency of $\omega = 1$.

As an example of the comparison between the two responses, we will consider the magnitude response of Fig. 3-17 [5]. Let us assume that the response can have a maximum variation of 10% in the passband ($A_1 = 0.9$) and also in the region above $\omega = 2$ ($A_2 = 0.1$).

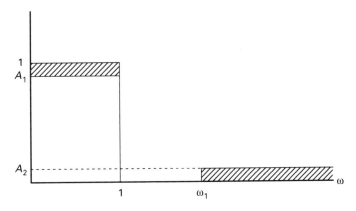

Figure 3-17 Attenuation requirements.

For a normalized Butterworth response, the curve is 3 dB down at $\omega = 1$. Obviously this curve is 0.92 dB down at this point. If we use the Butterworth response, we must then assume a higher cutoff frequency, ω_{cB}, which will be greater than unity. The Butterworth response can then be expressed as

$$|H(j\omega)|_B = \frac{1}{\left[1 + (\omega/\omega_{cB})^{2n}\right]^{1/2}}$$

At $\omega = 1$ this function must be greater than or equal to 0.9. We can express this as

$$\frac{1}{\left[1 + (1/\omega_{cB})^{2n}\right]^{1/2}} \geq 0.9$$

which implies that

$$(1/\omega_{cB})^{2n} \leq \frac{19}{81}$$

At $\omega = 2$, the function must be less than 0.1, or

$$\frac{1}{\sqrt{1 + (2/\omega_{cB})^{2n}}} \leq 0.1$$

which implies that

$$(1/\omega_{cB})^{2n} \geq \frac{99}{2^{2n}}$$

Since

$$\frac{99}{2^{2n}} \leq (1/\omega_{cB})^{2n} \leq \frac{19}{81}$$

then

$$\frac{99}{2^{2n}} \leq \frac{19}{81}$$

The minimum integer value of n that will satisfy this last condition is 5. For the case of $n = 5$, the two inequalities tell us that

$$1.15604 \leq \omega_{cB} \leq 1.26318$$

We will choose $\omega_{cB} = 1.2$ to complete the Butterworth problem.

Now we will consider the realization of the same specifications with a Chebyshev response. The value of the response at $\omega_c = 1$ must exceed 0.9, giving

$$\frac{1}{\left[1+\varepsilon^2\right]^{1/2}} \geq 0.9$$

At $\omega = 2$ the condition is

$$\frac{1}{\left[1+\varepsilon^2 C_n^2(2)\right]^{1/2}} \leq 0.1$$

We solve for ε from the first equality, resulting in

$$\varepsilon^2 \leq \frac{19}{81}$$

or $\varepsilon \leq 0.48432$.

Using the largest allowed value of ε in the second inequality gives

$$\left| C_n(2) \right| \geq 20.544$$

From Table 3-2 we find that the third-order polynomial will satisfy this requirement. Thus a third-order Chebyshev response or a fifth-order Butterworth will satisfy the specifications.

3.4.3 Linear Phase (Bessel) Filters

A waveform consisting of several frequency components can become distorted, that is experience a change of shape, if the relative phases of the components are modified. As a simple example of this, let us assume that a signal consisting of two components is applied to a phase-shifting network, as shown in Fig 3-18. The input waveform is given by $v = 10 \sin \omega_o t + 4 \sin 2\omega_o t$ and is shown in Fig. 3-18(b). If the phase shifter provides a phase shift of

$$\Theta = \frac{-0.045}{2\pi} \omega$$

in degrees, and we set the frequency of the fundamental component to 1 kHz, the output signal will be

$$v_o = 10 \sin(\omega_o t - 45) - 4 \sin(2\omega_o t - 90)$$

Figure 3-18(c) shows this waveform, which is delayed from the original but is not distorted in shape. The phase shift of the fundamental is $-45°$ referenced to the fundamental component. The phase shift of the second harmonic is $-90°$ referenced to the second harmonic component. The time delay of both components is the same, although

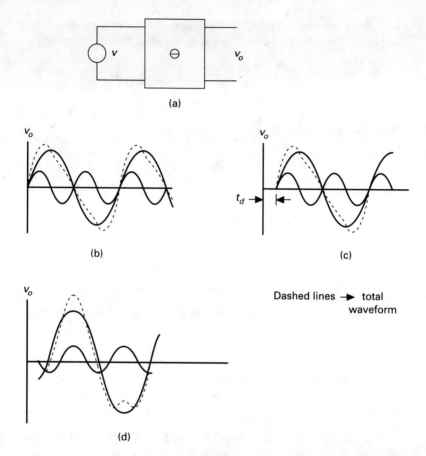

Figure 3-18 (a) A phase shift network. (b) A waveform with a fundamental plus second harmonic. (c) A delayed waveform with no phase distortion. (d) A delayed waveform with phase distortion.

the phase shifts are different. This results from the fact that each phase shift is referenced to a different period.

For a waveform to be delayed, but experience no waveshape distortion, the phase shift must be a linear function of frequency, or

$$\Theta = -k\omega \tag{3-20}$$

Now we will modify the phase shift network to introduce a phase shift of

$$\Theta = -1.14 \times 10^{-7}\,\omega^2$$

In this instance, Θ is a nonlinear function of ω.

The output signal is now

$$v_o = 10 \sin(\omega_o t - 45) + 4 \sin(2\omega_o t - 180)$$

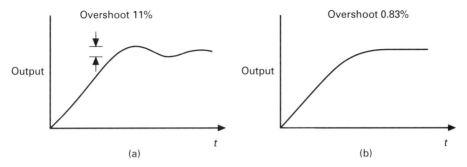

Figure 3-19 Step response of two 4-pole filters. (a) Butterworth. (b) Linear phase.

This waveform is shown in Fig. 3-18(d), and we see that the peak symmetry about the zero voltage line is destroyed. We emphasize that no change of amplitude of either component takes place in the phase shift network. The distortion is due only to the change of relative phase between the components.

We can prove the general result that a pure time delay requires a linear phase shift by the use of Laplace transforms. Given a function of time expressed by $f(t)$, the Laplace transform is

$$L[f(t)] = F(s)$$

A pure time delay of t_d in $f(t)$ can be expressed by $f(t-t_d)$. The Laplace transform of this delayed function for a sinusoid ($s = j\omega$) is

$$L\left[f(t-t_d)\right] = e^{-j\omega t_d} F(j\omega)$$

The additional phase shift due to the delay is given by

$$\Theta = -\omega t_d \tag{3-21}$$

In a filter designed to attenuate higher-frequency components, waveshape change may result from both amplitude and phase distortion. However, the attenuated components are often noise or undesired components that were inadvertently introduced into the system. Therefore, the desired signal is generally not distorted due to amplitude attenuation, which eliminates these undesired signals. On the other hand, phase distortion in the passband may result in waveshape change. This phase distortion can be demonstrated from the step responses of Fig. 3-19. The overshoot of the Butterworth filter is considerably greater than that of the linear phase filter. The Chebyshev response results in an even greater overshoot than does the Butterworth filter and is not used in applications requiring low overshoot.

The linear phase or maximally flat time delay filter is designed to attenuate high frequencies while providing for a phase shift that is linear with frequency within the passband. The transition region is not as sharp as the Chebyshev or even the Butterworth filter, but the waveshape change due to phase distortion is minimal.

TABLE 3-4 OVERSHOOT OF BUTTERWORTH AND LINEAR PHASE FILTERS

n	Butterworth overshoot, percent	Linear phase overshoot, percent
1	0.0	0.0
2	4.3	0.43
3	8.2	0.75
4	10.9	0.83
5	12.8	0.76

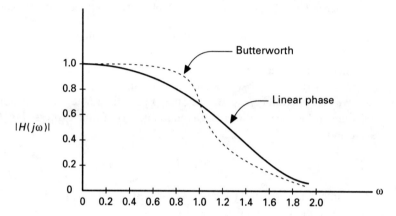

Figure 3-20 Magnitude response of 4-pole filters.

Table 3-4 compares the overshoot for the Butterworth and linear phase filters for filters up to fifth order.

Figure 3-20 shows a comparison of frequency responses of the Butterworth and linear phase filter for $n = 4$. The linear phase filter obviously is not as sharp in the transition region as the Butterworth filter.

Rather than derive the general expression for the linear phase transfer function, we will specifically consider the second-order case to demonstrate the method of obtaining linear phase. A general, second-order, low-pass transfer function can be expressed as

$$H(s) = \frac{1}{s^2 + a_1 s + a_0} \tag{3-22}$$

The phase of this function for sinusoidal excitation is

$$\Theta = -\tan^{-1} \frac{a_1 \omega}{a_0 - \omega^2} \tag{3-23}$$

Differentiating Θ with respect to ω in Eq. (3-21) gives

$$-\frac{d\Theta}{d\omega} = t_d \tag{3-24}$$

TABLE 3-5 DENOMINATORS OF LINEAR PHASE FILTERS

n	Denominator
1	$s+1$
2	$s^2 + 3s + 3$
3	$s^3 + 6s^2 + 15s + 15$
4	$s^4 + 10s^3 + 45s^2 + 105s + 105$
5	$s^5 + 15s^4 + 105s^3 + 420s^2 + 945s + 945$

Taking the derivative of Θ from Eq. (3-23) gives

$$-\frac{d\Theta}{d\omega} = \frac{a_1 a_0 + a_1 \omega^2}{a_0^2 + (a_1^2 - 2a_0)\omega^2 + \omega^4} \qquad (3\text{-}25)$$

In the ideal case, the time delay t_d is constant, but for the actual filter we see that the time delay is a function of ω. Dividing the denominator into the numerator of Eq. (3-25) gives a time delay of

$$t_d = \frac{a_1}{a_0} + \omega^2 \frac{(3a_1 - a_1^3/a_0)}{a_0^2} + \cdots \qquad (3\text{-}26)$$

For small values of ω, the lower-order terms are dominant, thus Eq. (3-26) can approximate a constant if the coefficient of ω^2 is set to zero. Actually, two conditions can be applied to Eq. (3-26). The first condition applied is

$$a_1 = a_0$$

to normalize the approximate time delay to 1 second. The second condition to approximate a constant delay is

$$a_1 = a_0 = 3$$

The second-order transfer function is then

$$H(s) = \frac{1}{s^2 + 3s + 3} \qquad (3\text{-}27)$$

The denominator in Eq. (3-27) is a Bessel function of the first kind. If this procedure is followed for other orders of filter, the resulting Bessel functions are those given in Table 3-5.

A recursion formula [6] allows higher-order Bessel functions to be calculated from

$$D_{n+1}(s) = (2n+1) D_n(s) + s^2 D_{n-1}(s) \qquad (3\text{-}28)$$

The normalization of the time delay to a delay of unity at $\omega = 0$ is essentially equal to a frequency normalization also. However, the normalized frequency does not relate to an amplitude condition as ω_c does in the Butterworth and Chebyshev responses.

We must consider the effect of denormalization on the time delay. The normalized frequency will be designated u, so that a third-order transfer function could be expressed as

$$H(ju) = \frac{1}{(ju)^3 + 6(ju)^2 + 15ju + 15}$$

The time delay is expressed in terms of the normalized frequency u as

$$t_d = -\frac{d\Theta}{du} \quad (3\text{-}29)$$

If the frequency is denormalized to a value of $\omega = u\omega_o$, we can write the actual time delay as

$$t_o = -\frac{d\theta}{d\omega} \quad (3\text{-}30)$$

where θ is the actual phase shift.

It can be shown that the phase shift of the normalized function is the same as that of the denormalized function, or $\Theta = \theta$. Since $d\omega = d(u\omega_o) = \omega_o du$, the actual delay can be written as

$$t_o = -\frac{1}{\omega_o}\frac{d\theta}{du} = \frac{t_d}{\omega_o} \quad (3\text{-}31)$$

Since the normalized delay, t_d, is equal to unity, we can write that the actual delay is

$$t_o = \frac{1}{\omega_o} \quad (3\text{-}32)$$

Thus the time delay of the denormalized function at $\omega = 0$ is a factor of ω_o less than the normalized value at $u = 0$.

As u is increased, the time delay varies from the zero frequency value, as indicated in Table 3-6 [6].

We note from Table 3-6 that the time delay deviates from its unity zero-frequency value by less than 1% up to $u = 1$ for n greater than 2.

TABLE 3-6 DEVIATION IN TIME DELAY FROM ZERO-FREQUENCY VALUE

n	u for 1% deviation	u for 10% deviation	u for 20% deviation	u for 50% deviation
1	0.10	0.34	0.50	1.00
2	0.56	1.09	1.39	2.20
3	1.21	1.94	2.29	3.40
4	1.93	2.84	3.31	4.60
5	2.71	3.76	4.20	5.78
6	3.52	4.69	5.95	6.97

TABLE 3-7 ATTENUATION AS A FUNCTION OF u

n	u for $\frac{1}{50}$ dB	u for $\frac{1}{10}$ dB	u for $\frac{1}{2}$ dB	u for 1 dB	u for 3 dB
1	0.08	0.14	0.35	0.51	1.00
2	0.11	0.26	0.57	0.80	1.36
3	0.14	0.34	0.75	1.05	1.75
4	0.17	0.40	0.89	1.25	2.13
5	0.20	0.45	1.01	1.43	2.42
6	0.22	0.50	1.12	1.58	2.70

A second set of values that is important in selecting the order of the linear phase filter is listed as Table 3-7 [6].

The use of these tables is demonstrated in the following two examples.

Example 3-4

Given a 3-pole linear phase filter with a 3-dB frequency of 1 kHz, find the normalizing factor ω_o.

Solution: For a 3-pole filter, Table 3-7 shows that the 3-dB point occurs at $u = 1.75$. Since $u = \omega/\omega_o$, and the desired 3-dB point is $\omega = 2\pi \times 10^3$, the value of ω_o is

$$\omega_o = \frac{2\pi \times 10^3}{1.75} = 3.59 \times 10^3$$

Once the normalized filter is synthesized, the factor ω_o can be used to denormalize the element values.

Example 3-5

A delay of 20 μs is required along with a 3-dB bandwidth of 1.06×10^5 rad/s (±5%). Find the value of n required.

Solution: The normalization factor is

$$\omega_o = \frac{1}{t_o} = \frac{1}{20 \times 10^{-6}} = 5 \times 10^4 \text{ rad/s}$$

Since

$$u = \frac{\omega}{\omega_o}$$

the 3-dB value of u must be

$$u_{3db} = \frac{\omega_{3db}}{\omega_o} = \frac{1.06 \times 10^5}{5 \times 10^4} = 2.125$$

From Table 3-7 we see that a fourth-order filter has a 3-dB point of $u = 2.13$, thus this value of n will satisfy the specifications within the 5% tolerance.

The time delay should be checked from Table 3-6. For $n = 4$ it is seen that the error in time delay is less than 1% up to a value of $u = 1.93$, which is near the 3-dB point. Therefore, this value of n will simultaneously satisfy both specifications.

Exercises

E3.4a A Butterworth filter with a 3-dB frequency of 10 kHz is to attenuate a 1-V, 25-kHz signal to a value of less than 0.1 V. Calculate the minimum required value of n (the number of poles).

E3.4b Repeat exercise E3.4a for a Chebyshev filter.

E3.4c Repeat exercise E3.4a for a linear phase filter.

3.5 PRESAMPLING FILTERS

In order to demonstrate the use of the theory developed in this chapter, we will consider the use of a presampling filter. This problem relates to the conversion of a triangular waveform to a digital sequence of values using ideal impulse sampling. The Fourier series for the waveform of Fig. 3-21 is given by

$$v(t) = \left[2.5 + \frac{20}{\pi^2} \sum_{n=1}^{\infty} (-1)^{n+1} \sin(2n-1)\omega t \right] V$$

$$= 2.5 + 2.0264 \sin \omega t - 0.2252 \sin 3\omega t + 0.0811 \sin 5\omega t$$

$$- 0.0414 \sin 7\omega t + 0.0250 \sin 9\omega t - 0.0167 \sin 11\omega t + \cdots V$$

In this expression, the fundamental frequency is $\omega = 2\pi$ krad/s. For our purposes we will approximate the waveform by the first four terms.

An 8-bit ADC is used to convert this signal to digital values. A 3-pole Butterworth response is selected for the prealiasing filter. Our problem consists of two parts:

1. Select the 3-dB bandwidth of the filter to lead to a maximum attenuation of any of the first four components (including the DC component) of 1 LSB or less.
2. Select a sampling frequency to limit any aliased component that falls within the filter bandwidth to 1/2 LSB or less.

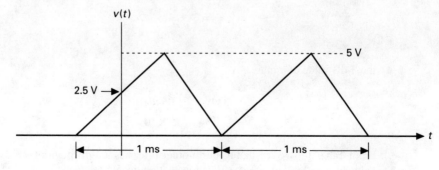

Figure 3-21 A triangular waveform.

Chap. 3   Filters

For an 8-bit ADC to convert a 5-V span, the quantum interval will be

$$q = \frac{5\text{ V}}{255} = 0.0196 \text{ V} = 1 \text{ LSB}$$

The attenuation amounting to 1 LSB must be found for each component. This can be calculated as follows:

1. Attenuation of 1-kHz component for 1 LSB error:

$$At_1 = \frac{2.0264 - 0.0196}{2.0264} = 0.990$$

2. Attenuation of 3-kHz component for 1 LSB error:

$$At_3 = \frac{0.2252 - 0.0196}{0.2252} = 0.913$$

3. Attenuation of 5-kHz component for 1 LSB error:

$$At_5 = \frac{0.0811 - 0.0196}{0.0811} = 0.758$$

Next, we select the 3-dB frequency of a 3-pole Butterworth filter to lead to an attenuation that is less than 0.990 at 1 kHz, less than 0.913 at 3 kHz, and less than 0.758 at 5 kHz. The denormalized response of the filter is given by

$$|H(j\omega)| = \frac{1}{\sqrt{1 + (\omega/\omega_c)^6}}$$

where ω_c is the 3-dB frequency.

The previous equation can be used to find ω_c by substituting the attenuation for $|H(j\omega)|$ and the corresponding frequency for ω. This is done for all three components, to find the largest required value of ω_c. This value is

$$f_c = 5.257 \text{ kHz}$$

To check this value, we can calculate the magnitude of each component at the filter output. These magnitudes are found by multiplying an input magnitude by the value of

$$\frac{1}{\sqrt{1 + (\omega/\omega_c)^6}}$$

using the corresponding value of ω. Table 3-8 contains the results of these calculations.

We note that the differences between input and output of the DC component, the 1-kHz component, and the 3-kHz component show less than 1 LSB or 0.0196 V. The

TABLE 3-8 ATTENUATION OF A THIRD-ORDER BUTTERWORTH FILTER

f, kHz	Input magnitude, V	Output magnitude, V
0	2.5000	2.5000
1	2.0264	2.0264
3	0.2252	0.2214
5	0.0811	0.0615
7	0.0414	0.0106 .0161
9	0.0250	0.0049
11	0.0167	0.0018

5-kHz component shows a difference of exactly 0.0196 V, thus the first requirement of the system is met.

The sampling frequency can now be selected by ensuring that any aliased component that falls below 5.257 kHz has a magnitude less than or equal to 1/2 LSB or 0.0098 V. From Table 3-8 we note that the 9-kHz component is the lowest-frequency component that satisfies this requirement. We must not allow the 7-kHz component to be aliased into the 0 to 5.257 kHz range. We can accomplish this by choosing a sampling frequency greater than 5.257 kHz plus 7 kHz. In this instance we might choose a sampling frequency of 14 kHz or $\omega_s = 88$ krad/s. For this choice, the 9-kHz, 11-kHz, and higher-frequency components would alias into the filter bandwidth, but each of these magnitudes is less than 0.0098 V. We note that the sample rate in this example is 2.8 times the frequency of the highest component of interest.

A more general consideration relates to the error introduced into the passband of the presampling filter. Most repetitive engineering waveforms include frequency components that diminish at higher frequencies. The highest frequency of interest may have an amplitude that is 0.01 to 0.1 times the span of the system. If the signal to be converted can consist of any frequency component in the passband, with an amplitude equal to one-half of the span, the error caused by the filter can be much greater. It should be emphasized that this is a rare situation and the following calculations are very conservative for most typical applications.

We will assume that an n-pole Butterworth presampling filter is used. The transfer function of such a filter is given by

$$|H(j\omega)| = \frac{1}{\left[1 + \left(\frac{f}{f_c}\right)^{2n}\right]^{1/2}}$$

where f_c is the 3-dB point.

We will also assume that the maximum magnitude of any frequency component can have a value of one-half the span, or

$$V = \frac{V^{span}}{2}$$

A point of interest is often the 1/2 LSB point, that is, the frequency at which the magnitude of the output voltage has dropped 1/2 LSB from the input value. At this

Chap. 3 Filters

frequency, the frequency error introduced by the filter equals the 1/2 LSB of the digital word.

If a voltage of

$$V \sin \omega t = \frac{V^{span}}{2} \sin \omega t$$

is applied to the filter, the output will have a magnitude of

$$\frac{V^{span}}{2} \times \frac{1}{\left[1 + \left(\frac{f}{f_c}\right)^{2n}\right]^{1/2}}$$

The filter error between input and output is

$$error = \frac{V^{span}}{2} \left(1 - \frac{1}{\left[1 + \left(\frac{f}{f_c}\right)^{2n}\right]^{1/2}}\right)$$

The value of 1/2 LSB depends on the number of bits N in the converted word and is given by

$$1/2 \text{ LSB} = \frac{V^{span}}{2^{N+1}} \qquad (3\text{-}33)$$

Equating the two previous equations and solving for the 1/2 LSB frequency, $f = f_{1/2 \text{LSB}}$, results in

$$f_{1/2 \text{LSB}} = f_c \left[\frac{2^{N+1} - 1}{2^{2N} - 2^{N+1} + 1}\right]^{1/2n} \qquad (3\text{-}34)$$

This equation can be accurately approximated (less than 2% error for $N \geq 6$) by

$$f_{1/2 \text{LSB}} = f_c \left(\frac{2^{N+1}}{2^{2N}}\right)^{1/2n} = f_c \times \frac{1}{2^{(N-1)/2n}} \qquad (3\text{-}35)$$

Equation (3-35) tells us that if the maximum span is used at all frequencies, the filter will introduce a 1/2 LSB error at a frequency of $f_{1/2 \text{LSB}}$.

For comparison purposes, let us consider a 10-bit conversion followed by either a 1-pole Butterworth filter or a 4-pole Butterworth filter. Both filters have 3-dB frequencies of 10 kHz.

For the 1-pole filter ($n = 1$), Eq. (3-35) yields, for $N = 10$,

$$f_{1/2 \text{LSB}} = 10^4 \left[\frac{2^{11}}{2^{20}}\right]^{1/2} = 442 \text{ Hz}$$

The frequency leading to a 1/2 LSB error is a factor of 22.6 times lower than the 3-dB frequency of the filter.

A 4-pole filter gives a value of

$$f_{1/2\,\text{LSB}} = 10^4 \left[\frac{2^{11}}{2^{20}}\right]^{1/8} = 4585\ \text{Hz}$$

In this case, the 1/2 LSB error frequency is 2.2 times lower than the 3-dB frequency of the filter.

These results show the improvement in flatness of response with increasing n for the Butterworth filter. If it is significant to extend the 1/2 LSB frequency without changing f_c, the order of the filter must be increased.

Exercise

E3.5a Calculate $f_{1/2\,\text{LSB}}$ for a 3-pole Butterworth filter with a corner frequency of 10 kHz.

GENERAL EXERCISES

3.1 A signal that can be expressed as
$$x(t) = 4 \sin \omega_1 t + 2 \sin 2\omega_1 t + \sin 3\omega_1 t + 0.2 \sin 4\omega_1 t$$
is sampled at a frequency of 10 kHz. What is the maximum value that ω_1 can have before an aliased component equals $4\omega_1$?

3.2 A signal is expressed as
$$x(t) = 4 \sin \omega_1 t + 2 \sin 2\omega_1 t + \sin 3\omega_1 t + K_4 \sin 4\omega_1 t$$
where $f_1 = \omega_1/2\pi = 2$ kHz. If the sampling frequency is 15.8 kHz, what must the value of K_4 be to limit the lowest-frequency aliased component magnitude to 0.1? Assume that the sampling process duplicates that of Fig. 3-3 with $b = 50\ \mu\text{s}$.

3.3 A normalized filter with a bandwidth of 1 rad/s and a 1-Ω load resistor has values of $L = 2.62\ H$ and $C = 1.14\ F$. If the actual filter to be used has a bandwidth of $2\pi \times 10^3$ rad/s and a 500-Ω load, find the denormalized values of L and C.

3.4 The filter shown in Fig. PR3.4 has a 1 rad/s bandwidth. Denormalize the element values to result in a filter with a 2-kHz bandwidth and a 200-Ω load.

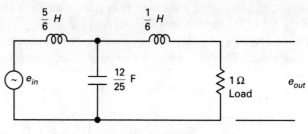

Figure PR3.4

3.5 If $|H(j\omega)|^2 = \dfrac{1}{1+\omega^6}$, find $H(s)$.

3.6 If $|H(J\omega)|^2 = \dfrac{1}{9+3\omega^2+\omega^4}$, find $H(s)$.

3.7 A signal contains seven frequency components, each having a peak amplitude of 2 V. The frequencies of the components are 2, 4, 6, 8, 10, 12, and 14 kHz. A third-order Butterworth filter is used, followed by ideal impulse sampling at 31.5 kHz. Calculate the highest 3-dB frequency that can be used for the filter to limit the total aliased value of rms voltage in the 0–20 kHz band to 0.4 V.

3.8 Repeat problem 3.7 assuming a Chebyshev filter is used. Use a ripple width of 0.1 dB.

3.9 Repeat problem 3.7 assuming a Bessel filter is used.

3.10 If the 3-dB frequency of the Butterworth filter in problem 3.7 is set at 10 kHz, what must the sampling frequency be to cause the rms voltage in the 0–20 kHz band to equal 0.2 V?

3.11 Find the smallest value of n and the largest value of ε to give a ripple width that is less than or equal to 0.1 and a 3-dB point less than or equal to 1.25 rad/s.

3.12 A second-order Butterworth filter and a second-order Chebyshev filter are used to satisfy the specifications shown in Fig. PR3.12. The response should lie between 0.9 and 1 in the passband and should be less than 0.05 above ω_1. Find ω_1 for each case.

3.13 Three-pole, low-pass Butterworth and Bessel filters are constructed to have a normalized 3-dB bandwidth of 1 rad/s.

    a. At what frequencies will each filter pass a signal that equals 90% of the value passed at DC?

    b. If the zero-frequency time delay of the denormalized Bessel filter is 80 µs, what is the denormalized 3-dB bandwidth?

3.14 A signal that can be expressed as

$$e(t) = 6 \sin 2000t + 2 \sin 6000t$$

is impulse sampled at $\omega_s = 18{,}000$ rad/s and then filtered. The filter used is a 3-pole Butterworth filter with a 3-dB frequency of 9,000 rad/s. Calculate the peak magnitudes and frequencies of the two lowest-frequency aliased components.

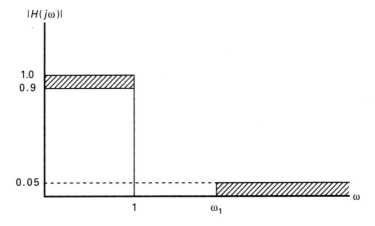

Figure PR3.12

3.15 Repeat problem 3.14 assuming a 3-pole Chebyshev filter is used with a 0.2-dB ripple.

3.16 Repeat problem 3.14 assuming a 3-pole Bessel filter is used.

REFERENCES

1. Bobrow, L. S., *Elementary Linear Circuit Analysis*, 2nd Ed., Holt, Rinehart and Winston, New York, 1987, Chapter 14.
2. Leigh, J. R., *Applied Digital Control*, Prentice Hall, Englewood Cliffs, NJ, 1985, Chapter 2.
3. Van Valkenburg, M. E., *Analog Filter Design*, Holt, Rinehart and Winston, New York, 1982, Chapter 6.
4. Zverev, A. I., *Handbook of Filter Synthesis*, John Wiley and Sons, New York, 1967.
5. Johnson, D. E., *Introduction to Filter Theory*, Prentice Hall, Englewood Cliffs, NJ, 1976, Chapter 3.
6. Weinberg, L., *Network Analysis and Synthesis*, McGraw-Hill, New York, 1962, Chapter 11.

4
Synthesis of Filters

There are two types of electronic filters in use today: passive and active. The passive filter is composed of passive elements only, that is, resistors, capacitors, and inductors. The active filter uses gain elements such as transistors or op amps along with capacitors and resistors to create the filter. The primary motivation behind active filters is the elimination of inductors from the circuit, since this element is the most nonideal and cannot be integrated. Another advantage of the active filter is that a wide variety of off-the-shelf filters are available to be used with a minimum of additional components. We will first consider passive synthesis methods and then proceed to a consideration of active synthesis.

Although filter synthesis is a very broad topic that cannot be treated in detail in a single chapter, we have included the subject in order to present an introduction to the filter design process. Consequently, this chapter will cover some fundamental concepts that will lead to a few practical synthesis methods. The material is not intended to be exhaustive, but it will allow a designer to synthesize actual filters when off-the-shelf filters are not available or when a simple discrete element filter is all that is required.

4.1 PASSIVE SYNTHESIS

In the few pages to be devoted to the topic of filter synthesis, only the pertinent points necessary to understand the basic ideas of design will be considered. Many statements will be made without proof, with a reference cited that contains the proof. The first point that should be rather obvious is that both inductors and capacitors are necessary to obtain complex poles. Since Butterworth, Chebyshev, and Bessel filters all require complex poles, both inductors and capacitors must be used to synthesize these filters with passive networks. Therefore, we will confine our discussion to LC or RLC networks.

4.1.1 Properties of LC Networks

There are several properties of LC networks that ultimately relate to synthesis of transfer functions. These are now listed.

1. LC impedances, admittances, or transfer functions have poles that lie along the $j\omega$ axis of the Laplace plane. This results from the fact that the real part of a pole of a driving point or transfer function leads to a dissipation of power [1]. The ideal LC network is lossless, and thus all poles lie on the $j\omega$ axis.
2. An LC impedance or admittance (called a driving point function) will never have repeated poles [1].
3. An LC driving point function will always be the ratio of two polynomials, either an even over an odd or an odd over an even, and the orders of the two polynomials differ by one [1].
4. The slope of the LC impedance as a function of frequency is positive. If $Z = jX(\omega)$ is the impedance of the network, then

$$\frac{dX(\omega)}{d\omega} > 0$$

This last condition leads to the important corollary that the poles and zeros alternate along the $j\omega$ axis [1]. Figure 4-1 shows a typical plot of $X(\omega)$ for an LC network. A driving point function is either an impedance or an admittance. These functions are determined by the ratio of voltage to current or current to voltage at a single pair of terminals. The types of transfer functions of interest are expressed as the ratio of an output quantity to an input quantity. A typical transfer function is the ratio of voltage at a pair of output terminals to the voltage applied across the pair of input terminals.

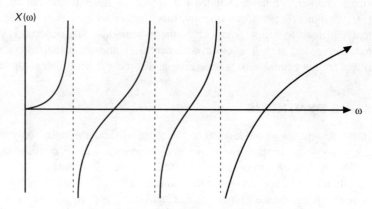

Figure 4-1 Reactance of an LC network.

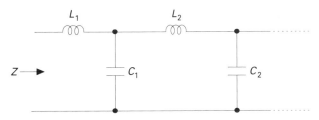

Figure 4-2 Cauer's first form.

4.1.2 Synthesis of LC Driving Point Functions

Rarely are we interested in creating a driving point function to use directly in a circuit. There are occasions when we may wish to shape or modify the frequency spectrum of an input voltage to an amplifier by using suitable LC networks as input or feedback elements, as shown in Fig. 2-13. In this chapter we study driving point synthesis primarily because it is used in creating a desired transfer function. Since we will direct our efforts toward the synthesis of ladder networks, we will be interested in what are referred to as the two Cauer forms. The Cauer 1 form involves the repeated removal of poles at $s = \infty$, and the Cauer 2 form involves the repeated removal of poles at $s = 0$. These procedures will be illustrated by examples.

Cauer's first form. Cauer's first form has the general configuration of Fig. 4-2. As s approaches infinity a dominant element will determine the behavior of the impedance. For example, L_1 in Fig. 4.2 determines the impedance as s approaches infinity because the impedance of C_1 goes to zero. With this form, the dominant element can be removed and the remaining impedance approaches zero.

In order to demonstrate this concept, let us consider the impedance

$$Z(s) = \frac{s^3 + 4s}{s^2 + 1}$$

As s approaches infinity, the higher-order terms of the numerator and denominator become dominant and

$$Z(s) \to \frac{s^3}{s^2} = s$$

We can remove this pole simply by subtracting s from $Z(s)$. This gives

$$\frac{s^3 + 4s}{s^2 + 1} - s = \frac{3s}{s^2 + 1}$$

The impedance $Z(s)$ can be expressed as

$$Z(s) = s + \frac{3s}{s^2 + 1}$$

If we associate the removal of the pole with the removal of a corresponding physical element, then we have synthesized a part of the impedance. Figure 4-3 demonstrates this concept.

Since the behavior of the dominant element exhibits itself as s approaches infinity, and the remaining impedance goes to zero as s approaches infinity, the removal of a dominant element from the remaining impedance cannot be accomplished. However, the remaining admittance exhibits a pole at $s = \infty$. Consequently, a dominant element can be removed from the admittance. The remainder admittance can be inverted and a pole again exists at $s = \infty$. This process can be repeated until the last element is removed.

Although the method of removing the dominant pole demonstrated previously could be used, a simpler method of pole removal exists. This method begins by arranging the polynomials of the numerator and denominator in descending order. The lower-order polynomial is then divided into the higher-order one, and the remainder forms the remaining function. For example, a pole can be removed from the previous impedance by the division

$$s + \frac{3s}{s^2+1}$$

$$s^2 + 1 \;\overline{\big)\; s^3 + 4s}$$
$$\underline{s^3 + s}$$
$$3s$$

The remainder impedance is inverted to create a pole at infinity and the process continues. Rather than go through these steps, it is more convenient to continue the division as follows:

$$\begin{array}{c} s \leftarrow Z \qquad \text{(an inductor)} \\ s^2+1 \;\overline{\big)\; s^3+4s} \\ \underline{s^3+s} \\ 3s \;\overline{\big)\; s^2+1} \qquad \tfrac{1}{3}s \leftarrow Y \quad \text{(a capacitor)} \\ \underline{s^2} \\ 1 \;\overline{\big)\; 3s} \qquad 3s \leftarrow Z \quad \text{(an inductor)} \\ \underline{3s} \\ 0 \end{array}$$

This procedure automatically inverts the remainder and removes the pole at infinity. If a pole removal first takes place from an impedance, the next removal is from an admittance, and the next is from an impedance. An impedance pole at infinity is equivalent to a series inductance, while a pole in the admittance at infinity is equivalent to a shunt capacitance. The network represented by the successive divisions is shown in Fig. 4-4.

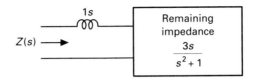

Figure 4-3 Removal of an element with a pole at $s = \infty$.

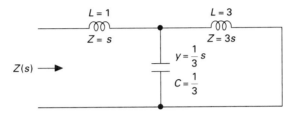

Figure 4-4 A synthesized impedance in Cauer's first form.

Example 4-1

Synthesize the impedance

$$Z(s) = \frac{s^2 + 4}{s^3 + 9s}$$

using Cauer's first form.

Solution: Since the impedance does not have a pole at $s = \infty$, the procedure must start with the admittance. The division proceeds as shown. Note that it is helpful to identify each element as an impedance or admittance to determine whether the element is placed in series (Z) or in parallel (Y).

```
              s ← Y           (a capacitor)
        ┌─────────────
s² + 4  │ s³ + 9s
        │ s³ + 4s           1/5 s ← Z        (an inductor)
        ─────────
              5s    ┌─────────
                    │ s² + 4
                    │ s²               5/4 s ← Y       (a capacitor)
                    ─────────
                          4   │ 5s
                              │ 5s
                              ─────
                                0
```

The synthesized impedance is shown in Fig. 4-5.

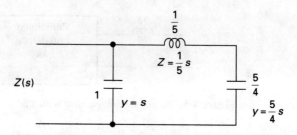

Figure 4-5 A Cauer network with a capacitor as the first element.

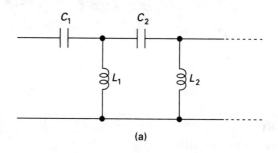

(a)

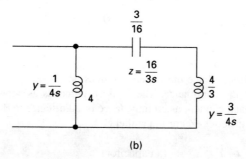

(b)

Figure 4-6 Cauer's second form. (a) General configuration. (b) An example.

Cauer's second form. Cauer's second form assumes a network as shown in Fig. 4-6(a), which has series impedance poles and shunt admittance poles at $s = 0$. In this case, poles are removed from $s = 0$. In order to accomplish this the numerator and denominator are arranged in ascending order. The two leading terms are examined to see if a pole exists at $s = 0$. If so, the division process begins.

If not, the function is inverted before beginning the division. We shall demonstrate this method with the function

$$Z(s) = \frac{s^3 + 4s}{s^2 + 1}$$

The numerator and denominator are arranged in ascending order and examined for a pole at $s = 0$. This results in

$$Z(s) = \frac{4s + s^3}{1 + s^2}$$

The impedance does not have the required pole, so the division process must start with $Y(s)$. This gives

$$
\begin{array}{c}
\dfrac{1}{4s} \leftarrow Y \\[4pt]
4s+s^3 \,\Big|\, \begin{array}{l} 1+s^2 \\ 1+\dfrac{1}{4}s^2 \end{array} \qquad \dfrac{16}{3s} \leftarrow Z \\[8pt]
\hline
\dfrac{3}{4}s^2 \,\Big|\, 4s+s^3 \\
\phantom{\dfrac{3}{4}s^2\,\big|\,} 4s \qquad\qquad \dfrac{3}{4s} \leftarrow Y \\[6pt]
\hline
 s^3 \,\Big|\, \begin{array}{l} \dfrac{3}{4}s^2 \\ \dfrac{3}{4}s^2 \end{array} \\[8pt]
\hline
 0
\end{array}
$$

The resulting network is shown in Fig. 4-6(b).

In summary, the Cauer forms can be implemented by the following procedure:

1. Arrange the numerator and denominator of the function in descending order for the Cauer 1 form or in ascending order for the Cauer 2 form.
2. Check to see if a pole exists at $s = \infty$ for the Cauer 1 form or at $s = 0$ for the Cauer 2 form. If not, invert the function.
3. Use repeated division to remove poles at $s = \infty$ for the Cauer 1 form or $s = 0$ for the Cauer 2 form.
4. Identify each pole removed as an impedance or admittance, and identify the appropriate element associated with each pole.

Exercises

    E4.1a Using Cauer's first form, synthesize the impedance

$$Z(s) = \frac{s^3 + 9s}{s^2 + 4}$$

    E4.1b Using Cauer's second form, synthesize the impedance

$$Z(s) = \frac{s^2 + 1}{s^3 + 4s}$$

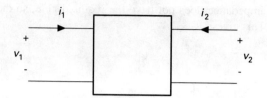

Figure 4-7 A two-port network.

4.2 SYNTHESIS OF TRANSFER FUNCTIONS

4.2.1 Two-Port Parameters

Most practical networks include an input port and an output port. Such a configuration is called a two-port network. In general, the input and output currents and voltages are defined in terms of Fig. 4-7. Note that both input and output currents are positive when flowing into the network.

Two of the variables can be written in terms of the remaining two variables. The variables used to express the equations are arbitrary and are selected to fit a given application. For example, the equations may be expressed in any of the following forms:

$$i_1 = y_{11}v_1 + y_{12}v_2 \tag{4-1a}$$
$$i_2 = y_{21}v_1 + y_{22}v_2 \tag{4-1b}$$

$$v_1 = z_{11}i_1 + z_{12}i_2 \tag{4-2a}$$
$$v_2 = z_{21}i_1 + z_{22}i_2 \tag{4-2b}$$

$$v_1 = h_{11}i_1 + h_{12}v_2 \tag{4-3a}$$
$$i_2 = h_{21}i_1 + h_{22}v_2 \tag{4-3b}$$

The first set of parameters are called the y parameters, or short-circuit admittances, of the network. These quantities are defined by

$$y_{11} = \left.\frac{i_1}{v_1}\right|_{v_2=0} \tag{4-4a}$$

$$y_{12} = \left.\frac{i_1}{v_2}\right|_{v_1=0} \tag{4-4b}$$

$$y_{21} = \left.\frac{i_2}{v_1}\right|_{v_2=0} \tag{4-4c}$$

$$y_{22} = \left.\frac{i_2}{v_2}\right|_{v_1=0} \tag{4-4d}$$

Physically, v_1 or v_2 can be forced to equal zero, by short-circuiting the corresponding port.

The z parameters, or open-circuit impedances, are defined by

Chap. 4 Synthesis of Filters

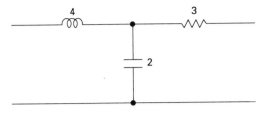

Figure 4-8 A tee network.

$$z_{11} = \left.\frac{v_1}{i_1}\right|_{i_2=0} \tag{4-5a}$$

$$z_{12} = \left.\frac{v_1}{i_2}\right|_{i_1=0} \tag{4-5b}$$

$$z_{21} = \left.\frac{v_2}{i_1}\right|_{i_2=0} \tag{4-5c}$$

$$z_{22} = \left.\frac{v_2}{i_2}\right|_{i_1=0} \tag{4-5d}$$

In measuring the z parameters, i_1 or i_2 can be forced to zero by opening the corresponding port.

The h parameters, or hybrid parameters, are often used in transistor circuit analysis and are defined by

$$h_{11} = \left.\frac{v_1}{i_1}\right|_{v_2=0} \tag{4-6a}$$

$$h_{12} = \left.\frac{v_1}{v_2}\right|_{i_1=0} \tag{4-6b}$$

$$h_{21} = \left.\frac{i_2}{i_1}\right|_{v_2=0} \tag{4-6c}$$

$$h_{22} = \left.\frac{i_2}{v_2}\right|_{i_1=0} \tag{4-6d}$$

We can demonstrate the method of evaluating these parameters for the network of Fig. 4-8.

The circuits used to evaluate the y parameters are shown in Fig. 4-9. The y parameters are found to be

$$y_{11} = \frac{6s+1}{24s^2+4s+3}$$

$$y_{12} = \frac{-1}{24s^2+4s+3}$$

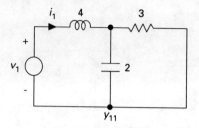
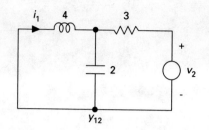
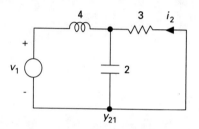
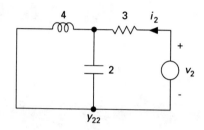

Figure 4-9 Circuits for evaluating y parameters.

$$y_{21} = \frac{-1}{24s^2 + 4s + 3}$$

$$y_{22} = \frac{8s^2 + 1}{24s^2 + 4s + 3}$$

For passive circuits, $y_{12} = y_{21}$.

Figure 4-10 shows the circuits used to evaluate the z parameters. The z parameters are

$$z_{11} = \frac{8s^2 + 1}{2s}$$

$$z_{12} = \frac{1}{2s}$$

$$z_{21} = \frac{1}{2s}$$

$$z_{22} = \frac{6s + 1}{2s}$$

For passive circuits, $z_{12} = z_{21}$. It should be noted that

Chap. 4 Synthesis of Filters

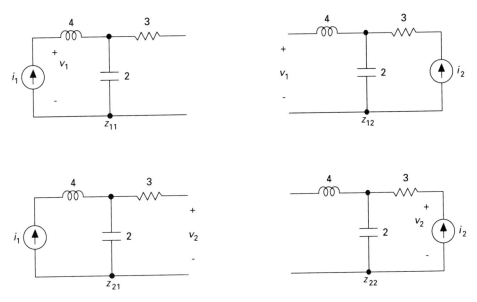

Figure 4-10 Circuits for evaluating z parameters.

$$y_{ij} \neq \frac{1}{z_{ij}}$$

We will not pursue the hybrid parameters, since they are of little interest in the filter synthesis method to be considered.

4.2.2 Transfer Function Synthesis

A significant configuration in a filter network is represented in Fig. 4-11. This model might represent an amplifier stage with low output impedance driving a filter that works into a second amplifier stage. The resistor would then represent the normalized input impedance of the second stage. If the output impedance of the first stage is considerably less than the input impedance of stage two, this model is then appropriate. This condition is often satisfied for practical circuits.

In the configuration shown, the following four equations are used;

$$i_1 = y_{11}v_1 + y_{12}v_2 \qquad (4\text{-}1a)$$
$$i_2 = y_{21}v_1 + y_{22}v_2 \qquad (4\text{-}1b)$$
$$i_2 = -v_2 \text{ (due to 1-}\Omega\text{ load)}$$
$$y_{12} = y_{21} \text{ (for all passive networks)}$$

For the configuration of Fig. 4-11, we are generally interested in the voltage transfer function, hence we solve these equations for the ratio v_2/v_1. This gives

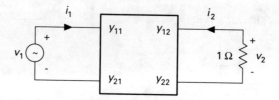

Figure 4-11 A filter configuration.

$$\frac{v_2}{v_1} = \frac{-y_{21}}{1 + y_{22}} \tag{4-7}$$

Equation (4-7) forms the basis of a synthesis procedure for LC networks. In addition to this equation, we apply the following three important facts:

1. y_{21} and y_{22} are expressible as the ratio of two polynomials in s, one of which is even while the other is odd.
2. The poles of y_{22} are always included as poles of y_{21}. (y_{21} may have poles in addition to those of y_{22}.) [1]
3. Zeros in the transfer function can be created by proper placement of elements.

The frequencies at which the transfer function equals zero are called transmission zeros. The key to the procedure is in the synthesis of the admittance y_{22}. As we synthesize this admittance using one of the Cauer forms, we create not only the poles and zeros of y_{22}, but we simultaneously create the poles of y_{21}. Then by examining the transmission zeros of the overall function, v_2/v_1, we can produce these values by proper placement of the elements.

We will demonstrate this three-step process by synthesizing a third-order Butterworth low-pass filter. In this case, using Table 3-1, we need a transfer function of the form

$$\frac{v_2}{v_1} = \frac{K}{s^3 + 2s^2 + 2s + 1}$$

where K is a constant. This function assumes a terminating impedance of 1 Ω and a corner frequency of 1 rad/s, as explained in Chapter 3.

Step 1. Create even over odd or odd over even functions for y_{21} and y_{22}. This can be done by equating the transfer function to Eq. (4-7), giving

$$\frac{-y_{21}}{1 + y_{22}} = \frac{K}{s^3 + 2s^2 + 2s + 1} = \frac{K}{(s^3 + 2s) + (2s^2 + 1)}$$

The numerator K, which is an even function of s, will be the numerator of $-y_{21}$. Therefore, we must divide both numerator and denominator of the transfer function by the odd part of the denominator polynomial to create

Chap. 4   Synthesis of Filters

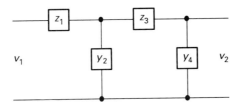

Figure 4-12 A general ladder network.

$$\frac{-y_{21}}{1+y_{22}} = \frac{\dfrac{K}{s^3+2s}}{1+\dfrac{2s^2+1}{s^3+2s}}$$

We can now identify y_{21} and y_{22} as

$$y_{21} = \frac{-K}{s^3+2s}$$

and

$$y_{22} = \frac{2s^2+1}{s^3+2s}$$

We note that both functions are ratios of even to odd polynomials and both have the same poles.

Step 2. Evaluate the locations of the transmission zeros of v_2/v_1 for the original transfer function. For the given function the three zeros are all at $s=\infty$. As we synthesize y_{22} we must place the elements properly to cause three transmission zeros at $s=\infty$.

Step 3. Synthesize y_{22} by removing poles at those values of s which produce transmission zeros. In a ladder network such as the Cauer forms, a series impedance or shunt admittance creates a transmission zero at the same value of s at which the series impedance or shunt admittance exhibits a pole. This can be seen from the general ladder network of Fig. 4-12.

If a series impedance becomes infinite at some Laplace frequency, it blocks current flow and causes a transmission zero at that value of s. If a shunt admittance becomes infinite at some value of s, a transmission zero is caused due to the existence of a short circuit path to ground. Thus if we want to create transmission zeros at some value of s, we simply remove poles in y_{22} at this value of s.

In this example, we want transmission zeros at $s=\infty$, therefore we will choose Cauer's first form, which removes poles at $s=\infty$. We start the synthesis of y_{22} in this example by noting that this admittance has no pole at $s=\infty$. Thus we must begin the process by inverting this admittance, leading to a pole removal from an impedance. The Cauer synthesis procedure results in the following values:

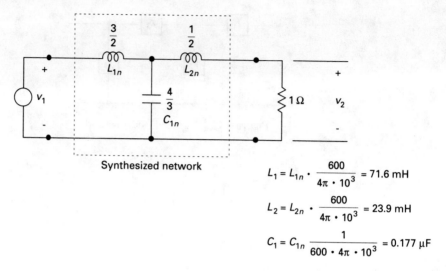

Figure 4-13 A filter circuit.

$$L_1 = L_{1n} \cdot \frac{600}{4\pi \cdot 10^3} = 71.6 \text{ mH}$$

$$L_2 = L_{2n} \cdot \frac{600}{4\pi \cdot 10^3} = 23.9 \text{ mH}$$

$$C_1 = C_{1n} \frac{1}{600 \cdot 4\pi \cdot 10^3} = 0.177 \text{ μF}$$

$$
\begin{array}{r|l}
 & \frac{1}{2}s \leftarrow Z \\
2s^2 + 1 & s^3 + 2s \\
 & s^3 + \frac{1}{2}s \\
\hline
 & \frac{3}{2}s
\end{array}
\quad
\begin{array}{r|l}
 & \frac{4}{3}s \leftarrow Y \\
\frac{3}{2}s & 2s^2 + 1 \\
 & 2s^2 \\
\hline
 & 1
\end{array}
\quad
\begin{array}{r|l}
 & \frac{3}{2}s \leftarrow Z \\
1 & \frac{3}{2}s \\
 & \frac{3}{2}s \\
\hline
 & 0
\end{array}
$$

The first element must result in an impedance of $s/2$. An inductor of value $1/2$ H is required for this element. The second element is an admittance of value $4s/3$, which is given by a $4/3$ F capacitance. The third element is implemented by a $3/2$ H inductance. The synthesized network for y_{22} is shown in Fig. 4-13. Note that since this admittance pertains to port 2 (y_{22}), the leading element (a $1/2$ H inductor) connects to port 2.

Chap. 4   Synthesis of Filters

We should also note that the series inductances cause poles in both y_{22} and y_{21} at $s = \infty$, as does the shunt capacitor. Furthermore, these elements create zeros in the transfer function (transmission zeros) at $s = \infty$.

Let us assume that the actual terminating resistance is 600 Ω rather than 1 Ω, and the desired cutoff frequency is $f_c = 2$ kHz. The denormalized values of elements are found from Eqs. (3-11c) and (3-11b) to be

$$L_1 = \frac{3}{2} \times \frac{600}{4\pi \times 10^3} = 71.6 \text{ mH}$$

$$L_2 = \frac{1}{2} \times \frac{600}{4\pi \times 10^3} = 23.9 \text{ mH}$$

$$C_1 = \frac{4}{3} \times \frac{1}{600 \times 4\pi \times 10^3} = 0.177 \text{ μF}$$

The next example demonstrates the synthesis of a transfer function with all transmission zeros at $s = 0$.

Example 4-2

Synthesize the following transfer function assuming a 1-Ω load:

$$\frac{v_2}{v_1} = \frac{-Ks^3}{s^3 + 2s^2 + 2s + 1}$$

Solution: The numerator is odd, therefore we must divide both numerator and denominator by the even part of the denominator to form

$$\frac{\dfrac{-Ks^3}{2s^2 + 1}}{1 + \dfrac{s^3 + 2s}{2s^2 + 1}}$$

This gives

$$y_{22} = \frac{s^3 + 2s}{2s^2 + 1}$$

The transmission zeros are all at $s = 0$ in this case, thus poles of y_{22} must be removed at $s = 0$ by Cauer's second form. The procedure is demonstrated as follows:

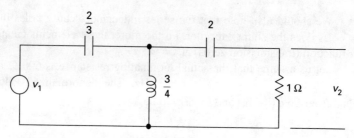

Figure 4-14 Filter for Example 4-2.

$$\frac{1}{2s} \leftarrow Z$$

$$
\begin{array}{r|l}
2s + s^3 & 1 + 2s^2 \\
& 1 + \frac{1}{2}s^2 \qquad \frac{4}{3s} \leftarrow Y \\
\hline
& \frac{3}{2}s^2 \quad \Big| \; 2s + s^3 \\
& \qquad\qquad 2s \qquad\qquad \frac{3}{2s} \leftarrow Z \\
& \qquad\qquad \overline{} \\
& \qquad\qquad s^3 \quad \Big| \; \frac{3}{2}s^2 \\
& \qquad\qquad\qquad\qquad \frac{3}{2}s^2 \\
& \qquad\qquad\qquad\qquad \overline{} \\
& \qquad\qquad\qquad\qquad 0
\end{array}
$$

The resulting filter is shown in Fig. 4-14.

The circuit of Example 4-2 is seen to be a high-pass filter. The transfer function has three zeros at $s = 0$ and approaches $-K$ as $s \to \infty$. The next example has a transfer function with zeros of transmission at both $s = 0$ and $s = \infty$.

Example 4-3

Synthesize the following transfer function assuming a 1-Ω load:

$$\frac{v_2}{v_1} = \frac{-Ks^3}{s^4 + s^3 + 4s^2 + 2s + 3}$$

Solution: Again, the numerator is odd, allowing us to arrange the transfer function as

Chap. 4   Synthesis of Filters

$$\frac{\dfrac{-Ks^3}{s^4+4s^2+3}}{1+\dfrac{s^3+2s}{s^4+4s^2+3}}$$

The original function has three transmission zeros at $s=0$ and one at $s=\infty$. In this situation we must remove one pole from y_{22} at $s=\infty$ using Cauer's first form and then remove three poles at $s=0$ using Cauer's second form. Starting the synthesis procedure with the reciprocal of y_{22} to obtain the proper pole, the Cauer first form is used. This gives

$$s^3+2s \overline{\smash{\big)}\, s^4+4s^2+3} \quad\Big|\; s+\dfrac{2s^2+3}{s^3+2s}$$

$$\underline{s^4+2s^2}$$

$$2s^2+3$$

The remainder impedance is now arranged to use Cauer's second form to remove three poles at $s=0$:

$$2s+s^3 \overline{\smash{\big)}\, 3+2s^2} \quad\Big|\; \dfrac{3}{2s} \leftarrow Z$$

$$\underline{3+\dfrac{3}{2}s^2}$$

$$\dfrac{1}{2}s^2 \overline{\smash{\big)}\, 2s+s^3} \quad\Big|\; \dfrac{4}{s} \leftarrow Y$$

$$\underline{2s}$$

$$s^3 \overline{\smash{\big)}\, \dfrac{1}{2}s^2} \quad\Big|\; \dfrac{1}{2s} \leftarrow Z$$

$$\underline{\dfrac{1}{2}s^2}$$

$$0$$

The finished circuit is shown in Fig. 4-15.

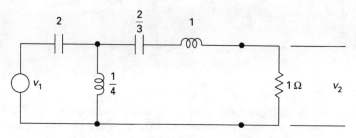

Figure 4-15 Filter for Example 4-3.

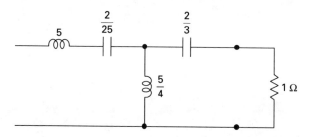

Figure 4-16 Alternate realization of filter for Example 4-3.

Had we chosen to remove the three poles at $s = 0$ before removing the one at $s = \infty$, the network would have appeared as shown in Fig. 4-16.

4.2.3 Zero Shifting to Obtain Finite Transmission Zeros

Many practical filters can be implemented by the method outlined in the preceding paragraphs. These filters have no transmission zeros at finite frequencies, which is typical of practical low-pass or high-pass filters. In some instances it might be advantageous to place one or more transmission zeros at finite frequencies. For example, it is sometimes useful to locate a transmission zero at 60 Hz to discriminate more effectively against 60-Hz hum.

Since removal of poles from a driving point function leads to transmission zeros for a ladder network, the two Cauer forms are used to create these nonfinite zeros at $s = 0$ or $s = \infty$. In order to create a finite transmission zero, a pole must be removed from the driving point function at the desired frequency. Since a pole may not exist at this frequency, the driving point function must be modified to create a pole at the desired frequency. This modification must not change the overall transfer function. A process called zero shifting allows this objective to be accomplished.

The overall procedure for synthesizing a transfer function with finite transmission zeros will next be outlined and demonstrated. The procedure consists of the following steps:

1. Create even over odd or odd over even functions for y_{21} and y_{22} in the usual way.
2. Identify the locations of the transmission zeros.
3. Remove all poles at $s = 0$ and $s = \infty$ by the normal synthesis process.

Chap. 4   Synthesis of Filters

4. Remove an element from the remainder driving point function to create a zero at the frequency of the transmission zero.
5. Remove a pole from the reciprocal of the remainder function at the desired frequency. This will lead to a transmission zero at this point.

As an example of this procedure we will consider the transfer function

$$\frac{v_2}{v_1} = \frac{s^2 + 4}{s^3 + 2s^2 + 2s + 1}$$

This equation can be rearranged to find that

$$y_{22} = \frac{2s^2 + 1}{s^3 + 2s}$$

The transmission zeros are seen to be at $s = \infty$, $+j2$, and $-j2$. We remove a pole at infinity by applying the procedure of Cauer's first form as follows:

$$\frac{1}{2}s + \frac{\frac{3}{2}s}{2s^2 + 1}$$

$$2s^2 + 1 \overline{\smash{\big)}\, s^3 + 2s}$$
$$\phantom{2s^2 + 1 \overline{\smash{\big)}\,}} s^3 + \frac{1}{2}s$$
$$\phantom{2s^2 + 1 \overline{\smash{\big)}\,}} \overline{ \frac{3}{2}s}$$

The pole removal results in an inductance having a value of $L = 1/2$. The remaining impedance does not have a pole at $s = j2$, although it does have a pole at $s = +j0.707$. It is difficult to shift poles from one finite frequency to another because we are dealing with infinite quantities. It is much easier to deal with shifting zeros to a desired point. This accomplishes the same thing as shifting a pole, because the reciprocal function of one having a zero at a finite frequency will have a pole at this point.

We note that the impedance remainder in the preceding division has the form of a parallel resonant circuit, as shown in Fig. 4-17(a). As we examine this remaining impedance, we see that inverting this function results in the admittance

$$Y = \frac{2s^2 + 1}{\frac{3s}{2}}$$

This function has a finite zero at $s = j0.707$. At $s = j2$, Y has a finite value of admittance. We now assume that this admittance can be broken into two parts: one part

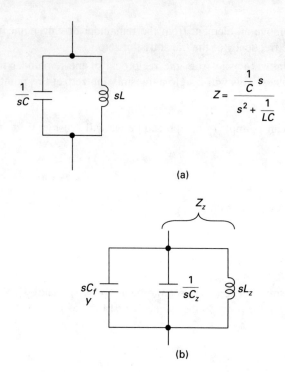

Figure 4-17 (a) A parallel circuit. (b) Shifting the pole component of the network in (a).

that has zero admittance at $s = j2$ and one part that leads to the finite admittance of Y at $s = j2$. This is shown in Fig. 4-17(b). We can write $Y = Y_z + sC_f$, where $Y_z = 0$ at $s = j2$.

The function sC_f must equal the value of Y at $s = j2$. This allows us to find the proper value of C_f since

$$j\omega C_f \bigg|_{s=j2} = \frac{2s^2 + 1}{\frac{3s}{2}} \bigg|_{s=j2}$$

Evaluating Y at $s = j2$ gives a value of

$$C_f = \frac{7}{6}$$

Knowing the value of C_f allows us to solve for Y_z. This gives

$$Y_z = Y - sC_f = \frac{2s^2 + 1}{\frac{3s}{2}} - \frac{7s}{6}$$

This admittance is found to be

Chap. 4 Synthesis of Filters

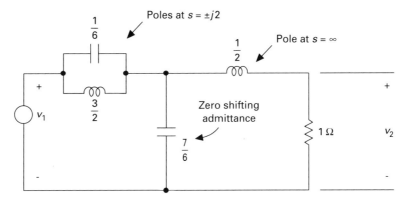

Figure 4-18 The finite zero filter of Example 4-4.

$$Y_z = \frac{s^2 + 4}{6s}$$

The reciprocal of Y_z has poles at $s = \pm j2$, which can be removed. This impedance is

$$Z_z = \frac{6s}{s^2 + 4}$$

and can be synthesized by a parallel LC circuit, as shown in Fig. 4-17.

For this particular function, values of

$$C_z = \frac{1}{6} \text{ and } L_z = 1.5$$

realizes the impedance Z_z. The resulting network is shown in Fig. 4-18.

We note that the shunt capacitor would normally cause a second zero at $s = \infty$, but due to the capacitor in the tank circuit, this is no longer the case. We further note that the circuit is not canonic. That is, there are four elements required to realize a third-order function. The zero-shifting capacitor leads to this extra element. An inductor shifts a zero to the left for an admittance and to the right for an impedance, that is, an inductor will lower the zero frequency for an admittance and raise it for an impedance. A capacitance shifts a zero left for an impedance and right for an admittance.

While this section on passive synthesis has included only a small fraction of the information on this topic, useful filters can be constructed using the methods discussed. This material is valid only if the source impedance is much smaller than the terminating resistance. This condition often results in practice if a low-impedance stage such as an emitter follower or voltage-feedback op amp drives the filter input.

Exercises

E4.2a Synthesize a voltage transfer function with a second-order Butterworth response. Assume the filter is loaded with a 1-kΩ resistor and the corner frequency is to be 20 kHz.

E4.2b Repeat exercise E4.2a for a second-order Chebyshev response. Assume a maximum ripple width in the passband of 1 dB.

4.3 LOW-PASS TO HIGH-PASS TRANSFORMATION

The low-pass filter is used more often than other types of filters in data acquisition systems. In situations that call for a high-pass or a band-pass response, it is possible to transform the low-pass transfer function into a high-pass or a band-pass function [2]. The filter can then be synthesized to obtain the desired response.

The key to transforming a low-pass function to a high-pass function is the variable transformation $s_l \rightarrow 1/s_h$, that is, the Laplace variable s in the low-pass function is replaced by $1/s$ to produce the high-pass transfer function. In the Cauer forms that include only a single element in each leg of the ladder, this transformation changes poles at $s = \infty$ to poles at $s = 0$. Since poles in the ladder network cause zeros of transmission, zeros of transmission move from $s = \infty$ to $s = 0$.

As a simple example of this transformation, the third-order low-pass Butterworth function will be transformed to a high-pass function. The low-pass function is given by

$$H(s_l) = \frac{-K}{s_l^3 + 2s_l^2 + 2s_l + 1}$$

Substituting $1/s_h$ for s_l leads to

$$H(s_h) = \frac{-K s_h^3}{1 + 2s_h + 2s_h^2 + s_h^3}$$

which is a high-pass response with a 3-dB point at $\omega = 1$ and a maximally flat response about $s_h = \infty$. This function was synthesized in Example 4-2 and shown in Fig. 4-14. The element values of the filter can be denormalized to create the desired 3-dB frequency.

The denormalized high-pass transfer function can also be found by using the transformation

$$s_l = \frac{\omega_c^2}{s_h}$$

on the denormalized low-pass function where ω_c is the 3-dB frequency or the cutoff frequency of both filters. This produces a high-pass transfer function with a 3-dB frequency or cutoff frequency of ω_c. This transfer function is called the denormalized transfer function. This method will be used in the following section.

Exercises

E4.3 Transform the low-pass filter realized in exercise E4.2a to a high-pass filter with the same corner frequency.

E4.3 Transform the low-pass filter realized in exercise E4.2b to a high-pass filter with the same corner frequency.

4.4 ACTIVE SYNTHESIS

Although a mild interest in active filters existed prior to the development of the integrated circuit, the major advancements in active filters have been motivated by a desire to construct integrated circuit filters. The results of that work have given the engineer more choices in selecting the type of filter to be used in the system. Completely integrated filters such as the switched capacitor filter might be used, a hybrid filter combining IC chips with discrete capacitors may be used, or a discrete component, active filter with transistors, resistors, and capacitors may be selected.

There are several reasons why the active filter is used. An active filter can eliminate inductors, decrease the size of required capacitors, obtain high values of Q in narrow-band applications, and occupy very little physical volume.

Of course, these advantages are accompanied by disadvantages. The active filter requires DC power, may experience stability problems, and may be sensitive to temperature change.

Elimination of inductors has been a goal of active synthesis for many years. Of the three passive elements used in filters, the inductor is the most nonideal element. It is used as an energy storage element, but it exhibits relatively high resistive energy loss compared to the capacitor. The inductor cannot be integrated on a silicon ship and must be eliminated from any design that will be synthesized as an integrated circuit. Even before the field of integrated circuits emerged, however, methods of avoiding inductors, especially those of large value, were developed.

One approach to eliminating conventional inductors from circuits is to simulate the inductor by means of integrated circuit components. This approach has not met with great success. A second approach is to simulate the response of circuits that use inductors. This is the more popular method of creating filter circuits.

Any passive circuit that has complex poles requires both inductors and capacitors. This means that all of the popular response types such as Butterworth, Chebyshev, Bessel, or elliptic can be achieved with passive circuits by using these two elements. In active filter design, a considerably different configuration might be used to create a resonant circuit without using an inductor. Several other configurations can be used to obtain these types of response. The following pages will only summarize a small number of approaches.

The idea of an active circuit can be understood by examining the RLC circuit shown in Fig. 4-19. If R is very large so that v_1 and R simulate a current source of value

$$i = \frac{v_1}{R}$$

the output voltage is

$$v_2 = i \times Z_{LC}$$

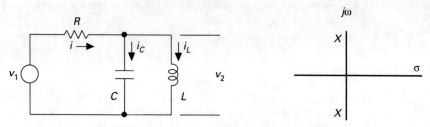

Figure 4-19 A parallel LC circuit and pole locations.

The current i consists of the sum of i_C and i_L. If

$$i_L = -i_C$$

then i equals zero, that is, no external current enters the LC combination, but the voltage v_2 is present. If a voltage appears across an impedance that has no entering current, the impedance is infinite.

At resonance, this ideal situation is only approximated due to losses in the inductor, capacitor, and source resistance, but a very high impedance results. We can write that

$$i_c = \frac{v_2}{1/sC} = sCv_2$$

and

$$i_L = \frac{v_2}{sL}$$

Substituting $j\omega$ for s and equating i_C to $-i_L$ allows the resonant frequency to be found. The result is

$$j\omega C v_2 = -\frac{v_2}{j\omega L}$$

at resonance, for which the resonant frequency is

$$\omega_o = \frac{1}{\sqrt{LC}}$$

The significant point here is that the current through the inductor is ideally 180° different from the current through the capacitor. It is impossible to get currents in adjacent elements of a passive RC or RL circuit to have a 180° phase difference. However, if positive feedback, implying a gain or active circuit, is used, this phase shift can be achieved. The circuit of Fig. 4-20 demonstrates this point.

The transfer function of this circuit can be found to be

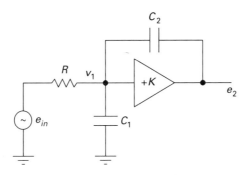

Figure 4-20 A positive feedback circuit.

$$\frac{e_2}{e_{in}} = \frac{K}{s\left[C_1 R + C_2 R - K C_2 R \right] + 1}$$

The negative sign in the denominator arises because a positive voltage v_1 can result in current entering this node through C_2 if K is greater than unity. There are three conditions on K that lead to three different results. If $KC_2 < (C_1 + C_2)$, then the transfer function appears to be that of a normal RC circuit. If $KC_2 = (C_1 + C_2)$, the function appears to be a purely resistive function. If $KC_2 > (C_1 + C_2)$, then the circuit can appear to be inductive. This circuit does not simulate an inductor, because the impedance value does not vary with frequency in the same way that an inductor does, but the current flow through C_2 is in the same direction. Although the circuit shown may have some stability problems, this basic idea can be used to create active circuits.

4.4.1 Bach Filters

The Bach filter [3] is not the most efficient method of creating a Butterworth or Chebyshev response due to the number of op amps required, but it is a method that allows easy extension to any number of poles desired. This method was originally developed to be used with vacuum tube circuits in the 1940s but has been extended to op amp stages. One op amp is required for each pole of the response if more than one pole is to be used.

Before we look at the Bach method, one point deserves emphasis. One might ask why a 4-pole circuit cannot be realized with four identical poles created by RC circuits and isolated by buffer stages such as those of Fig. 4-21. This circuit can be used in some cases, but it is not nearly as sharp as a 4-pole Butterworth filter, as shown in Fig. 4-22. This filter is not very appropriate for applications requiring a sharp transition region. Of course, the 4-pole Chebyshev function results in a sharper transition than does the Butterworth function.

Low-pass transfer function. Figure 4-23 indicates a network with a transfer function $T(s) = e_o/e_i$, where s is the Laplace variable. For sinusoidal excitation, this variable reduces to $s = j\omega$. There may be other transfer functions of interest, such as e_o/i_i or

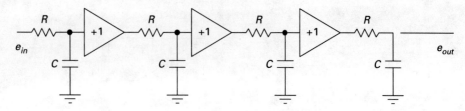

Figure 4-21 A 4-pole RC filter.

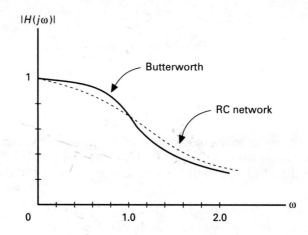

Figure 4-22 Comparison of Butterworth to RC response (4 poles).

Figure 4-23 Filter network.

i_o/e_i, but we will restrict the discussion here to the voltage transfer function. This function is useful in active filter design, since impedance levels are conveniently controlled by op amps.

For the Bach filter it is convenient to use the denormalized transfer functions. The transfer functions for maximally flat magnitude responses up to fourth order (four poles) are listed below. These are also called the denormalized Butterworth polynomials.

$$T(s) = \frac{K_1}{s + \omega_c} \tag{4-8a}$$

$$T(s) = \frac{K_2}{s^2 + 1.414\,\omega_c s + \omega_c^2} \tag{4-8b}$$

$$T(s) = \frac{K_3}{s^3 + 2\omega_c s^2 + 2\omega_c^2 s + \omega_c^3} \tag{4-8c}$$

$$T(s) = \frac{K_4}{s^4 + 2.613\omega_c s^3 + 3.414\omega_c^2 s^2 + 2.613\omega_c^3 s + \omega_c^4} \tag{4-8d}$$

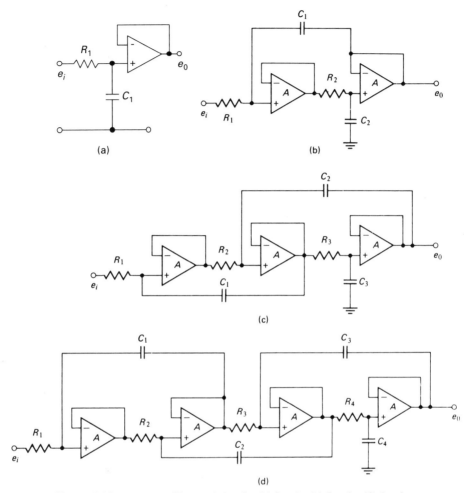

Figure 4-24 Low-pass filters: (a) 1-pole; (b) 2-pole; (c) 3-pole; (d) 4-pole.

The 3-dB cutoff frequency is designated by ω_c. Due to the availability of tabulated transfer functions it is generally necessary only to refer to a table to select the desired function. We must of course assume that the designer can relate the desired characteristics to the tabulated functions. The first part of the filter synthesis problem, that of selecting an appropriate transfer function, is therefore solved, and we can now proceed to the second part. The filter must now be designed to implement the selected function.

There are always several possible solutions to a given synthesis problem. One method of synthesis is based on the circuits of Fig. 4-24 [3]. An n-pole low-pass filter can be constructed by simply extending the pattern of Fig. 4-24 to n stages. If we define the parameters $\omega_1 = 1/R_1C_1$, $\omega_2 = 1/R_2C_2$, $\omega_3 = 1/R_3C_3$, and $\omega_4 = 1/R_4C_4$, the transfer functions for the four circuits become

$$T(s) = \frac{\omega_1}{s + \omega_1} \qquad (4\text{-}9a)$$

$$T(s) = \frac{\omega_1\omega_2}{s^2 + \omega_1 s + \omega_1\omega_2} \tag{4-9b}$$

$$T(s) = \frac{\omega_1\omega_2\omega_3}{s^3 + \omega_1 s^2 + \omega_1\omega_2 s + \omega_1\omega_2\omega_3} \tag{4-9c}$$

$$T(s) = \frac{\omega_1\omega_2\omega_3\omega_4}{s^4 + \omega_1 s^3 + \omega_1\omega_2 s^2 + \omega_1\omega_2\omega_3 s + \omega_1\omega_2\omega_3\omega_4} \tag{4-9d}$$

The transfer function for an n-pole filter can be found by extending these formulas.

If a maximally flat magnitude response is desired, the transfer function corresponding to the appropriate number of poles is selected from Eq. (4-8). The corresponding equation from Eq. (4-9) is used, and circuit element values are chosen to make Eq. (4-9) equal Eq. (4-8). An example will demonstrate this method.

Example 4-4

Design a low-pass, maximally flat magnitude response filter with an asymptotic rolloff of 18 dB/octave. The corner frequency should be 2 kHz. The source resistance is 100 Ω while the load resistance is 10 kΩ.

Solution. For Butterworth filters the rolloff rate is $6n$ dB/octave, where n is the number of poles of the transfer function. This design requires three stages to reach an 18 dB/octave rolloff. Equation (4-8c) corresponds to the transfer function for this case. For a 2-kHz bandwidth the equation becomes

$$T(s) = \frac{K_3}{s^3 + 8\pi \times 10^3 s^2 + 32\pi^2 \times 10^6 s + 64\pi^3 \times 10^9}$$

Equation (4-9c) can be made to equal this equation if $\omega_1 = 8\pi \times 10^3$, $\omega_2 = 4\pi \times 10^3$, and $\omega_3 = 2\pi \times 10^3$ and we assume that $K_3 = \omega_1\omega_2\omega_3 = 64\pi^3 \times 10^9$. The circuit of Fig. 4-24(c) has the appropriate transfer function if the element values are properly selected. Let us choose $R_1 = 10$ kΩ to create a relatively high input impedance compared to the 100-Ω source resistance. The required value of C_1 is

$$C_1 = \frac{1}{\omega_1 R_1} = 0.0398 \;\mu\text{F} \approx 0.04 \;\mu\text{F}$$

Selecting R_2 and R_3 as 10-kΩ resistances leads to values of $C_2 = 0.08$ μF and $C_3 = 0.16$ μF. The transfer function of this circuit is

$$T(s) = \frac{64\pi^3 \times 10^9}{s^3 + 8\pi \times 10^3 s^2 + 32\pi^2 \times 10^6 s + 64\pi^3 \times 10^9}$$

Figure 4-25 shows the frequency response of this filter.

Chap. 4   Synthesis of Filters

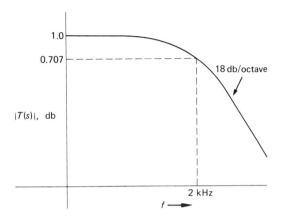

Figure 4-25 Frequency response of 3-pole Butterworth filter.

A buffer stage can be used to obtain more favorable loading conditions if necessary. However, there is a great deal of flexibility in impedance level with a circuit of this type. The input impedance can be made to have a high value by selecting R_1 to be large. The output impedance is determined by the output impedance of the last stage, and this value can be quite small. Low-frequency filters having corner frequencies of a few hertz can be constructed with practical values of capacitance by this method.

Starting with Bessel or Chebyshev polynomials results in the realization of maximally flat phase or equal-ripple response filters by this same method.

High-pass transfer function.   Given the low-pass transfer function for a desired response, it is a simple matter to obtain the high-pass transfer function yielding the same type of response using the transformation of Section 4.3. A frequency transformation that preserves the same bandwidth can be used to convert from the low-pass function to the high-pass function. The transformation introduced in the previous section requires that the variable s be replaced with ω_c^2/s, where ω_c is the bandwidth of both low-pass and high-pass functions. For example, the two-pole Butterworth response of Eq. (4-8b) will transform as follows upon applying the preceding frequency transformation:

$$\frac{K_2}{s^2 + 1.414\,\omega_c s + \omega_c^2} \rightarrow \frac{K_2}{\dfrac{\omega_c^4}{s^2} + 1.414\,\dfrac{\omega_c^3}{s} + \omega_c^2}$$

$$= \frac{(K_2/\omega_c^2)s^2}{s^2 + 1.414\,\omega_c s + \omega_c^2} = \frac{K_2' s^2}{s^2 + 1.414\,\omega_c s + \omega_c^2}$$

The Butterworth high-pass transfer functions are maximally flat about $s = \infty$, whereas the low-pass functions are maximally flat about $s = 0$. These high-pass functions up to the fourth order are as follows:

$$T(s) = \frac{sK'_1}{s + \omega_c} \tag{4-10a}$$

$$T(s) = \frac{s^2 K'_2}{s^2 + 1.414\,\omega_c s + \omega_c^2} \tag{4-10b}$$

$$T(s) = \frac{s^3 K'_3}{s^3 + 2\omega_c s^2 + 2\omega_c^2 s + \omega_c^3} \tag{4-10c}$$

$$T(s) = \frac{s^4 K'_4}{s^4 + 2.613\omega_c s^3 + 3.414\omega_c^2 s^2 + 2.613\omega_c^3 s + \omega_c^4} \tag{4-10d}$$

Again the 3-dB frequency is given by ω_c.

The circuits of Fig. 4-24 can be converted to high-pass filters simply by interchanging the positions of the resistors and capacitors. The transfer functions then become

$$T(s) = \frac{s}{s + \omega_1} \tag{4-11a}$$

$$T(s) = \frac{s^2}{s^2 + \omega_2 s + \omega_2 \omega_1} \tag{4-11b}$$

$$T(s) = \frac{s^3}{s^3 + \omega_2 s^2 + \omega_3 \omega_2 s + \omega_3 \omega_2 \omega_1} \tag{4-11c}$$

$$T(s) = \frac{s^4}{s^4 + \omega_4 s^3 + \omega_4 \omega_3 s^2 + \omega_4 \omega_3 \omega_2 s + \omega_4 \omega_3 \omega_2 \omega_1} \tag{4-11d}$$

The high-pass filters can be constructed by choosing the parameters $\omega_1, \omega_2, \ldots, \omega_n$ such that the applicable equation from Eq. (4-11) equals the appropriate equation from Eq. (4-10).

Bessel, Chebyshev, and other responses can be realized for the high-pass filter also, requiring the appropriate starting polynomial.

There are many other methods that can be used to synthesize active filters. Some methods require fewer op amps than the method discussed, and these approaches can be found in the literature [4, 5].

4.4.2 Band-Pass Filters

High-Q band-pass filters have traditionally used resonant circuits containing both capacitance and inductance. Pressure to develop inductorless band-pass filters has come not only from the integrated circuit area, but also from the low-frequency communication field. Sharp, low-frequency filters are very difficult to synthesize with inductors due to the low values of coil Q at low frequencies. Research on active band-pass filters predates the development of the integrated circuit, but this development has certainly led to intensified efforts in the active filter area.

There are several theoretical methods of synthesizing general transfer functions containing complex poles near the $j\omega$-axis. So much information is available in the litera-

Chap. 4   Synthesis of Filters

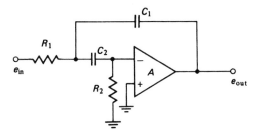

Figure 4-26 Sallen and Key resonator.

ture that even a brief review of this area cannot be attempted here. Thus the ensuing discussion will be limited to a consideration of second-order, highly selective band-pass filters that include the op amp as the basic component.

Sallen and Key resonator. The circuit shown in Fig. 4-26 was originally described by Sallen and Key [6]. The transfer function of this circuit is found to be

$$\frac{e_{out}}{e_{in}} = -\frac{\dfrac{A}{(A+1)R_1C_1}s}{s^2 + \dfrac{R_1C_1 + R_1C_2 + R_2C_2}{(A+1)R_1C_1R_2C_2}s + \dfrac{1}{(A+1)R_1C_1R_2C_2}} \quad (4\text{-}12)$$

where $s = j\omega$. For a second-order system of reasonably high Q, the transfer function can be expressed as

$$\frac{e_{out}}{e_{in}} = \frac{N(s)}{s^2 + s(\omega_n/Q) + \omega_n^2} \quad (4\text{-}13)$$

where ω_n is the resonant frequency. This parameter is actually the undamped natural frequency, but it differs little from the resonant frequency for sharply tuned circuits.

Equating coefficients of like powers of s in the denominator polynomials of Eqs. (4-12) and (4-13) results in

$$\omega_n = \frac{1}{\left[(A+1)R_1C_1R_2C_2\right]^{1/2}} \quad (4\text{-}14)$$

and

$$Q = \frac{\left[(A+1)R_1C_1R_2C_2\right]^{1/2}}{R_1C_1 + R_1C_2 + R_2C_2} \quad (4\text{-}15)$$

The gain of the circuit at resonance is

$$A_R = \frac{-AR_2C_2}{R_1C_1 + R_1C_2 + R_2C_2} \quad (4\text{-}16)$$

If $R_1 = R_2 = R$ and $C_1 = C_2 = C$, the preceding three equations simplify to the following:

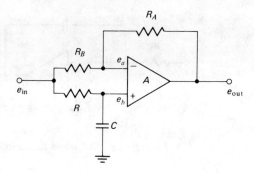

Figure 4-27 Op amp all-pass network.

$$\omega_n = \frac{1}{\sqrt{A+1}\,RC} \tag{4-17}$$

$$Q = \frac{\sqrt{A+1}}{3} \tag{4-18}$$

$$A_R = \frac{A}{3} \tag{4-19}$$

While it can be shown that the sensitivity of Q to any passive component is quite low, this circuit suffers from frequency limitations. If a resonant frequency of 1 kHz is required, good design practice dictates an amplifier bandwidth of at least 10 kHz. A gain of 40,000 with this bandwidth is rather typical of a high-quality op amp; thus the maximum Q obtainable is

$$Q = \frac{\sqrt{40{,}001}}{3} = 67$$

It can be shown that this circuit is limited to frequencies such that

$$100\omega_n Q^2 < \text{gain bandwidth of op amp} \tag{4-20}$$

Phase-shift filters. The last second-order system to be discussed is based on the all-pass network shown in Fig. 4-27. If loading effects of the op amp and capacitive losses are neglected, the output voltage is

$$e_{out} = A(e_b - e_a)$$

where

$$e_a = e_{in}\frac{R_A}{R_A + R_B} + e_{out}\frac{R_B}{R_A + R_B}$$

and

$$e_b = \frac{e_{in}}{1 + sCR}$$

Chap. 4   Synthesis of Filters   315

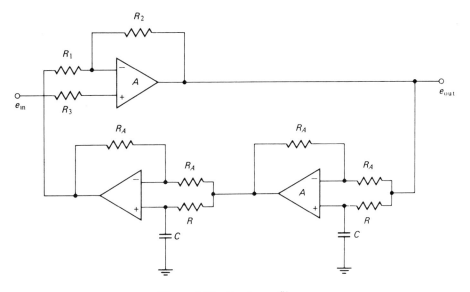

Figure 4-28   Band-pass filter.

The overall gain is

$$\frac{e_{out}}{e_{in}} = \frac{1 - \dfrac{sCRR_A}{R_B}}{1 + sCR} \frac{AR_B}{R_A + (A+1)R_B}$$

If R_A and R_B are selected to be equal, the gain expression reduces to

$$\frac{e_{out}}{e_{in}} = \frac{A}{A+2} \frac{1 - sCR}{1 + sCR} \tag{4-21}$$

The magnitude of the gain is $A/(A+2)$, which approximates unity for typical values of A, while the phase difference between the output and input signals is

$$\theta = -2 \tan^{-1} \omega CR \tag{4-22}$$

The phase shift varies from zero to $-180°$ as frequency increases. All magnitudes are passed by this network with unity gain, but phase shift is a function of frequency.

The all-pass network is an important element in several applications requiring phase control of signals. At this point, however, we are primarily interested in those features of this network that relate to active filters. Figure 4-28 shows a narrow-band filter that utilizes this all-pass element [1]. The transfer function for the filter is

$$\frac{e_{out}}{e_{in}} = \frac{s^2 + \dfrac{2}{RC} s + \dfrac{1}{R^2C^2}}{s^2 + \dfrac{2}{RC}\left(\dfrac{R_1 - R_2}{R_1 + R_2}\right) s + \dfrac{1}{R^2C^2}} \tag{4-23}$$

where we have assumed that $A \gg 2$. Comparing this transfer function to Eq. (4-13) results in

$$\omega_n = \frac{1}{RC} \qquad (4\text{-}24)$$

$$Q = \frac{R_1 + R_2}{2(R_1 - R_2)} \qquad (4\text{-}25)$$

and

$$A_R = \frac{R_1 + R_2}{R_1 - R_2} = 2Q \qquad (4\text{-}26)$$

where A_R is the overall resonant gain.

The selectivity and resonant gain of this filter depend strongly on the difference $R_1 - R_2$. For a Q of 100, the difference must be 1 percent. Resistances can be controlled closely enough to establish practical values of Q in the range of 100 to 200 for lumped or integrated circuits. If R_1 and R_2 are physically constructed from the same material, the temperature coefficients will be equal, minimizing temperature-sensitivity problems.

In the preceding analysis we have neglected loading of the input impedances of the op amp, the quality factor of the capacitor, and temperature variations of A. While it is necessary to select high-quality components, this circuit leads to a reasonably sharp band-pass filter.

4.4.3 Switched-Capacitor Filters

The switched-capacitor filter (SCF) is available on an IC chip. This device uses on-chip op amps and is thus limited to low-frequency filter applications. The SCF implements Butterworth, Chebyshev, and other responses. A typical off-the-shelf SCF might realize a seventh-order response. Present SCFs are limited to about 50 kHz; however BJT-CMOS versions will up this limit by a factor of 10. Experimental versions have realized narrow-band responses having a center frequency of several MHz, so the future of SCFs appears promising.

Exercises

E4.4a Design a 3-pole, low-pass, Butterworth filter with a 3-dB frequency of 6 kHz using the Bach method. The source resistance is 2 kΩ, while the load resistance is 6 kΩ.

E4.4b Design a 3-pole, high-pass, Chebyshev filter with a corner frequency of 10 kHz using the Bach method. The source resistance is 2 kΩ, while the load resistance is 6 kΩ.

GENERAL EXERCISES

4.1 Synthesize, using Cauer's first form, the impedance

$$Z(s) = \frac{s^2 + 1}{s^3 + 4s}$$

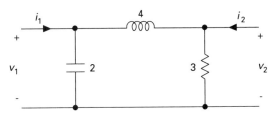

Figure PR4.5

4.2 Synthesize, using Cauer's second form, the impedance
$$Z(s) = \frac{s^2 + 4}{s^3 + 9s}$$

4.3 An LC impedance is
$$Z(s) = \frac{(s^2 + 1)(s^2 + 9)}{s(s^2 + 4)}$$

Realize this impedance using Cauer's first form.

4.4 Repeat problem 4.3 using Cauer's second form.

4.5 Evaluate the y parameters for the circuit of Fig. PR4.5.

4.6 Evaluate the z parameters for the circuit of Fig. PR4.5.

4.7 Realize an LC two-port network terminated in a 1-Ω resistor having a voltage transfer function of
$$\frac{V_2}{V_1} = \frac{Ks^2}{s^3 + 2s^2 + 2s + 1}$$

Plot the frequency response of V_2/V_1 over a range that includes $\omega = 0.1$ rad/s and $\omega = 10$ rad/s.

4.8 Realize an LC two-port network terminated in a 1-Ω resistor having a voltage transfer function of
$$\frac{V_2}{V_1} = \frac{K}{s^4 + s^3 + 34s^2 + 16s + 225}$$

4.9 Repeat problem 4.8 for
$$\frac{V_2}{V_1} = \frac{Ks^4}{s^4 + s^3 + 34s^2 + 16s + 225}.$$

4.10 a. Synthesize a 3-pole, low-pass Bessel filter having a 5-kHz, 3-dB frequency and a 2-kΩ load resistance.
b. Calculate the zero-frequency time delay of the synthesized filter.

4.11 Synthesize a 4-pole, low-pass Chebyshev filter having a 5-kHz, 3-dB frequency and a 2-kΩ load resistance. Assume a maximum ripple width in the passband of 0.5 dB.

4.12 A 3-pole, low-pass Butterworth filter and a 3-pole, low-pass Bessel filter are constructed both having a 3-dB bandwidth of 1 rad/s.
a. At what frequencies in the passband will each filter attenuate the input signal to a value of 0.9?

b. If the time delay of the denormalized Bessel filter is 80 µs, what is the denormalized 3-dB bandwidth?

4.13 Synthesize an LC filter terminated in a 1-Ω resistance to give
$$\frac{V_2}{V_1} = \frac{s^2}{s^4 + 10s^3 + 45s^2 + 105s + 105}$$

4.14 Synthesize an LC filter terminated in a 1-Ω resistance to give
$$\frac{V_2}{V_1} = \frac{s^2 + 9}{s^3 + 2s^2 + 2s + 1}$$

4.15 Synthesize an LC filter terminated in a 1-Ω resistance to give
$$\frac{V_2}{V_1} = \frac{s(s^2 + 4)}{s^4 + 3s^3 + 4s^2 + 3s + 1}$$

4.16 An LC filter terminated in a 1-Ω resistor has a transfer function of
$$\frac{V_2}{V_1} = \frac{N(s)}{2s^3 + s^2 + 3s + 1}$$

a. Choose $N(s)$ to create a high-pass transfer function.
b. Synthesize the high-pass filter.

4.17 Design a 4-pole, low-pass, Butterworth filter with a 3-dB point of 14 kHz using unity gain stages. The source resistance is 1 kΩ, while the load resistance is 10 kΩ.

4.18 Design a 4-pole, high-pass, Butterworth filter with a 3-dB point of 20 kHz using unity gain stages. The source resistance is 1 kΩ, while the load resistance is 10 kΩ.

REFERENCES

1. Johnson, D. E., *Introduction to Filter Theory*, Prentice Hall, Englewood Cliffs, NJ, 1976.
2. Van Valkenburg, M. E., *Analog Filter Design*, Holt, Rinehart and Winston, New York, 1982.
3. Comer, D. J., *Modern Electronic Circuit Design*, Addison-Wesley, Reading, MA, 1976.
4. Bruton, L. T., *RC-Active Circuits Theory and Design*, Prentice Hall, Englewood Cliffs, NJ, 1980.
5. Kennedy, E. J., *Operational Amplifier Circuits Theory and Applications*, Holt, Rinehart and Winston, New York, 1988.
6. Sallen, R. P. and E. L. Key, "A practical method of designing RC active filters." *IRE Transactions on Circuit Theory*, CT-2, March 1955.

5
Basic Error Considerations in Data Acquisition Systems

5.1 SIMPLE QUANTIZATION

Earlier chapters have considered the overall data acquisition system and its first three elements: the analog transducer, the amplifier, and the filter. We could now proceed to the device used to convert an analog to a digital signal (the analog-to-digital converter) or the device used to convert from a digital to an analog signal (the digital-to-analog converter). Rather than proceed directly to these topics, some general concepts relating to the conversion between analog and digital signals will be treated in this chapter. These concepts form the basis for the more specific coverage of converters in the following two chapters.

5.1.1 Quantum Interval and Error

A finite-length digital code can represent only a finite number of levels. On the other hand, an analog signal theoretically has an infinite number of values. When an analog signal is to be represented by a digital code, the range of the analog signal is divided into a finite number of levels. This division of an analog range into levels is called *quantization*. Figure 5-1 depicts an analog voltage quantized into equal width intervals.

The quantization can be defined in terms of the graph of Fig. 5-2(a). As the input varies from 0 to V_1, this signal is at level 0. Between V_1 and V_2, the signal is at level 1. Level 2 is entered as the input reaches V_2 and exited as this value reaches V_3. If the widths of the levels are equal, Fig. 5-2(b) can also be used to describe the relationship between input signal and level number. The width of the level is referred to as the quantum interval, q. This quantum interval is also equal to the LSB value. In Fig. 5-2(a)

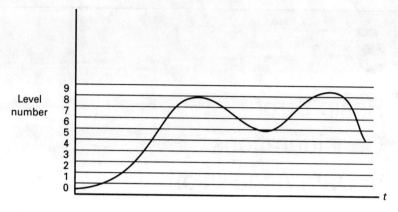

Figure 5-1 Quantization levels of an analog signal.

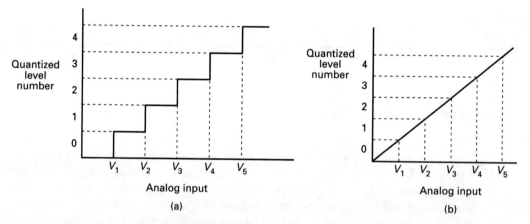

Figure 5-2 Analog input versus level number.

and (b), the quantum interval is $q = V_1$. Figure 5-3 shows a signal that has a quantum interval that increases with analog signal magnitude.

In converting the analog signal to a digital code, the digital code may simply represent the level at which the analog signal exists. For example, if a conversion is made at a time when the analog signal resides between levels 3 and 4, the output digital code should represent the number 3. The use of a level number to represent the analog signal leads to an inaccuracy called quantization error. This error can be evaluated from the graphs of Fig. 5-4.

The error is calculated as the difference between the actual analog signal and the value represented by the level number. The value represented by level number m is given by mq. In Fig. 5-4(a), level 0 represents an analog value of zero, and the error increases from 0 to q as the analog signal ranges from 0 to q. When the analog signal reaches q, the level number is 1. The error at this point again drops to zero, but it increases to q as the signal increases from q to $2q$. The greatest possible error is equal to q.

Chap. 5 Basic Error Considerations in Data Acquisition Systems

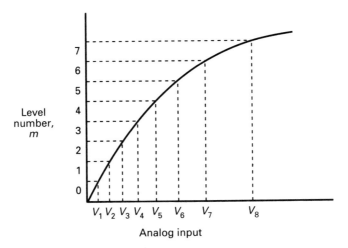

Figure 5-3 Variable quantum interval.

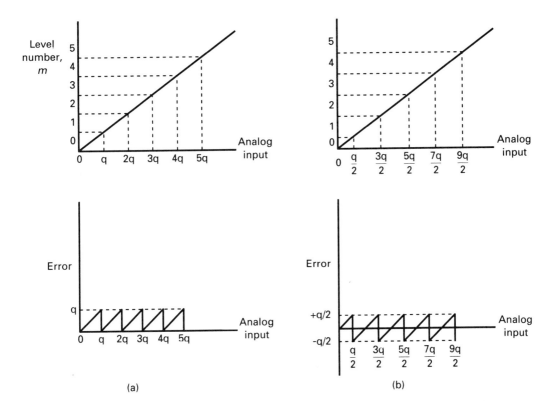

Figure 5-4 Analog input versus error. (a) Quantization error for equal quantum intervals. (b) Quantization error for first interval of $q/2$ or 1/2 LSB.

In Fig. 5-4(b) the first quantum interval is reduced to $q/2$ or 1/2 LSB, while all other intervals remain at a value of q [1]. For this case, the analog signal corresponds to level 1 at a value of $q/2$. Since level 1 represents a value of q, the error at this point is given by $-q/2$. As the analog signal increases from $q/2$ to $3q/2$, the error ranges from $-q/2$ to $+q/2$. In this situation, the maximum absolute error is $q/2$ rather than q. Most practical systems implement the scheme of Fig. 5-4(b). We can readily see that the quantization error in either of the methods depicted in Fig. 5-4 is minimized by minimizing the quantum interval q. The following paragraphs indicate the practical limitations relating to this minimization.

5.1.2 Span of an Analog Signal

The span of the analog signal to be converted is defined in terms of the minimum and maximum values of this signal. In a data acquisition system the transducer span may be quite different from the span of the analog signal to be converted. This results from the fact that the signal conditioning circuitry, that is, the amplifier and filter, may add gain and offset to the transducer output. We will assume in this section that the span of the analog voltage extends from 0 to v_{max}, that is, a unipolar signal system.

If we choose to represent the analog signal with an N-bit code, a maximum of 2^N levels can be numbered (including the zero level). The minimum quantum interval is calculated to be

$$q = \frac{v_{max}}{2^N} \tag{5-1}$$

This can be seen from Fig. 5-5. If v_{max} is made to correspond to the 2^Nth level, the total span is divided into 2^N levels, leading to Eq. (5-1).

The quantum interval, and thus the quantization error, can be minimized by increasing N to an appropriately high number. There are some obvious limitations to this approach. For example, the maximum number of bits may be limited by the computer or processor that follows the converter or by the converter itself. Furthermore, the lower

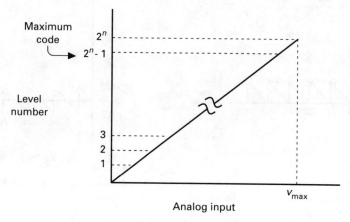

Figure 5-5 Output code as a function of analog input.

Chap. 5   Basic Error Considerations in Data Acquisition Systems

limit on the quantization error is influenced by the transducer and conditioning errors. It is unreasonable and expensive to make the quantization error much smaller than the other errors of the system.

Exercises

E5.1a An 8-bit ADC has an input range of 0 to 8 V. Calculate the quantization error for the converter.

E5.1b Repeat exercise E5.1a for a 10-bit ADC.

5.2 EFFECT OF ERRORS ON WORD SIZE

The number of bits used to represent an analog value influences the quantization error of the converter. The other errors of the system along with the quantization error determine the total system error. Too few bits in the conversion leads to a dominant quantization error. Too many bits leads to an expensive converter. This section considers the guidelines used to select a proper number or bits.

5.2.1 Quantization Error

Figure 5-6(a) indicates an analog signal with a constant error band, resulting from transducer and conditioning errors. The total error including quantization error is shown in Fig. 5-6(b). The number of bits used to represent the analog signal is the size of the digital word, and word size can be related to the relative error magnitudes. This can be seen from a consideration of the number of significant bits included in the digital word as the total number of bits is varied.

In this consideration we will assume that the transducer and conditioning errors are symmetrical about the true analog values. From Fig. 5-6(a) this assumption would lead to $\delta_2 = \delta_1 = \delta$. We will further assume that the quantization error can be set to equal δ by choosing the word size to be N_o. The quantization error, QE, of an N-bit system is given by

$$QE = q/2 = \frac{V_{max}}{2^{N+1}} \qquad (5\text{-}2)$$

For a word size of N_o, the quantization error is equated to δ, that is,

$$QE_o = q_o/2 = \frac{V_{max}}{2^{N_o+1}} = \delta \qquad (5\text{-}3)$$

If the word size is chosen to be less than N_o, the value of QE will be greater than that of δ. Since total error is the sum of δ and QE, the total error will be dominated by QE and approaches a value of $\pm q/2$ as N decreases. The LSB of the digital word will be inaccurate for this system. The number of meaningful bits will then be $N - 1$.

When N is increased to a value of N_o, the total error is

$$QE + \delta = \pm q/2 \pm q/2 = \pm q$$

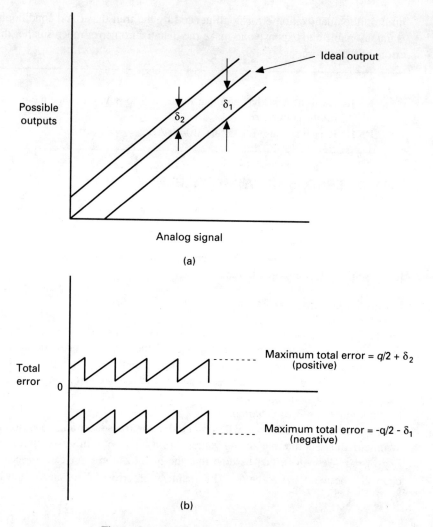

Figure 5-6 Maximum error after quantization.

As N increases above N_o, the error can be expressed as

$$QE + \delta = \pm \left[\frac{1 + 2^{N-N_o}}{2} \right] q$$

$$= \pm \left[\frac{1}{2} + 2^{N-N_o-1} \right] q \qquad (5\text{-}4)$$

The weighting of the N-bit code representing the analog signal as a function of bit position is

Bit position	N	...	i	...	3	2	1
Weighting	2^{N-1}	...	2^{i-1}	...	2^2	2^1	2^0

Chap. 5   Basic Error Considerations in Data Acquisition Systems

A 1 bit in the ith position represents a signal value of $2^{i-1}q$. Thus the error given in Eq. (5-4) will slightly exceed that given by the $N - N_o$ position. The error will cause $N - N_o + 1$ bits to be inaccurate. The number of meaningful bits in this case will be

$$N - (N - N_o + 1) = N_o - 1$$

When N is smaller than N_o, and the quantization error dominates the total error, the number of meaningful or significant bits varies with N. However as N reaches and exceeds N_o, the number of meaningful bits is a constant, equal to $N_o - 1$, determined primarily by transducer and conditioning error, δ. The accuracy of the converted word does not become higher as N increases. Since larger converters are more expensive, in general, a lower value of N is more desirable. The lower limit on N is N_o, and this value is often chosen for the word size. Solving Eq. (5-3) for N_o gives

$$N_o = \log_2 \left(\frac{v_{max}}{\delta} \right) - 1 \tag{5-5a}$$

$$= \ln \frac{\left(\frac{v_{max}}{\delta} \right)}{\ln 2} - 1 = 1.44 \ln \left(\frac{v_{max}}{\delta} \right) - 1 \tag{5-5b}$$

Of course, Eq. (5-5) will generally lead to a noninteger value of N_o and will be rounded up to the next-highest integer to obtain N.

Example 5-1

A transducer is specified by the manufacturer to have a $\pm 0.2\%$ least-squares linearity based on a lower end point of zero. The maximum output of the transducer is 10 V. Assuming that the linearity error of the transducer is the only system error other than quantization error, calculate a reasonable number of bits for the ADC.

Solution: We first compute the maximum transducer error from Eq. (1-17) to be

$$\Delta x_{max} = L_{ls} \, x_{max} = 0.002 \times 10 \text{ V} = 0.02 \text{ V}$$

From Eq. (5-5) and assuming an amplifier gain of A, N_o is found to be

$$N_o = 1.44 \ln \left(\frac{10A}{0.02A} \right) - 1 = 1.44 \ln (500) - 1 = 7.9$$

A value of $N = 8$ would be a reasonable choice for this transducer. We note that amplifier gain does not affect the calculation if both transducer span and error are amplified.

Example 5-2

Assume an amplifier with a gain of 4 follows the transducer of Example 5-1. If the output range of the amplifier is 0 to 15 V, calculate a reasonable word size.

Solution: In this system, the overall span of the transducer is not used. The error after amplification will be $4 \times 0.02 = 0.08$ V and v_{max} is 15 V. Equation (5-5) now gives

$$N_o = 1.44 \ln\left(\frac{15}{0.08}\right) - 1 = 6.54$$

In this case $N = 7$ would be a reasonable number of bits for the system.

In a typical system the transducer will exhibit other types of error in addition to linearity error. Noise sources and amplifier error will also degrade the accuracy of the analog signal to be converted. The concept of an error band was discussed in Chapter 1 and can be used to specify the total, worst-case error of the transducer. Amplifier errors were discussed in Chapter 2. Errors due to noise sources are more difficult to handle and will be discussed in the following paragraphs.

5.2.2 Noise Considerations

Several noise components, such as resistor noise and certain types of semiconductor noise, can be assumed to have a white noise characteristic. This means that the voltage and power spectra of these sources are flat with frequency. While a constant rms voltage can be found for a single source or several sources, as described in Section 1.6, this value reveals little about the instantaneous noise value. The random nature of the noise voltage leads to difficulty in evaluating the error voltage at the specific times of the analog-to-digital conversions. We now give the details of noise amplitude properties which were stated without proof in Chapter 1.

This uncertainty problem can be emphasized by considering the frequency spectrum of a unit impulse of voltage, $\delta(t)$. The Fourier integral is expressed as

$$F(j\omega) = \int_{-\infty}^{\infty} f(t)\, e^{-j\omega t} dt \tag{5-6}$$

The transform of $\delta(t)$ is given by

$$F(j\omega) = \int_{-\infty}^{\infty} \delta(t)\, e^{-j\omega t} dt = 1$$

The continuous amplitude spectrum for the unit impulse is shown in Fig. 5-7. This impulse function contains all frequencies with equal amplitudes.

The inverse transform integral is expressed as

$$f(t) = \frac{1}{2\pi} \int_{-\infty}^{\infty} F(j\omega) e^{j\omega t} d\omega \tag{5-7}$$

For the unit impulse, the inverse transform integral becomes

$$\delta(t) = \frac{1}{2\pi} \int_{-\infty}^{\infty} e^{j\omega t} d\omega$$

A constant continuous amplitude spectrum transforms to a unit impulse.

It was mentioned that the continuous amplitude spectrum of white noise is also constant, thus one might reasonably ask if the time response of noise is an impulse. From

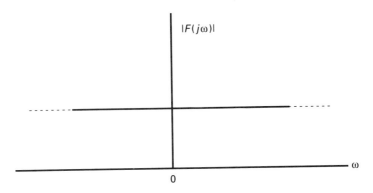

Figure 5-7 Amplitude spectrum for the unit impulse.

a theoretical standpoint, white noise components do not add together to create a single impulse function. The value of $F(j\omega)$ is constant for the impulse function, and the phase of $F(j\omega)$ is zero. Implied in the operation of forming the inverse transform of the constant spectrum is this zero phase for $F(j\omega)$ at all values of ω. The result of the inverse transform is the ideal impulse function.

Band-limited white noise from several sources will not exhibit this phase relationship, and an ideal impulse does not result. However, the random phases of the components will vary with time in such a way that white noise will generate random amplitude spikes. As the phases of many components approach a zero value, or a value that leads to a linear phase shift with frequency, a spike that grossly approximates an impulse occurs in the white noise time function. This phenomenon is quite visible when white noise is observed on an oscilloscope.

The random nature of amplitude spikes in white noise leads to a consideration that is different from that of most errors. For many error sources, a maximum limit can be used to execute worst-case design. The analog-to-digital conversion process does not convert the rms or any other fixed value of noise. Rather it converts the instantaneous values that exist at the times of conversion. Consequently, a statistical approach to the noise problem is necessary. Therefore, we will briefly review some pertinent facts relative to probability and white noise.

The probability density function describes the probability that a signal will assume a value within some defined range at any instant of time [2]. The waveform of Fig. 5-8 demonstrates this concept.

The probability that $x(t)$ falls within the range x to $(x+\Delta x)$ can be found by summing all values of Δt over a total time T and dividing by T. As T becomes large, the probability calculated this way becomes more accurate. The total time T can be allowed to approach infinity in the limit, giving

$$Prob\,[\,x < x(t) \le (x+\Delta x)\,] = \lim_{T \to \infty} \frac{\sum_i \Delta t_i}{T}$$

For small values of Δx, a probability density function $p(x)$ can be defined in terms of this probability as

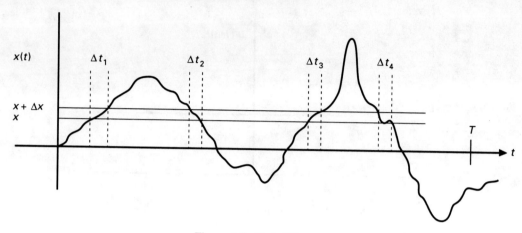

Figure 5-8 Probability measurement.

$$p(x)\,\Delta x = Prob\,[\,x < x(t) \le (x + \Delta x)\,]$$

As Δx is allowed to approach zero, $p(x)$ is accurately given by

$$p(x) = \lim_{\Delta x \to 0} \frac{Prob\,[\,x < x(t) \le (x + \Delta x)\,]}{\Delta x} \tag{5-8}$$

The probability distribution function, $P(x)$, is used to calculate the probability that a signal exists above or below some particular constant value. For example, the probability that $x(t)$ has a value less than or equal to x_1 is

$$P(x_1) = Prob\left[\,x(t) \le x_1\,\right] = \int_{-\infty}^{x_1} p(x)dx \tag{5-9}$$

The probability distribution function has a maximum value of 1. The probability that $x(t)$ falls within a range of x_1 to x_2 is found from

$$P(x_2) - P(x_1) = Prob\left[\,x_1 < x(t) \le x_2\,\right] = \int_{x_1}^{x_2} p(x)dx \tag{5-10}$$

For white noise, the probability density function is given by the Gaussian equation

$$p(x) = \frac{1}{\sigma\sqrt{2\pi}}\,e^{-x^2/2\sigma^2} \tag{5-11}$$

where σ is the rms value of the waveform. The probability that the function exists outside the limits of $\pm x_p$ is expressed by

$$Prob\left[\,|x(t)| > x_p\,\right] = 1 - Prob\left[\,-x_p < x(t) \le x_p\,\right]$$

$$= 1 - \frac{1}{\sigma\sqrt{2\pi}}\int_{-x_p}^{x_p} e^{-x^2/2\sigma^2}\,dx \tag{5-12}$$

Chap. 5 Basic Error Considerations in Data Acquisition Systems

TABLE 5-1 PROBABILITY THAT $|x(t)|$ EXCEEDS x_p FOR VARIOUS VALUES OF x_p

x_p	Prob. that $\|x(t)\|$ exceeds x_p
0.253 σ	0.8
0.842 σ	0.4
1.645 σ	0.1
2.054 σ	0.04
2.576 σ	0.01
3.090 σ	0.002
3.719 σ	0.0002
4.265 σ	0.00002
4.753 σ	0.000002

From a statistical table [3] the value of the area under the normal curve (the integral term) is evaluated. Table 5-1 shows a few values for the probability as a function of x_p.

The probability of $|x(t)|$ exceeding a value of 3.72 σ is 0.0002. From a practical standpoint a value of 4 σ is exceeded by $|x(t)|$ for less than 0.0001 or 0.01% of the total signal duration. This level of uncertainty is generally considered acceptable in most engineering applications. The worst-case error due to white noise is taken to be four times the rms value. This worst-case error value can then be added to the other worst-case errors of the system to arrive at a total error value. The following example demonstrates the use of the noise error in calculating an appropriate word size.

Example 5-3

A unipolar transducer with a full-scale output of 5 V is used in a data acquisition system. The maximum end-point linearity error is 8 mV, and the equivalent resistance of the transducer is 200 kΩ. If the circuit bandwidth is 100 MHz with a 1-pole response, what is a reasonable word size? Assume that $T = 297°$ K and neglect other possible error sources.

Solution: The resistance noise is first calculated from Eq. (1-56), using $B = 1.57 \times 100$ MHz. The rms noise voltage is

$$e_n = \sqrt{4 \times 1.38 \times 10^{-23} \times 297 \times 2 \times 10^5 \times 1.57 \times 10^8} = 7.17 \times 10^{-4} \text{ V}$$

The worst-case noise error is then taken to be $4e_n$, which is 2.87 mV.

The total error is the sum of linearity error and worst-case noise error, or $\delta = 10.87$ mV.

Equation (5-5) is now used to find N. Since

$$N_o = 1.44 \ln\left(\frac{5}{10.87 \times 10^{-3}}\right) - 1 = 7.83$$

then N is selected to be 8.

5.2.3 Signal-to-Noise Ratio

Although the signal-to-noise ratio (S/N) is not always used to specify data acquisition systems, some insight can be gained from a consideration of this ratio. We will first look at the S/N caused solely by quantization error.

For a maximum input signal of v_{max} the quantization error is

$$QE = \frac{v_{max}}{2^{N+1}} \qquad (5\text{-}2)$$

The maximum value of S/N is then

$$S/N = \frac{v_{max}}{QE} = 2^{N+1} \qquad (5\text{-}13)$$

When the signal is smaller than v_{max}, the S/N is smaller than the value given by Eq. (5-13), since the quantization error remains constant as the signal decreases.

Certain digital systems in the telecommunications field use a variable quantum interval to maintain a constant S/N over a 60-dB range of amplitudes. In this case each quantum is selected to be [4]

$$q_i = k \log i$$

The value of QE is now less for smaller signals than it is for large signals. The S/N is then approximately constant.

Rather than build a converter with a variable quantum interval, a method called *companding* is used to accomplish the same effect by modifying the input voltage signal. A signal compressor circuit has a nonlinear gain characteristic, as shown in Fig. 5-9. The transducer output is applied to the compressor input. The change in output voltage for a given input change is greater at low input levels than for high input levels.

The relationship between v_{out} and v_{in} is relatively complex in practical companding systems. In North America, the relationship is called the µ-law and can be expressed as

$$v_{out} = V_1 \frac{\ln(1 + \mu\, v_{in}/V_1)}{\ln(1 + \mu)}$$

where $\mu = 255$ and V_1 is the maximum input and also the maximum output compressed voltage [4]. A constant quantum interval for v_{out} now leads to an effective quantum interval for v_{in} that is nonlinear. If a signal is to be processed and then restored to its original waveshape, an expander with a gain characteristic that is the inverse of the µ-law is used following the digital-to-analog conversion.

The concept of companding is particularly useful for signals that are to be transmitted over phone lines or other transmission lines. The long lines are susceptible to picking up noise which could obscure small quantum intervals. On the other hand, if a large quantum interval is selected to minimize noise error, the high amplitude signals are difficult to convert. The nonlinear quantum interval solves this problem using larger values for conversion of smaller signals and smaller intervals for conversion of larger signals. We will consider only those systems with fixed or constant quantum intervals in the remainder of this textbook.

Chap. 5   Basic Error Considerations in Data Acquisition Systems   331

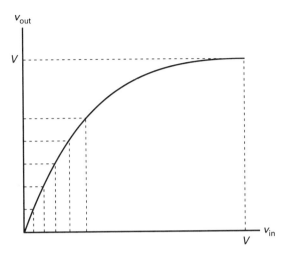

Figure 5-9 Nonlinear amplifier with compression characteristics.

With a constant quantum interval, Eq. (5-13) implies that the S/N can become arbitrarily high as N is increased. This would be true if quantization error were the only error of the system. Of course, this is rarely true for a practical system. If the worst-case value of transducer and conditioning errors is given by δ, and QE_o is set to equal δ by Eq. (5-3), the total error in the system is

$$2\delta = \frac{V_{max}}{2^N}$$

The S/N is now

$$S/N = \frac{V_{max}}{2\delta} = 2^N$$

where N is limited to the value given by Eq. (5-5b). Thus the S/N is limited by the transducer and conditioning errors which also determine an upper limit on N.

Exercises

E5.2a A transducer has an output range of 0 to 10 V and a maximum error of 0.02 V. Specify a reasonable number of bits for the ADC that converts this signal.

E5.2b The rms value of a white noise voltage is 1 mV. In a 1-second duration of this voltage, how many milliseconds will this voltage exceed a value of 3.09 mV?

5.3 SELECTION OF CONVERTER SIZE

Every data acquisition system may present different problems to the designer, due to the different specifications that serve as the starting point of the design. There are two general specifications that have a major effect on the design. The first specification is the number of bits needed for the ADC. In many systems, this number is fixed before the system design begins. For example, the pulse-code modulation scheme used to transmit long distance conversations over phone lines is an 8-bit system. Most digital music synthesizers use 8- to 12-bit words. Compact disk (CD) systems have standardized on 16-bit digital words.

When the number of bits to be used is known, the transducer selection, amplifier design, and filter design can be based on the criterion that the total system error should be less than or equal to 1/2 LSB. This topic has been discussed extensively in the preceding chapters.

In other systems the transducer specification may serve as the design starting point. A particular transducer may be the only one with the desired sensitivity or desired cost, and the remainder of the system must be designed around the transducer. If it is important that N be as large as possible, for accuracy purposes, care must be taken to minimize system errors so that N_o, calculated from Eq. (5-5), is large.

A third possibility is that N and the transducer are both specified. The allowable system error that can be introduced by signal conditioning circuitry can then be calculated, based on the 1/2 LSB criterion.

5.4 OVERVIEW OF DATA ACQUISITION SYSTEMS

The overall problem of acquiring digital data from a continuous input signal was discussed in the first chapter of this textbook. With the background of Chapters 1 through 4 in mind, it is appropriate to return now to a general system to emphasize certain key points. Figure 5-10 shows the block diagram of a general system.

The amplifier and first filter make up the signal conditioning section of the system. This section typically modifies the DC level of the transducer signal, filters noise from the analog signal, and limits the bandwidth to an appropriate value before sampling takes place. The input to the ADC consists of the desired transducer signal (properly conditioned) plus transducer errors, amplifier errors, and frequency response errors. Examples of transducer errors are nonlinearities, noise, and hysteresis errors. Examples of amplifier errors are offset, common mode, slew rate, and noise errors. The filter may introduce frequency response errors, phase distortion, and noise errors.

The ADC may have an analog multiplexer at the input allowing more than one channel to be converted to digital values with a single ADC. Many systems convert only one value and have no need of a multiplexer. In some systems a sample-and-hold circuit precedes the actual conversion from analog to digital. This conversion introduces quantization error, aperture or conversion error, and multiplexer error.

The sample rate is related to the highest frequency of importance contained in the transducer signal and to the desired accuracy. The first filter bandwidth relates to these parameters, as discussed in Chapters 3 and 4. The lowest possible sampling rate should be chosen to minimize storage needs, but for accuracy purposes, a sampling rate of 2.5 to 5 times the highest desired frequency often results.

The output of the ADC is then stored and processed by some type of digital system. Most often this is a computer with both RAM and disk storage capabilities. The acquired signal could be used in a variety of ways to fulfill the purposes of the system. It may be compressed and stored, as in voice or data "store and forward" systems. In such systems, the voice or data information is used to reconstruct the original analog input at some later time. In other systems, the digital processor might simply delay and attenuate the input,

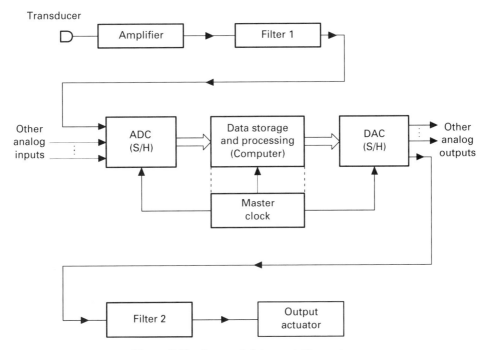

Figure 5-10 A general data acquisition system.

then add the delayed signal to the original. The reconstructed analog signal is an echo or reverberation signal. Still other systems may analyze the digital information to identify certain features. In some cases, the output may consist of printed values. In other instances, the data is processed or manipulated in order to produce a desired output signal. In these types of system the digital output code must often be converted back to an analog signal to drive an actuator.

The most popular system for reconstruction of a signal is the zero-order or sample-and-hold circuit followed by a low-pass filter. The DAC converts the digital output from the computer and holds the analog output value until the next digital value is converted. This operation represents the sample-and-hold function. The low-pass filter eliminates rapid transient changes in this reconstructed signal. In some cases, the actuator to be driven exhibits some filtering capability. A DC motor attenuates or eliminates high-frequency armature voltage components simply because the relatively high shaft inertia cannot respond to high frequencies.

There may be more than one output channel using some sort of multiplexing scheme in a small percentage of applications. More complex reconstruction schemes may also be used by the computer to achieve a more accurate output signal. These methods will be considered in a later chapter.

5.5 WAVEFORM RECONSTRUCTION

All previous material of this textbook has been concerned with acquiring analog signals and converting them to digital form. After such a signal is processed, the resulting digital words must be converted back to an analog signal. This process is called *waveform*

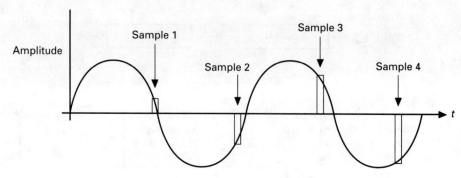

Figure 5-11 Reconstruction with a low sample rate.

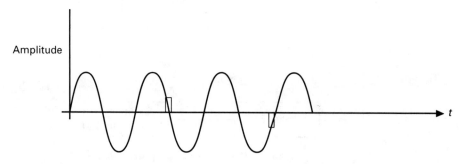

Figure 5-12 One possible waveform fitting two samples.

reconstruction. This section will consider a few methods of reconstructing the analog output signal.

The Nyquist sampling theorem, presented in Chapter 4, tells us that a waveform can be accurately reconstructed from a sequence of sampled values, if the sampling rate exceeds $2f_{max}$ where f_{max} is the highest frequency present in the signal.

In practice, it is exceptionally difficult to achieve an accurate reconstruction of a component of frequency f_{max} with a sampling rate that slightly exceeds $2f_{max}$. A very long delay is required to accurately reconstruct the waveform. For example, consider the waveform of Fig. 5-11. After only two samples, the first period of the waveform cannot be reconstructed, because several different sinusoids could pass through these two points, as shown in Fig. 5-12.

As more sample values are considered, the incorrect higher-frequency waveforms cannot fit all of the sample points. After many samples, the first, second, third, and other cycles of the original waveform can be constructed.

In addition to the fact that many sample points must be considered before accurate reconstruction can be done with low sampling rates, the system that performs the reconstruction is quite complex. Most practical systems accept errors in reconstruction but are relatively simple and lead to little or no delay.

Another consideration to be made relates to the necessity of band-limiting the frequency of an input signal. An ideal filter with no attenuation in the passband and 100% attenuation in the stopband is never available. The filtering function must be performed by a practical filter. As the cutoff frequency is approached from the passband

Chap. 5 Basic Error Considerations in Data Acquisition Systems   335

region, finite attenuation takes place. We recall from Chapter 3 that the 3-dB frequency of a Butterworth filter relates to the 1/2 LSB error frequency by

$$f_{1/2\,\text{LSB}} = f_c \left[\frac{2^{N+1} - 1}{2^{2N} - 2^{N+1} + 1} \right]^{1/2n} \tag{3-34}$$

$$\approx f_c \times \frac{1}{2^{(N-1)/2n}} \tag{3-35}$$

where N is the number of bits used for the digital word, n is the order of the filter, and f_c is the 3-dB frequency of the filter. For a 1-pole filter and a 10-bit word size, the 3-dB frequency must be 22.6 times the frequency $f_{1/2\,\text{LSB}}$. If a fourth-order Butterworth filter is used, f_c need only be 2.2 times $f_{1/2\,\text{LSB}}$. The frequency $f_{1/2\,\text{LSB}}$ is generally taken to be the highest frequency that we want to be within 1/2 LSB accuracy. We will emphasize that Eqs. (3-34) and (3-35) assume a high-frequency component that equals one-half of the total span. Generally, this component is much smaller, making these calculations very conservative.

We see that if a low-performance filter, such as a 1-pole filter, is used, the sampling frequency must be many times the frequency $f_{1/2\,\text{LSB}}$. A typical sampling frequency for this case might be 2.5 to 3 times f_c or 45 to 70 times $f_{1/2\,\text{LSB}}$.

We will find that the reconstruction or recovery sampling frequency must also exceed the highest frequency by a large factor to be accurately reconstructed, if a low-performance filter or no filter is used.

One of the simplest and most popular methods of reconstruction is shown in Fig. 5-13. The DAC output may contain transient glitches which can be eliminated by the S/H module. This method is called zero-order hold reconstruction. Typically the conversion is made by the DAC on the positive clock transition while the S/H module holds on the alternate or negative clock transition. This allows the DAC to settle before the sample is taken.

If $V(t) = A \sin \omega t$ is a component to be recovered using zero-order hold with a recovery sample time of T_{rs}, the diagram of Fig. 5-14 can be used to evaluate the maximum error.

The maximum slope of the curve occurs at $t = 0$ and is $A\omega$. Assuming T_{rs} is much smaller than T, the signal period, the error just prior to the first sample is approximately

$$V_{Emax} = T_{rs} A \omega$$

The error in terms of a percentage of A is

$$\text{percent error} = 100\omega\, T_{rs} = 200\, \pi f T_{rs} = 200\, \pi \frac{f}{f_{rs}} \tag{5-14}$$

where f is the signal frequency and f_{rs} is the recovery sample frequency. Solving for f_{rs} gives

$$f_{rs} = \frac{200\, \pi f}{\text{percent error}} \tag{5-15}$$

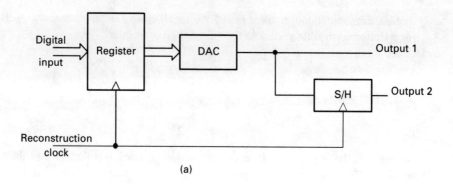

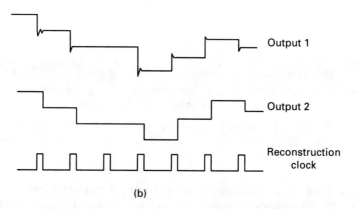

Figure 5-13 (a) Zero-order reconstruction circuit. (b) Outputs.

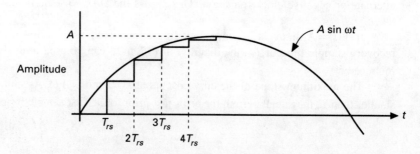

Figure 5-14 Waveform used to evaluate error.

Using only the zero-order hold circuit to reconstruct a signal component of $f = 20$ kHz, a 5% maximum error in this component will result for a recovery sample rate of

$$f_{rs} = \frac{200\,\pi \times 20{,}000}{5} = 2.51 \text{ MHz}.$$

This recovery sample frequency must be a factor of 125 greater than the signal frequency for a 5% maximum error.

Chap. 5   Basic Error Considerations in Data Acquisition Systems

Again we should recognize that Eq. (5-15) gives worst-case results because the high-frequency components of most signals have magnitudes that are small percentages of the total signal span. Furthermore, this equation uses a maximum error rather than an average or rms error. In music, a 20-kHz component may have an amplitude that is 1/1000 of the total span of the signal. A 5% maximum error in this component amounts to only 1/20,000 of the span, which is generally insignificant in sound reproduction.

A more realistic calculation defines the fraction of the total span that is made up by the maximum magnitude of the highest-frequency component. This value is given by

$$x = \frac{A_{max}}{V_{span}/2} \tag{5-16}$$

The maximum error for this signal using zero-order reconstruction is

$$V_{Emax} = T_{rs} A_{max} \omega = \omega T_{rs} x V_{span}/2$$

The error level of normal significance is the 1/2 LSB level, which can be expressed as

$$e_{1/2\,LSB} = \frac{V_{span}}{2^{N+1}} \tag{5-17}$$

Equating the maximum error to $e_{1/2\,LSB}$ allows the recovery sample rate to be found as

$$f_{rs} = 2^{N+1} \pi f x \tag{5-18}$$

Example 5-4

A signal with a maximum frequency of 20 kHz is to be reconstructed by a zero-order hold circuit. The 20-kHz component has a maximum magnitude of 0.1 V, while the span is 10 V. Calculate the recovery sample rate for 1/2 LSB error, assuming a 12-bit system.

Solution: The value of x is

$$x = \frac{0.1}{5} = 0.02$$

Substituting this value into Eq. (5-18) results in

$$f_{rs} = 2^{13} \times \pi \times 20 \times 10^3 \times 0.02 = 10.3 \text{ MHz}$$

This sample rate is 515 times the highest signal frequency.

Obviously, if great accuracy in reproducing high-frequency components is necessary, the zero-order hold circuit is somewhat impractical. The required high sample recovery frequency can be lowered considerably by following the DAC-sample/hold system by a low-pass filter. In general, the higher the order of the filter, the lower will be the required recovery sample frequency.

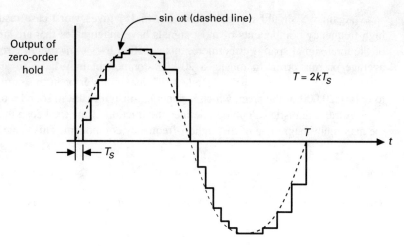

Figure 5-15 Signal used for Fourier analysis.

The error resulting from a filtered zero-order hold system can be found by using Fourier theory to evaluate the magnitude of the components of the zero-order signal. The effects of the filter on these components can then be evaluated to find the expression for the reconstructed signal. Figure 5-15 shows the zero-order hold waveform before filtering.

In order to simplify the analysis, an even number of samples per cycle of the signal waveform is assumed. This number of samples in a full cycle is $2k$, consequently $T = 2k\,T_s$. With this assumption, the zero-order signal is an odd function that will consist of only sinusoidal terms in the Fourier series [5]. The coefficients of these terms can be found from the well-known formula [5]

$$b_n = \frac{4}{T}\int_0^{T/2} f(t)\sin n\omega t\, dt \qquad (5\text{-}19)$$

where $f(t)$ is the signal to be expressed in Fourier series form. This signal is constant in each sampling interval and can be written as

$$f(t) = 0]_0^{T_s} + \sin\omega T_s]_{T_s}^{2T_s} + \ldots + \sin(k-1)\omega T_s]_{(k-1)T_s}^{kT_s}$$

$$= \sum_{i=1}^{k} \sin(i-1)\,\omega T_s]_{(i-1)T_s}^{iT_s}$$

The square bracket indicates the interval over which each term is valid.

The coefficient b_n can be evaluated from Eq. (5-19) as

$$b_n = \frac{4}{T}\int_0^{T/2}\left(\sum_{i=1}^{k}\sin(i-1)\,\omega T_s]_{(i-1)T_s}^{iT_s}\right)\sin n\omega t\, dt$$

TABLE 5-2 COEFFICIENTS OF FIRST FOUR COMPONENTS IN FOURIER SERIES

k	b_1	b_2	b_3	b_4
2	0.637	−0.637	0.212	1.43×10^{-10}
3	0.827	−0.413	-2.57×10^{-6}	−0.207
4	0.900	−0.318	-1.28×10^{-6}	−0.132
5	0.935	−0.259	-7.93×10^{-7}	−0.105
6	0.955	−0.217	-5.4×10^{-7}	−0.087
8	0.974	−0.165	-3.01×10^{-7}	−0.066
10	0.984	−0.132	-1.92×10^{-7}	−0.053
20	0.996	−0.067	-4.7×10^{-8}	−0.027
40	0.999	−0.033	-1.18×10^{-8}	−0.013
100	1.000	−0.013	-1.95×10^{-8}	−0.005

Since each term in the summation has a constant value only when t is in a specific interval and is zero when t falls outside this interval, the expression for b_n can also be written as

$$b_n = \frac{4}{T}\left(\sum_{i=1}^{k} \sin(i-1)\omega t \int_{iT_s}^{(i-1)T_s} \sin n\omega t\, dt\right)$$

$$= \frac{4}{T}\left(\sum_{i=1}^{k} \sin(i-1)\omega t \left[-\frac{\cos n\omega t}{n\omega}\right]_{(i-1)T_s}^{iT_s}\right)$$

$$= \frac{2}{n\pi}\left(\sum_{i=1}^{k} \sin(i-1)\omega t\, [\cos(i-1)n\pi/k - \cos ni\pi/k\,]\right)$$

Table 5-2 shows the coefficients of the fundamental, second, third, and fourth harmonics for several values of k.

This table allows us to write the first four components for the corresponding value of k. These components approximate the zero-order hold waveform for $\sin \omega t$. For example, a sample rate that is four times the signal frequency ($f_{rs} = 4f$, $k = 2$) results in an approximate waveform of

$$f_a(t) = 0.637 \sin \omega t - 0.637 \sin 2\omega t + 0.212 \sin 3\omega t$$

It should be understood that this approximate waveform may not accurately represent the actual zero-order hold waveform, especially for lower values of k. In such cases, the higher-order components are not negligible. This approximation becomes more accurate when the waveform is applied to a low-pass filter that greatly attenuates the higher harmonics, rendering all but the first few components negligible. In some cases, only the first two components are needed to calculate the error.

In order to calculate the error, we will assume that the highest-frequency component of interest has a frequency of f and a magnitude of A. We can again define x (bipolar) as

$$x = \frac{A}{v_{span}/2}$$

The perfectly constructed waveform would be

$$f(t) = \frac{x \, v_{span}}{2} \sin \omega t$$

The zero-order signal for $k = 4$ ($f_{rs} = 8f$) can be represented by

$$f_a(t) = 0.900 \sin \omega t - 0.318 \sin 2\omega t - 0.132 \sin 4\omega t$$

We note that the third harmonic component is negligible compared to the other three components. If this signal is now filtered by an n-pole Butterworth low-pass filter, the output can be expressed as

$$f_{af}(t) = \frac{x \, v_{span}}{2} \left[\frac{0.900}{\sqrt{1 + (f/f_c)^{2n}}} \sin(\omega t + \Theta_1) - \frac{0.318}{\sqrt{1 + (2f/f_c)^{2n}}} \sin(2\omega t + \Theta_2) \right.$$

$$\left. - \frac{0.132}{\sqrt{1 + (4f/f_c)^{2n}}} \sin(4\omega t + \Theta_4) \right]$$

where f_c is the 3-dB frequency of the filter. The phase shifts introduced by the filter are represented by $\Theta_1, \Theta_2,$ and Θ_4.

If f_c is selected to be $2f$ and a 4-pole filter is used, the filter output for $k = 4$ becomes

$$f_{af}(t) = \frac{x \, v_{span}}{2} \left[0.898 \sin(\omega t + \Theta_1) - 0.225 \sin(2\omega t + \Theta_2) - 0.008 \sin(4\omega t + \Theta_4) \right]$$

The magnitude of the third component is very small compared to the magnitude of either of the other components and can be neglected. A conservative estimate of the error can be found by taking the difference of the two remaining magnitudes as the maximum magnitude of $f_{af}(t)$. This yields

TABLE 5-3 MAXIMUM PERCENT ERROR IN ZERO-ORDER HOLD RECONSTRUCTION WITH AND WITHOUT FILTERING

		Zero-order hold followed by n-pole Butterworth filter								
		$f_c = f$			$f_c = 2f$			$f_c = 4f$		
k	Zero-order hold	$n=1$	$n=4$	$n=8$	$n=1$	$n=4$	$n=8$	$n=1$	$n=4$	$n=8$
2	100	90	59	38	100	86	82	90	85	85
4	71	54	38	36	48	33	33	50	51	51
10	31	38	31	30	24	11	11	20	19	19
20	16	33	30	30	17	6	5	11	9	9
40	8	31	30	29	14	3	2	7	4	4
100	3	30	29	29	12	1	1	5	2	2

$$\left| f_{af}(t) \right|_{max} = \frac{0.673 \times v_{span}}{2}$$

Comparing this value to the actual waveform to be reconstructed, that is,

$$\frac{x \, v_{span}}{2} \sin \omega t$$

results in a maximum error of 0.327 or approximately 33% of the original component.

The results of these calculations depend on the recovery sample rate, the filter cutoff frequency, the number of poles of the filter, and the type of filter used. Table 5-3 indicates the error percentages for several different situations. It should be emphasized that these are worst-case calculations for the maximum errors assuming a sinusoidal input of frequency f.

$$\frac{f_{rs}}{f} = 2k$$

Several interesting points result from an examination of Table 5-3. For the zero-order hold circuit with no filtering, a sample rate of $80f$ is required for an error of 8%. A 3% error occurs when the sample rate is $200f$. We note that if the 3-dB frequency of the filter equals f, the errors at the higher sampling rates are increased over the zero-order hold circuit alone. This is due to the 70.7% attenuation factor of the fundamental component by the filter.

When the 3-dB point of the filter equals $2f$, the attenuation of the fundamental component is much less and so also is the corresponding error. A sampling rate of $8f$ results in a 33% error for a 4-pole Butterworth filter. Raising the sample rate reduces the maximum error, reaching 1% for a $200f$ sampling rate.

As the 3-dB frequency is raised, the filter does not effectively eliminate the second and fourth harmonics of the zero-order hold signal. Typically this corner frequency is selected to be somewhere between f and $2f$.

While the errors in the table seem rather large for reasonable sample rates, we recognize that these calculations were based on the error in the highest-frequency component. A typical signal consists of many lower-frequency components that will be converted more accurately.

Example 5-5

A triangular waveform that varies from 0 to 10 V is shown in Fig. 5-16. The Fourier series for this signal is given by

$$f(t) = 5 + \frac{40}{\pi}\left[\sin \omega t - \frac{1}{9}\sin 3\omega t + \frac{1}{25}\sin 5\omega t - \ldots\right]$$

Assuming the data pertaining to the DC component plus the first three components are stored in a computer, calculate the erroneous components for a zero-order hold circuit followed by a 4-pole Butterworth filter. The fundamental frequency is $f = 2$ kHz, the filter bandwidth is 20 kHz, and the recovery sampling rate is $15f$.

Solution: We treat each component separately to calculate the effect of the zero-order hold. After the zero-order hold operation, each of the three components can be represented by its fundamental, second, and fourth harmonic. For the $\sin \omega t$ component of the ramp, the value of k is 30; for the $\sin 3\omega t$ term, the value of k is 10; and for the $\sin 5\omega t$ term, the value of k is 6. Table 5-2 can be used to find the approximate expression for $f(t)$.

The fundamental component of the signal, $4.05 \sin \omega t$, when reconstructed leads to three components of

$$4.05\,(0.998 \sin \omega t - 0.044 \sin 2\omega t - 0.0178 \sin 4\omega t)$$

Similarly, the third harmonic component of the signal, $-0.450 \sin 3\omega t$, leads to

$$-0.450\,(0.984 \sin 3\omega t - 0.132 \sin 6\omega t - 0.053 \sin 12\omega t)$$

and the fifth harmonic, $0.162 \sin 5\omega t$, leads to

$$0.162\,(0.955 \sin 5\omega t - 0.217 \sin 10\omega t - 0.087 \sin 20\omega t)$$

After filtering, the signal can be written as

$$f_{af}(t) = 5 + 4.042 \sin(\omega t + \Theta_1) - 0.178 \sin(2\omega t + \Theta_2) - 0.072 \sin(4\omega t + \Theta_4)$$
$$- 0.443 \sin(3\omega t + \Theta_3) + 0.59 \sin(6\omega t + \Theta_6) + 0.010 \sin(12\omega t + \Theta_{12})$$
$$+ 0.154 \sin(5\omega t + \Theta_5) - 0.025 \sin(10\omega t + \Theta_{10}) - 0.006 \sin(20\omega t + \Theta_{20})$$

The frequency components at $2f$, $4f$, $6f$, $10f$, $12f$, and $20f$ are error signals that detract from the accuracy of the actual signal.

For highly accurate signal reconstruction a first-order hold method can be used, although slightly more computing power is required. This method adds a slope to the reconstructed waveform between samples. Two popular techniques used are the predictive first-order hold and the linear first-order hold shown in Fig. 5-17.

Chap. 5   Basic Error Considerations in Data Acquisition Systems   343

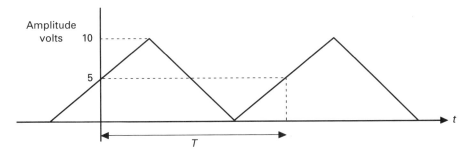

Figure 5-16 A triangular waveform.

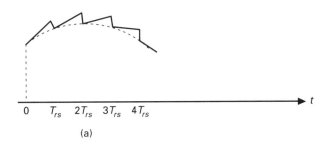

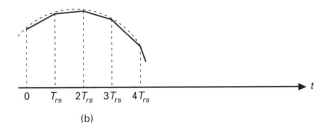

Figure 5-17 First-order hold waveforms: (a) predictive; (b) linear.

The value of the waveform between samples in the predictive method is estimated by assuming the slope of the waveform equals that slope existing between the last two known sample values. The waveform value is corrected at each sample time. This method allows the waveform to be reconstructed immediately as discrete values are calculated by the processing algorithms.

The linear first-order hold scheme must wait for a sample value before estimating the waveform between this most recent value and the previous value. The estimate is based on a straight line connecting the two points. In this method, the system output can appear only after a delay of one sample time, T_{rs}.

The computer outputs intermediate values between sample points based on predicted values. The values at the sample points result from the digital processing algorithms used, while the values between samples are calculated very easily based on either the predictive or linear first-order hold methods.

Exercises

E5.5a Rework Example 5-4 assuming the 20-kHz component has a maximum amplitude of 0.005 V.

E5.5b Rework Example 5-4 assuming the highest frequency of interest is 6 kHz and has a maximum amplitude of 0.03 V.

GENERAL EXERCISES

5.1 A range of 0–5 V is to be converted to 6-bit digital words. What is the minimum quantum error for the system?

5.2 Repeat problem 5.1 for a 12-bit system.

5.3 A transducer operates over a 0–6 V range. The overall transducer accuracy is specified as ±0.6% of FSO. Neglecting all other error sources, specify a reasonable number of bits for the digital representation of this signal.

5.4 Repeat problem 5.3 with a transducer range of 0–15 V.

5.5 Repeat problem 5.3 with an accuracy of ±0.2% of FSO.

5.6 Repeat problem 5.3 with a white noise voltage of 0.04 V rms present at the transducer output.

5.7 If N is selected to be 8 in problem 5.3, what is the maximum total error in the system? If N is selected to be 12, what is the maximum total error in the system?

5.8 A 5-pole Butterworth filter with $f_c = 100$ kHz is used as a presampling filter. What is the highest frequency that can be converted with 1/2 LSB frequency error in an 8-bit system? Assume the peak-to-peak magnitude of the signal equals the system span.

5.9 Using the system of problem 5.8, how many bits of accuracy result in converting a signal of full span and frequency of 40 kHz?

5.10 A 12-bit system with a range of 0–12 V is used in reconstructing a signal. If the highest-frequency component is $0.05 \sin \omega t$ where $\omega = 2\pi \times 16{,}000$ rad/s, calculate the minimum required recovery sample rate for 1/2 LSB error.

5.11 A 12-bit system with a range of 0–12 V is used in reconstructing a signal. If the recovery sample rate is 200 kHz and the maximum component magnitude is 1/20 of the span (for higher-frequency components), what is the maximum frequency that can be converted with 1/2 LSB maximum error?

REFERENCES

1. Loriferne, B., *Analog-Digital and Digital-Analog Conversion*, Heyden and Son, London, 1982, Chapter 3.
2. Bendat, J. S., and A. G. Piersol, *Measurement and Analysis of Random Data*, Wiley and Sons, New York, 1966, Chapter 1.
3. Kelley, T. L., *The Kelley Statistical Tables*, Harvard University Press, Cambridge, MA, 1948.
4. Taub, H., and D. L. Schilling, *Principles of Communication Systems,* 2nd ed., McGraw-Hill, New York, 1986.
5. Van Valkenburg, M. E., *Network Analysis*, 3rd ed., Prentice Hall, Englewood Cliffs, NJ, 1974.

6

The Digital-to-Analog Converter

The key element that allows interfacing between analog transducers and digital systems is the converter. The analog-to-digital converter, or ADC, accepts the band-limited analog input signal and converts this signal to a time series of digital codes at the output. The output code at a particular instant represents the strength of the input signal at the time of the conversion. The digital system uses the string of ADC output codes as the digital representation of the input signal.

After the digital system has processed the input signal, the processed signal is also represented by a series of digital codes. In many applications, it is necessary to convert this processed signal to an analog or continuous output signal. The conversion of the series of digital codes to a continuous signal can be done with a digital-to-analog converter (DAC). The DAC accepts a digital input code and converts this signal to a corresponding analog output signal. The strength of the output signal depends on the value of the input code at the time of conversion.

A digital reverberation system, used to simulate an echo chamber in musical applications, is an example of a system that requires an ADC and a DAC. The input music is converted to a digital series of codes by the ADC. This signal is then delayed and attenuated digitally to create a simulated echo signal. A second echo signal is produced by delaying and attenuating the signal even further. The three digital signals are added together to produce a composite signal. This signal is then converted back to an analog waveform by a DAC, amplified and perhaps filtered, and applied to a speaker. The result is a signal containing an echo. The amount of delay of the two additional components and the attenuation of these signals can be controlled by the digital processor to determine the quality of the echoed signal.

Since the DAC is generally less complex than the ADC and is also used as a component of some types of ADC, this chapter discusses the DAC. The ADC is then considered in the next chapter.

6.1 PERFORMANCE PARAMETERS OF THE DAC

This section defines some of the major parameters that determine DAC performance.

6.1.1 Number of Bits

There are several parameters that influence the selection of the DAC. The number of bits that the DAC can convert is a very important consideration. The selection of the DAC size is generally governed by the width of the output code. An 8-bit output code requires a DAC that converts at least 8 bits. In some instances a DAC that is larger than the output code may be used due to the availability of an off-the-shelf DAC of a certain size. A 7-bit code may be converted by an 8-bit DAC simply because this DAC size is common. The LSB of the DAC would be left unused in this case.

At this point it should be mentioned that the output word size is determined by two factors: the ADC size and the digital signal processing methods applied. In many systems, the ADC size, the number of bits used in processing, and the DAC size are equal. Other processing tasks may require $N + 2$ bits to achieve sufficient accuracy with the processing method used, but the final results may only be accurate to N bits. In such a case, the two least significant bits of the code can be ignored and the DAC will convert the remaining N bits. The ADC size will then be 2 bits larger than the DAC size.

For our purposes we will assume a DAC size that is equal to the ADC size. Chapter 5 has considered several factors that influence the ADC size.

6.1.2 Settling Time

The settling time, t_s, measures the interval between an abrupt change of input code and the time that the output has reached the correct analog value to within some specified accuracy. Often the accuracy is ±1/2 LSB. That is, when the output approximates the correct value within one-half of a quantum interval, settling time is completed.

Settling time determines the upper limit on frequency of conversion for the DAC. The DAC should never exceed a conversion frequency of

$$f_{\max} = \frac{1}{t_s}$$

A conversion frequency that exceeds this value results in output voltages that are inaccurate. Figure 6-1 demonstrates settling time for a DAC that includes an input register. The parameter is measured as the time interval between the 50% point of the latch enable signal that sets the input data into the register and the time for the output to settle to ±1/2 LSB.

Chap. 6 The Digital-to-Analog Converter

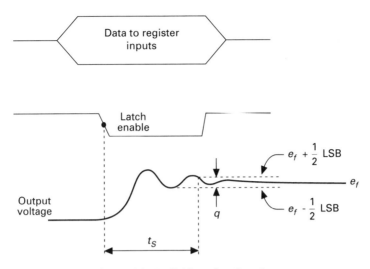

Figure 6-1 Definition of settling time.

6.1.3 Conversion Factor

The conversion factor, *CF*, which is the same as the quantum interval, *q*, is defined as the change in output signal for a unit change of input code. For a voltage output, the units of *CF* are volts/unit or simply volts. Some off-the-shelf DACs have a fixed *CF*, while others allow this parameter to be varied by an external resistor or by a power supply voltage. The largest voltage that can be produced by an *N*-bit, binary DAC is

$$v_{\max} = \left(2^N - 1\right) \times CF \tag{6-1}$$

6.1.4 Accuracy or Linearity

As the input digital code increases, the analog output voltage should increase proportionally. However, due to various error sources, the increase in e_{out} may not be perfectly linear with input code change. This departure from the ideal is often specified as linearity. An 8-bit linearity implies an error within ±1/2 LSB. This parameter was discussed in detail in Chapter 1 and will not be considered further at this point.

6.1.5 Resolution

The resolution quantity measures the smallest change in input signal that is guaranteed to produce a discernible output change. Often it is expressed in terms of a number of bits. For example, 8-bit resolution indicates that a change in the 8th-least significant bit will cause a discernible output change. While the resolution specification generally corresponds to the total number of bits of the DAC, this may not be true for larger DACs. An 8-bit DAC usually has 8-bit resolution, but a 14-bit DAC may have only 12-bit resolution. This means that a change in the 13th or 14th bit of the input code may lead to an output change that is smaller than the change due to nonlinearity or to a temperature drift.

> **Exercises**
>
> **E6.1a** Calculate the quantum interval of a 10-bit DAC that has a maximum output of 8 V.
>
> **E6.1b** Calculate the maximum output of a 12-bit DAC with a quantum interval of 0.0040 V.

6.2 DESIGN OF THE DAC

This section discusses several configurations that can be used for a DAC. The two most important circuits, the binary-weighted DAC and the ladder DAC, are emphasized.

A DAC converts a digital code to an analog signal that has a value that represents the value of the digital code. The most popular DAC converts straight binary code to analog, but other codes such as BCD, offset binary, or 2's complement are sometimes converted.

One problem that must be considered in converting a digital input code is that of voltage level variation from one bit of the code to another. For TTL, the high voltage level may range from 2.4 V to 5 V, while the low level might vary from 0 V to 0.7 V for different circuits. If an 8-bit register drives the 8 inputs of a DAC, all inputs may be driven with slightly different voltages, even when the register contains all 1s.

These inaccurate values cannot be converted directly to analog levels. There are two popular approaches to the solution of this problem. One approach uses this input voltage to activate a semiconductor switch. This switch then controls a precise current that can now be accurately converted. The second method sets the digital code into a register that is part of the DAC. The outputs of this register are carefully regulated to precise voltage levels allowing accurate conversion.

Figure 6-2 shows one possible block diagram for the DAC. If a register is included as part of the DAC, a load strobe input will be present. The voltages X_1 and $-X_1$ are reference voltages. At least one reference voltage is necessary for accurate conversion, however, this voltage is often derived from the power supply voltage. The voltages V_1 and $-V_1$ are power supply voltages. Some DACs require only one power supply voltage. The output analog voltage is designated V_A.

With the increased importance of the microprocessor (µP) in control applications, a double-buffered DAC has been developed. Such an arrangement is shown in Fig. 6-3. This system allows the µP to write to any of the three buffer registers at any time without changing the DAC output voltages. At an appropriate time, the DAC register strobe can be applied to update the DAC registers. In some situations, it is necessary to control the time of updating to avoid transients.

The µP-compatible DAC can also be used in systems that do not utilize a µP. In such a case, the input code should be capable of transfer directly from the input data lines to the DAC register. This is referred to as the flow-through mode. The latch enable lines

Chap. 6   The Digital-to-Analog Converter

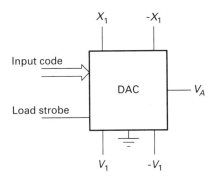

Figure 6-2 Block diagram of a DAC.

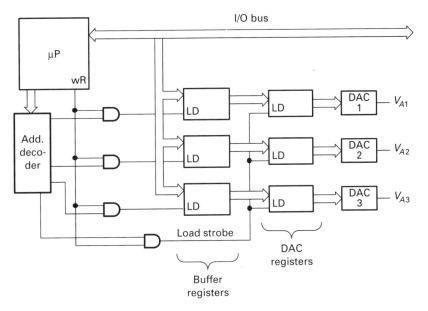

Figure 6-3 Double-buffered DACs.

of both the buffer register and DAC register are asserted and remain asserted in this case. Typically, the registers are transparent D latches to allow operation in the flow-through mode.

If only one transparent register were used with a μP, invalid data would be indicated at all times except during the short time when a write operation guarantees valid data on the bus. If the write strobe enabled the single latch, incorrect data would enter the DAC when this strobe is first asserted. Toward the end of the strobe, valid data appears and the correct word would be latched into the register as the write strobe becomes deasserted. The DAC would generate incorrect output values during the strobe until the valid data appears. The use of two latches avoids this problem. At the end of the first write strobe, the correct data resides in the buffer register. A second write strobe transfers this data to the DAC register for conversion.

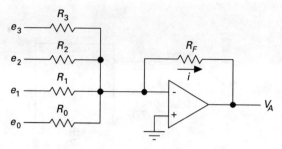

Figure 6-4 A binary-weighted DAC.

6.2.1 Small Word Size, Binary-Weighted DACs

The simplest converter from a conceptual viewpoint is the binary-weighted DAC. This converter creates currents that are proportional to the weights of each code bit. These currents are summed and used to produce a proportional voltage. Figure 6-4 shows a 4-bit, binary-weighted DAC. The voltage V_A can be expressed as

$$V_A = -\left(\frac{e_3}{R_3} + \frac{e_2}{R_2} + \frac{e_1}{R_1} + \frac{e_0}{R_0}\right)R_F$$

We will assume the input code is represented by $a_3\, a_2\, a_1\, a_0$ where a_i can be either 0 or 1. If the voltage level associated with these two values is 0 and V_B, and we choose the resistor values to be

$$R_0 = 2R_1 = 4R_2 = 8R_3$$

the output voltage becomes

$$V_A = -\frac{R_F}{R_0} V_B \left(a_3 8 + a_2 4 + a_1 2 + a_0\right) \qquad (6\text{-}2)$$

Equation (6-2) assumes that a_3, the most significant bit (MSB), is connected to the input resistor R_3.

The absolute value of the quantum interval or conversion factor for the DAC is the voltage per LSB. In this case,

$$CF = -\frac{R_F V_B}{R_0} \qquad (6\text{-}3)$$

The value of CF is determined by the designer. Often, this factor will be set to the maximum allowable value, which is limited by the maximum output of the op amp, V_{OM}. The largest conversion factor is calculated from Eq. (6-1) to be

$$CF_{max} = \frac{V_{OM}}{2^N - 1}$$

Chap. 6   The Digital-to-Analog Converter

If this conversion factor were used, the largest input code would develop an output equal to V_{OM}. In order to allow this input to generate an output voltage within the active region of the amplifier, the maximum conversion factor is often taken as

$$CF_{max} = \frac{V_{OM}}{2^N} = q_{max} \qquad (6\text{-}4)$$

This value allows one quantum interval between the maximum output and the edge of the active region. We will apply Eq. (6-4) in succeeding material.

Example 6-1

The binary-weighted DAC of Fig. 6-4 has values of $R_F = 10$ kΩ and $R_0 = 360$ kΩ.

a. Calculate the conversion factor for the converter when $V_B = -8$ V.
b. If V_{OM} is 8 V, modify R_F to result in CF_{max}.

Solution:
a. The conversion factor for this case is calculated from Eq. (6-3) to be

$$CF = \frac{10 \times 8}{360} = 0.222 \text{ V}$$

The maximum output of this system would be

$$e_{max} = 15 \times 0.222 = 3.33 \text{ V}$$

The other values of resistance would be $R_1 = 180$ kΩ, $R_2 = 90$ kΩ, and $R_3 = 45$ kΩ.

b. The maximum conversion factor is

$$CF_{max} = \frac{8}{2^4} = 0.5 \text{ V}$$

Using this value in Eq. (6-3) allows R_F to be written as

$$R_F = \frac{CF_{max} R_0}{V_B} = 22.5 \text{ k}\Omega$$

With this conversion factor the maximum output of the system is 7.5 V.

A practical realization of the binary-weighted DAC is shown in Fig. 6-5. The voltage at the inverting node is $e_1 = V_{REF}$. When $a_0 = a_1 = a_2 = a_3 = 0$, all transistors are cut off and no current flows from the inverting node. The output voltage of the first op amp equals V_{REF} in this case. When a transistor is saturated by applying a high level to an input, a current of V_{REF}/R, where R is the corresponding collector resistor, flows from

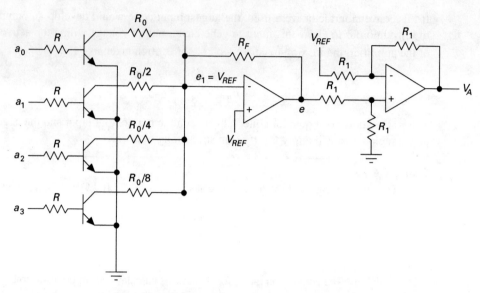

Figure 6-5 A practical 4-bit, binary-weighted DAC.

the inverting node. This assumes that $V_{CE\,(sat)} = 0$. The collector current also flows through R_F. In general, the output voltage of the first op amp can be written as

$$e = V_{REF}\left(1 + \frac{8a_3 R_F}{R_0} + \frac{4a_2 R_F}{R_0} + \frac{2a_1 R_F}{R_0} + \frac{a_0 R_F}{R_0}\right)$$

The second op amp is a difference circuit with an output of

$$V_A = e - V_{REF}$$

thus we can write

$$V_A = \frac{V_{REF} R_F}{R_0}(8a_3 + 4a_2 + 2a_1 + a_0) \tag{6-5}$$

The binary-weighted DAC is used for small word size systems. The practical upper limit on N is 6 for reasons to be discussed next.

The two major problems with this DAC are the large resistor spread required for large word sizes and the speed limitation due to voltage mode switching. Of course, this latter problem does not preclude use of the binary-weighted DAC in lower frequency switching.

The parasitic capacitance at the collector of each switching transistor slows the conversion time of the DAC. When a transistor has been saturated and then is switched off due to an input bit change, there is a charging time constant at the collector. For the LSB, this time constant is maximum and is determined by the product of R_0 and the parasitic capacitance. A 5-pF capacitance and a 20-kΩ resistance results in a time constant of 0.1 μs. If three time constants are allowed for this transition, the maximum switching speed for this bit is about 3.3 MHz. This figure would be lowered further by the saturation storage time of the transistors.

Chap. 6   The Digital-to-Analog Converter

One problem related to the resistor spread is that the smallest input resistor, associated with the MSB, leads to an output that is a factor of 2^{N-1} times larger than the output for the LSB. Thus if the value of this resistor is inaccurate or drifts slightly due to temperature change, a change of 1 part in 2^{N-1} may lead to an output error that is larger than 1 LSB.

The error in output voltage can be related to the change in the MSB resistor by writing

$$e_{MSB} = \frac{V_{REF} R_F}{R_M} \qquad (6\text{-}6)$$

where R_M is the MSB resistor. Nominally, the MSB resistor is related to the LSB resistor R_L by

$$R_M = \frac{R_L}{2^{N-1}} \qquad (6\text{-}7)$$

The conversion factor is

$$CF = \frac{V_{REF} R_F}{R_L} \qquad (6\text{-}8)$$

If we assume that R_M increases to R_M^+ or decreases to R_M^-, the value of e_{MSB} changes to

$$e_{MSB}^+ = \frac{V_{REF} R_F}{R_M^-}$$

or

$$e_{MSB}^- = \frac{V_{REF} R_F}{R_M^+}$$

We can equate the possible changes in e_{MSB} to $CF/2$ or $LSB/2$ to calculate the necessary changes in R_M. Solving for fractional changes in R_M in the manner outlined gives

$$\frac{\Delta R_M^-}{R_M} = \frac{1}{2^N - 1} \approx \frac{1}{2^N} \qquad (6\text{-}9a)$$

and

$$\frac{\Delta R_M^+}{R_M} = \frac{1}{2^N - 1} \approx \frac{1}{2^N} \qquad (6\text{-}9b)$$

Table 6-1 indicates the change in MSB resistance leading to an output change of LSB/2, assuming all other parameters remain constant.

Of course, Table 6-1 gives a worst-case situation by assuming no change in R_F. If the weighting resistors and R_F are made of the same material, temperature tracking of these resistors will decrease the overall drift. However, we see that a 10-bit converter requires only a 1 part in 1024 change in resistance to cause an output change of LSB/2.

TABLE 6-1 FRACTIONAL CHANGE IN R_M TO CAUSE 1/2 LSB OUTPUT ERROR

n	$\dfrac{\Delta R}{R_M}$
4	$\dfrac{1}{16}$
6	$\dfrac{1}{64}$
8	$\dfrac{1}{256}$
10	$\dfrac{1}{1024}$
12	$\dfrac{1}{4096}$

A second problem caused by the large resistance spread is difficulty of IC fabrication. If the DAC is to be integrated, a large difference in resistor values cannot be tolerated, so an alternate configuration must be considered for N above 6.

6.2.2 The Ladder Configuration

The ladder network is used to solve the problem of resistor spread and minimize drift problems in DACs with larger values of N. Virtually all IC DACs use some variation of the ladder configuration. For IC DACs of 12 bits and above, laser trimming of resistors is used to achieve the necessary accuracy.

The basic ladder configuration is shown in Fig. 6-6. The semiconductor switches used in the circuit of Fig. 6-6(a) connect the resistor to ground or to virtual ground. Thus regardless of switch position, the voltage at the common terminal of the switch is always zero. This is called a current switching ladder, since current through the ladder resistors never changes. This current is directed through one of two paths by the switches. The network of Fig. 6-6(b) can then be used for analysis.

Beginning at the right-hand side of the ladder we see that the two parallel resistors of value $2R$ reduce to an equivalent resistance of R. This equivalent resistance adds in series with R_3 to present a total resistance of $2R$, looking right from node 2, not including R_4. This resistance appears in parallel with R_4 giving a resistance of R looking right from node 2 and including the effect of R_4. We see that this network is iterative in that the resistance seen at a given node, looking to the right, equals the value seen at the next node.

The current I_1, given by

$$I_1 = \frac{V_{REF}}{2R}$$

is seen to split in half at the second node. A value of

$$I_2 = \frac{I_1}{2}$$

Chap. 6   The Digital-to-Analog Converter

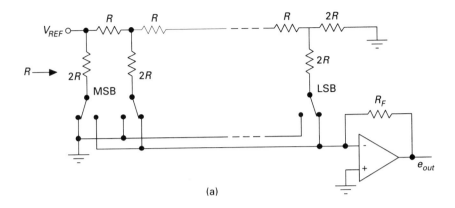

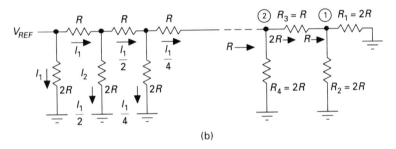

Figure 6-6 (a) A ladder DAC. (b) The equivalent ladder circuit.

flows through the second vertical resistance. Each succeeding vertical resistor has a value of current flow equal to one-half that of the previous resistor. The Nth vertical resistor has a current flow of

$$I_N = \frac{I_1}{2^{N-1}}$$

This is the LSB current, while I_1 is the MSB current.

When a bit is 1, the switch will divert the current to the inverting input of the op amp. This current flows through the resistor R_F, allowing the output voltage to be expressed as

$$e_{out} = -\left(a_{N-1} + \ldots + \frac{a_2}{2^{N-3}} + \frac{a_1}{2^{N-2}} + \frac{a_0}{2^{N-1}}\right)\frac{V_{REF}R_F}{2R} \qquad (6\text{-}10)$$

The voltage V_{REF} can be reversed to create a positive value of output voltage.

The conversion factor for the ladder DAC is

$$CF = \frac{V_{REF}R_F}{2^N R} \qquad (6\text{-}11)$$

The spread of resistance values for the ladder is now only a 2:1 spread, and a change in value of any single resistance does not affect the output voltage as drastically as it can in the binary-weighted DAC of Fig. 6-4. Furthermore, if fabricated on a chip, the resistors have relatively well-matched temperature coefficients to minimize drift.

The DAC of Fig. 6-6 exhibits a speed advantage over the DAC of Fig. 6-5, since current mode switching takes place. The voltage level at the common terminal of the switch remains constant as the switch position changes. The current at this point also remains constant, with only the path changing. Since no voltage change takes place, the charge on the parasitic capacitances at the input does not change, nor do the transistors move in and out of saturation.

Exercises

E6.2a Rework Example 6-1 with $V_B = 6$ V.

E6.2b Assume the DAC of Fig. 6-6(a) is an 8-bit device. If V_{REF} is -12 V and $R_F = 20$ kΩ, calculate R to result in a conversion factor of 0.022 V.

6.3 A PRACTICAL DAC

This section looks at a specific, practical μP-compatible DAC and relates its specifications and configuration to those discussed in the previous two sections.

The National Semiconductor DAC, model DAC0830, is a μP-compatible DAC that has been available for over a decade. This device is a double-buffered DAC that also allows flow-through operation. It produces a current output designed to drive an op amp in the inverting configuration.

The major specifications are as follows:

1. Converter size, $N = 8$ bits
2. Resolution, 8 bits
3. Output current settling time, 1 μs
4. Conversion factor = $V_{REF}/256$ (typical $V_{REF} = 15$ V)
5. Linearity, 8 bits (end point linearity)

The conversion factor can be varied by changing V_{REF} or by changing the gain of the external op amp. The reference voltage can be varied in both the positive or negative directions. This allows four-quadrant multiplication. A multiplying DAC produces an output that is proportional to the input code multiplied by the reference voltage. The equation for output current for this DAC is

$$I_{out} = \frac{V_{REF}}{15 \text{ k}\Omega} \times \frac{\text{Digital input}}{256}$$

Since V_{REF} can be positive or negative and a second op amp can be added to allow the MSB to carry sign information, the DAC can function as a four-quadrant multiplier.

The block diagram of the DAC is shown in Fig. 6-7. The input register and the DAC register are loaded separately when used with a μP. The multiplying converter uses

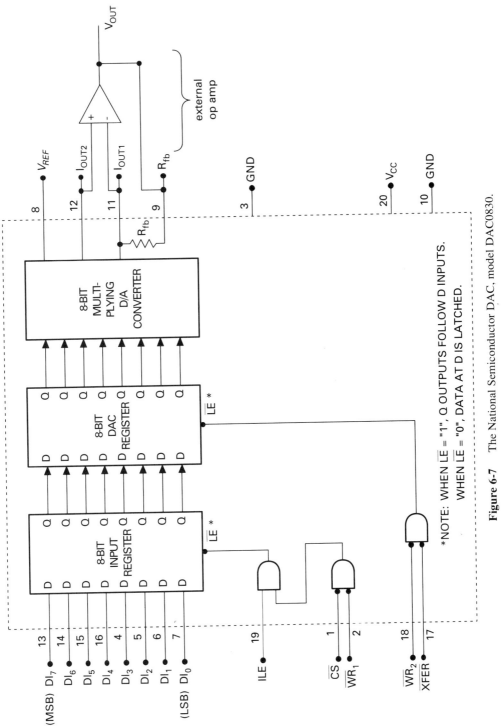

Figure 6-7 The National Semiconductor DAC, model DAC0830.

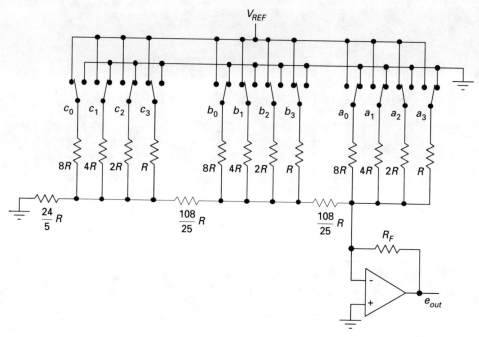

Figure 6-8 A weighted ladder BCD DAC.

an R–$2R$ ladder as discussed earlier. The output current must flow into a virtual ground mode and thus the necessity of the op amp.

6.4 OTHER TYPES OF DAC

Converters are available that accept digital codes other than binary fractional codes. The BCD DAC and bipolar DAC are discussed in this section.

6.4.1 BCD DACs

BCD DACs can be constructed in one of two rather simple ways. A ladder network can be used to create the appropriate weighting of digits. The circuit of Fig. 6-8 can be shown to create the output

$$e_{out} = \frac{-V_{REF}R_F}{R}\left[\left(a_3 + \frac{a_2}{2} + \frac{a_1}{4} + \frac{a_0}{8}\right) + \frac{1}{10}\left(b_3 + \frac{b_2}{2} + \frac{b_1}{4} + \frac{b_0}{8}\right) \right.$$
$$\left. + \frac{1}{100}\left(c_3 + \frac{c_2}{2} + \frac{c_1}{4} + \frac{c_0}{8}\right)\right] \quad (6\text{-}12)$$

It is a trivial exercise, left to the student, to show that Eq. (6-12) is correct (see problem 6.11).

Chap. 6 The Digital-to-Analog Converter

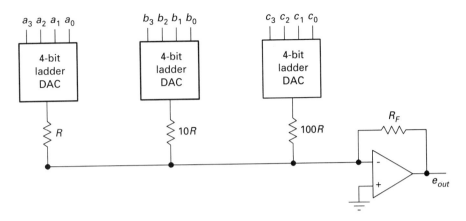

Figure 6-9 A BCD DAC.

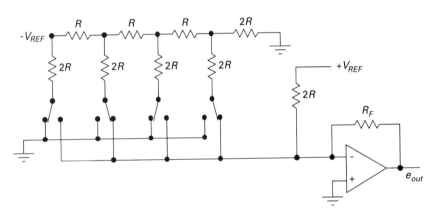

Figure 6-10 An offset binary DAC.

A second method of creating a BCD DAC is to use 4-bit binary DACs for each BCD digit and sum these DAC outputs with proper weighting. Such a scheme is shown in Fig. 6-9. Some commercially available binary DACs produce a current output that is proportional to the input code. If these DACs are used for digit conversion, they can be summed directly into the output op amp. The correct weighting of each current must be done with the individual DACs. This method minimizes the switching time of the converter by eliminating large voltage excursions at the op amp input.

6.4.2 Bipolar DACs

A bipolar DAC accepts an input code that can represent both positive and negative numbers. The output can generate positive or negative voltages that represent these codes. The simplest bipolar DAC uses the offset binary code. In this case a binary DAC simply sums the appropriate offset current into the op amp stage to create the offset binary output. A 4-bit ladder DAC for offset binary output is shown in Fig. 6-10.

TABLE 6-2 OUTPUT OF OFFSET BINARY DAC

Input code	Number	Output $\times V_{REF} R_F/2R$
0000	−8	−1
0001	−7	−7/8
0010	−6	−6/8
0011	−5	−5/8
0100	−4	−4/8
0101	−3	−3/8
0110	−2	−2/8
0111	−1	−1/8
1000	0	0
1001	1	1/8
1010	2	2/8
1011	3	3/8
1100	4	4/8
1101	5	5/8
1110	6	6/8
1111	7	7/8

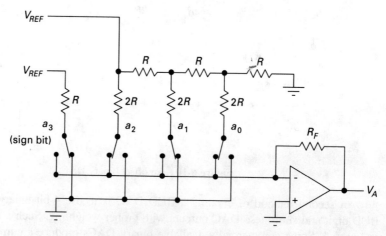

Figure 6-11 A 2's complement DAC.

With the additional offset current, the output of this DAC is given by

$$e_{out} = \frac{V_{REF}R_F}{2R}\left(a_3 + \frac{a_2}{2} + \frac{a_1}{4} + \frac{a_0}{8}\right) - \frac{V_{REF}R_F}{2R} \qquad (6\text{-}13)$$

From Eq. (6-13), Table 6-2 can be generated.

A bipolar 2's complement DAC is very similar to the binary DAC, except the current through the MSB resistor is opposite to that through the other bit resistors. This circuit is shown in Fig. 6-11. The output voltage is now

Chap. 6   The Digital-to-Analog Converter

TABLE 6-3 OUTPUT OF 2'S COMPLEMENT DAC

Input code	Number	Output $\times V_{REF} R_F/2R$
1000	−8	−1
1001	−7	−7/8
1010	−6	−6/8
1011	−5	−5/8
1100	−4	−4/8
1101	−3	−3/8
1110	−2	−2/8
1111	−1	−1/8
0000	0	0
0001	1	1/8
0010	2	2/8
0011	3	3/8
0100	4	4/8
0101	5	5/8
0110	6	6/8
0111	7	7/8

$$e_{out} = \frac{V_{REF} R_F}{R} \left(-a_3 + \frac{a_2}{2} + \frac{a_2}{4} + \frac{a_0}{8} \right) \qquad (6\text{-}14)$$

The a_3 bit carries the sign information. If $a_3 = 0$, the sign is positive and Eq. (6-14) leads to positive output voltages. When $a_3 = 1$, the sign is negative and a negative voltage is produced. Table 6-3 shows the outputs for the various possible inputs.

6.5 THE GLITCH PROBLEM IN DACs

A significant problem occurring in DACs is that of glitching. When a digital code is set into a DAC register, the unequal switching times of register flip-flops or the digital bit switches can cause large transient spikes at the DAC output. As a code changes from one value to be converted to the next value, all bits of the code do not change simultaneously. This problem can be demonstrated by considering the change of a 4-bit code from binary 3 to binary 4. As the code of 0100 is written into the register that previously contained 0011, several transient states may exist for short periods of time before the register output equals 0100. The possible states can be found from the Karnaugh map of Fig. 6-12. The transient states existing along these paths are tabulated in Table 6-4.

As the bits cause the transient states to be reached, the DAC may respond quickly enough to convert these incorrect values. For example, as the code changes from 0011 to 0001 to 0000 to 0100, the DAC may generate outputs proportional to 3, 1, 0, and 4. The output is quite unstable during the transient states. Whenever a code in the register is to be replaced by a code that is logically nonadjacent, a glitch can occur. The magnitude of the glitch depends on the significance of the bit that is changing, as indicated in Fig. 6-13(a).

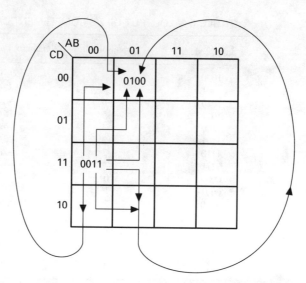

Figure 6-12 Possible paths from 0011 to 0100.

TABLE 6-4 TRANSITION PATHS FROM 0011 TO 0100

Initial state	Transient state	Transient state	Final state
0011	0010	0000	0100
0011	0001	0000	0100
0011	0010	0110	0100
0011	0001	0101	0100
0011	0111	0110	0100
0011	0111	0101	0100

Certain systems that are driven by a DAC ignore the glitches that appear at the DAC output. DC motors, for example, cannot respond to these sharp transients due to the inertia of the motor and are thus ignored. In applications that cannot tolerate these glitches, a sample-and-hold circuit is typically used. This circuit is driven by a signal that is out of phase with the signal that drives the input code. In this way, sampling is done after the transient has disappeared, as shown in Fig. 6-13(b). The details of the sample-and-hold circuit will be covered in a later chapter.

GENERAL EXERCISES

6.1 Explain why two registers are needed rather than a single edge-triggered register to make a DAC function with a µP and also in flow-through mode.

6.2 The binary-weighted DAC of Fig. 6-4 has a maximum output voltage of +10 V. If $R_F = 40$ kΩ and $V_B = -4$ V, select values of R_3, R_2, R_1, and R_0 to achieve the maximum possible conversion factor.

6.3 Repeat problem 6.2 for a 6-bit DAC. In this case, find values of R_5, R_4, R_3, R_2, R_1, and R_0. What is the CF in this circuit?

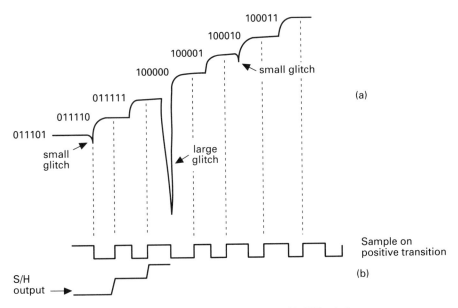

Figure 6-13 (a) DAC glitch problem. (b) S/H solution.

6.4 In Fig. 6-5, assume that $R_F = 4$ kΩ, $R_0 = 20$ kΩ, and $V_{REF} = 8$ V. What is V_A for an input code of 1010?

6.5 In problem 6.4, select R_F to create a conversion factor of 0.6 V.

6.6 The LSB resistor in a 6-bit binary-weighted DAC is 320 kΩ. What is the value of the MSB resistor? What is the change in absolute value of MSB resistance to cause a worst-case change in V_A of $\pm 1/2$ LSB?

6.7 The ladder DAC of Fig. 6-6(a) has $R_F = 20$ kΩ, $V_{REF} = -12$ V, and is an 8-bit DAC. Calculate R to result in $CF = 0.04$ V. Calculate the output voltage for an input code of 10011101.

6.8 If the conversion factor of an 8-bit ladder DAC is 0.04 V and $V_{REF} = -12$ V, calculate the change in V_{REF} to cause a $\pm 1/2$ LSB change in V_A.

6.9 If $R_F = 20$ kΩ and $R = 20$ kΩ, select V_{REF} to give a $CF = 0.8$ V in the circuit of Fig. 6-10.

6.10 Suppose one has an 8-bit converter and a 7-bit converter. The analog output ranges of both converters are 0–10 V. What is the LSB value of each converter? Now explain how applying 7 bits to the 8-bit converter (LSB of the 8-bit converter unused) results in the same quantum interval as the original 7-bit converter.

6.11 Derive Eq. (6-12) for the BCD DAC. Show that Fig. 6-8 performs the correct conversion for the code 100101010111.

REFERENCES

1. TRW Electronics Component Group, *TRW LSI Products Division*, TRW, La Jolla, CA, 1985.
2. Engineering Staff, *Data Acquisition Databook Update and Selection Guide*, Analog Devices, Norwood, MA, 1986.

3. Loriferne, B., *Analog-Digital and Digital-Analog Conversion*, Heyden and Sons, London, 1982.
4. Comer, D. J., *Electronic Design with Integrated Circuits*, Addison-Wesley, Reading, MA, 1981.

7

The Analog-to-Digital Converter

The analog-to-digital converter, or ADC, is driven by a continuous voltage or current signal and generates an output code that is proportional to the input signal. This device provides the digital system, often a computer, with the sequential digital codes that represent the analog input signal. The ADC generally has two control lines that allow handshaking with external devices. The first of these is a start conversion input that instructs the ADC to begin converting the analog signal to a corresponding code. When the conversion is completed, an end-of-conversion output is asserted to inform the waiting device that a valid output code is available from the ADC. If the device is designed for use with a μP or other computer, an output enable line must be asserted to enable the digital code onto the μP bus. When the output enable line is not asserted, the digital code output lines are in the high-impedance state. A clock input line to the ADC is driven by a clock signal to provide a time reference.

If the ADC is μP-compatible, it may also provide several multiplexed input channels. An address must then be provided to a set of encoded address lines and an address enable strobe applied to select the proper channel. Whichever channel is selected will remain connected to the ADC input until the address lines are changed and another address latch enable strobe is applied. A block diagram of a μP-compatible ADC is shown in Fig. 7-1.

7.1 PERFORMANCE PARAMETERS OF THE ADC

In order to select an ADC it is important to understand the definitions of the converter specifications and how they apply to the analog signal being converted.

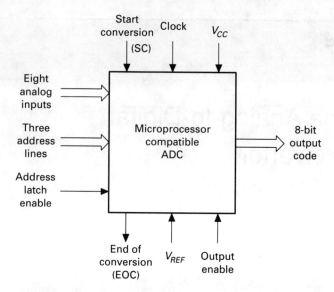

Figure 7-1 A microprocessor-compatible ADC block diagram.

7.1.1 Number of Bits

The number of bits in the output code of the ADC is one of the first considerations made in selecting an ADC. This number, N, is chosen so that the resulting ADC quantization error is slightly smaller than the total system error. These considerations were presented in detail in Chapter 5 and will not be discussed further here. One or two bits more than required might be chosen to gain the benefits of using an off-the-shelf ADC.

7.1.2 Conversion Time

The time from one transition of the start conversion pulse (SC) to the assertion of the end-of-conversion (EOC) line indicates the conversion time of the ADC. Often, a high-to-low transition of the SC signal initiates the conversion. The 50% point of this transition signifies the beginning of conversion time. The 50% point of the EOC line indicates the end-of-conversion time.

7.1.3 Quantum Interval or LSB Value

The quantum interval is determined by the magnitude of the analog input required to result in a unit change in the output code. This interval should not be measured between the output codes for 0 and 1, because this first interval is adjusted to be $q/2$. All succeeding intervals are q. This smaller initial interval is selected to minimize the absolute value of quantization error, as explained in Chapter 5.

7.1.4 Accuracy

The quantization error in an ADC is generally $\pm q/2$. Additional sources of error are linearity, offset, multiplexer, and full-scale errors. The total absolute error is a significant

Chap. 7   The Analog-to-Digital Converter

specification that adds all error sources, including quantization error. A typical absolute error specification is ±1 LSB or ±q.

Exercises

E7.1a An analog system monitors temperature. The output of the amplifier has a span of 10 V with a total system error of 0.014 V. Select a reasonable number of bits for the ADC.

E7.1b Calculate the size of the quantum interval for the converter selected in E7.1a.

7.2 TYPES OF ADC

There are several methods of converting an analog signal to a digital code. The following list indicates several possible types of ADC:

1. Parallel, simultaneous, or flash converter
2. Successive approximation converter
3. Iterative converter
4. Ramp or staircase converter
5. Tracking converter
6. Integrating or dual slope converter

The process of analog-to-digital conversion is more complex than the process of digital-to-analog conversion. An ADC either requires much more circuitry (parallel converter) or much more time for conversion than does the DAC. The following paragraphs will discuss each of these conversion schemes along with the advantages and disadvantages of each.

7.2.1 Parallel Converter

The block diagram of a simple parallel converter is shown in Fig. 7-2. This device is also called a simultaneous converter or a flash converter. These names correctly imply a very rapid analog-to-digital conversion for this ADC. No other method used for conventional ADCs results in conversion speeds approaching those obtained by this method.

The parallel converter establishes $2^N - 1$ voltage levels, each differing from the next by the quantum interval q. The lowest voltage level is established at $q/2$ with succeeding levels of $3q/2, 5q/2, 7q/2, \cdots, (2^{N+1} - 3)q/2$. These levels are easily established by a string of series resistors connected to a reference voltage.

One comparator circuit is required for each quantum level resulting in $2^N - 1$ comparators for an N-bit ADC. Circuit complexity increases rapidly with N, but VLSI techniques now allow 8-bit parallel converters to be fabricated on a single chip.

The first comparator switches to an output of $+V_H$, the high logic level, when the analog input exceeds a voltage of $q/2$. Each succeeding comparator switches from a low output to the high level as the analog input increases by q over the switching point of the previous comparator. If $V_{A\max}$ is the largest input voltage to be converted, the quantum interval is calculated from

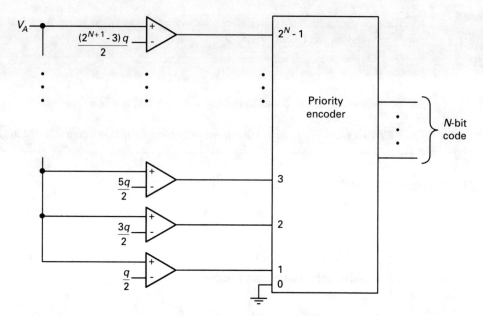

Figure 7-2 A parallel converter.

$$q = \frac{V_{Amax}}{2^N - 3/2} \tag{7-1}$$

The comparators drive the input of a priority encoder logic circuit. This device has 2^N input lines, numbered successively from 0 to $2^N - 1$. If a single input line is asserted, while all other lines are deasserted, the encoder generates a binary output code corresponding to the number of the input line. A priority encoder allows several input lines to be asserted simultaneously and produces an output code corresponding to the highest-numbered asserted input line.

For each value of V_A applied to the input, several adjacent comparators will be asserting input lines to the comparator. The reference voltages driving each of these comparators will be less than V_A. All other comparators with reference voltages exceeding V_A will not assert any encoder inputs. The priority encoder will generate a binary code corresponding to the highest-number comparator that asserts an input line. This code is proportional to the value of V_A.

As the input voltage changes to a different level, the output code responds after a short period of time. This response time is determined by the propagation delay time of the comparators plus the propagation delay time of the priority encoder. Since the precise time that the input signal crosses a quantum level is undetermined, a convert signal or start conversion signal is normally included to reference the time of conversion. This still allows the output code to be changing at the time some following device is reading the code. A transient, incorrect code could be registered under these circumstances.

In order to eliminate problems of transient output codes, two methods may be used. The first method is to insert a sample-and-hold circuit between the analog signal V_A and the ADC input. These circuits will later be discussed in detail. For the present time we need only to recognize that when a sample pulse is applied to the sample-and-hold circuit, the output will move to the same voltage as the input voltage. As the input voltage changes, the output remains constant. This value changes only when another sample pulse is applied, at which time the output will change to the input value that is now present.

If the sample pulse is applied to the sample-and-hold circuit, its output, which is also the input to the ADC, remains constant. The output code that results will not change until the sample pulse is again changed. The ADC output is then read a sufficient time after the sample pulse occurs to guarantee an accurate code.

A second method to avoid transient output codes uses latching comparators. The leading edge of the start conversion input latches the outputs of the comparators at the values present at that time. The encoder inputs will not change until the next assertion of the start conversion input. The output code remains constant between assertions of the start conversion input, except for the short period required for conversion, that is, the propagation delay of the encoder.

This method is similar to the sample-and-hold method, except the output of the comparators are sampled and held rather than the analog input. The sampled values are now digital levels and can be stored in digital latches incorporated into the comparator.

For N greater than 6, the design of the priority encoder becomes rather complex. In order to implement this circuit in a straightforward manner, each pair of adjacent comparators drives an exclusive OR gate, as shown in Fig. 7-3. The only OR gate that has an output of logic 1 will be that gate driven by an input of 1 and an input of 0. All gates below this one will be driven with two inputs of 1, while all gates above this one have 0 inputs. Only a single input now drives the encoder at any given time, and this device can be a simple encoder rather than a priority encoder. The comparator outputs are often referred to as a thermometer code, since all lower outputs are 1 while all upper outputs are 0.

As an example of a parallel converter we will cite the TRW TDC1007 monolithic video ADC. This device is a 64-pin, 8-bit ADC that operates at very high sampling rates. The converter has 8-bit resolution and allows sample rates to reach 20 MHz. The voltage references for the 255 comparators are created by a resistor string. The chip is implemented with ECL (emitter-coupled logic) technology but has TTL-compatible inputs and outputs.

The comparators have latched outputs that store the output values approximately 10 ns after the rising edge of a convert signal. The encoder is enabled on the falling edge of the convert signal, and the encoded values are latched into an output register on the next rising edge of this signal. The convert line must be driven high for a minimum of 15 ns and must be low for a minimum of 25 ns. The total minimum conversion time is then 40 ns. The specifications indicate a 20-MHz sampling rate, but selected devices can approach a 30-MHz rate.

The TDC1029 converter, also made by TRW, is a 6-bit converter that has a 100-MHz sample rate. This ECL device is similar to that of the TDC1007, but it allows a higher sample rate in exchange for a smaller number of bits converted.

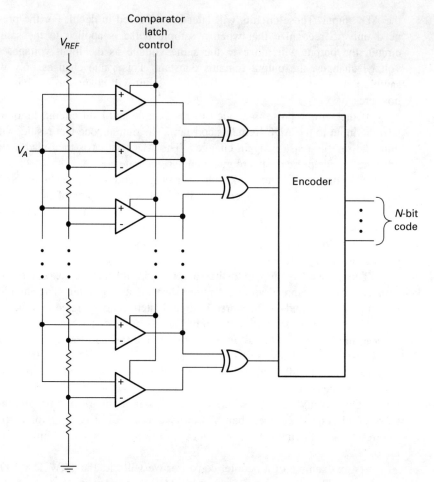

Figure 7-3 A parallel converter ADC using latching comparators and exclusive OR gates.

7.2.2 Successive Approximation ADC

The successive approximation converter is based on an intelligent trial-and-error method. If a person must guess some fixed number in a given range and has feedback to know whether each guess is high or low, each guess should split the known range. Given a range of 0 to 100, the first guess is 50. If this guess is higher than the fixed number, the new range is known to be 0 to 50. The second guess should be 25. If this guess is low, the new range is 25 to 50. The third guess should be 37 or 38. By successively limiting the range by a factor of one-half, the final answer is reached after a reasonably small number of guesses. This scheme is applied in the successive approximation ADC shown in Fig. 7-4.

When the SC input is asserted, the control unit instructs the sequencer to place a 1 in the MSB position of the holding register. All other bits remain at the zero level. This is the first guess or approximation made. The largest possible number would correspond to all bits in the holding register being set to the 1 level. The DAC converts the digital

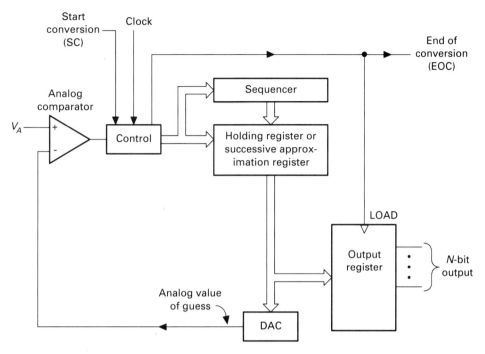

Figure 7-4 Successive approximation ADC.

guess to an analog value. Since the largest digital value would be 2^N-1, the largest analog value generated is

$$V_{Gmax} = (2^N - 1) \times q_D$$

where q_D is the conversion factor of the DAC. This largest value equals the range of the analog input. The first guess is approximately one-half of this range, with an analog value of

$$V_G = 2^{N-1} \times q_D$$

The comparator output reflects the relative values of the input V_A and the approximation V_G. If V_A is larger than V_G, a logic 1 drives the control circuit. A value of V_G larger than V_A results in a comparator output of logic 0. With the knowledge of whether the approximation is too large or too small, the control circuit instructs the sequencer to modify the guess. The second approximation adds a 1 to the next-most-significant bit. If the first guess was too small, a 1 remains in the MSB; if the first guess was too large, the MSB is reset to 0. The result of the second guess is compared to the input to determine the third guess. This procedure is repeated until N approximations have been made.

Each approximation can be checked in a clock period, thus an N-bit conversion requires approximately N clock periods. A small additional time period is required to transfer the data in the holding register to the output register. As this transfer is made, the EOC line is asserted. This line can be used by an external device to indicate that the

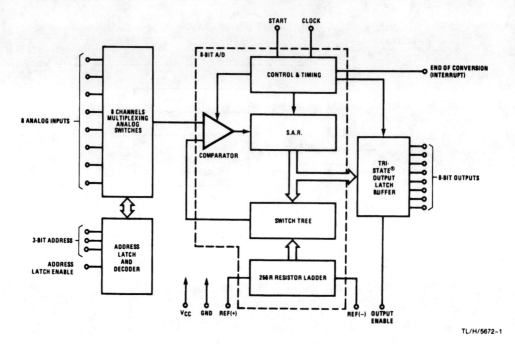

Figure 7-5 Block diagram of the National Semiconductor ADC0809. Reprinted with permission of National Semiconductor.

conversion has been completed. Some ADC output registers have three-state outputs for use with microprocessors. In this case an output enable line is used to enable data from the register onto the microprocessor bus.

The successive approximation ADC is not as fast as the parallel converter, but it is faster than other types such as the ramp, staircase, or tracking converter. It also exhibits a fixed conversion time, which can be an advantage in some applications. This type of converter has been very popular over the years for applications requiring relatively fast conversion. Only in recent years has the parallel converter begun to challenge the dominance of the successive approximation circuit. Even now, for conversions greater than 8 bits, the successive approximation method is used more than any other type.

The National Semiconductor ADC0809 device will serve as an example of the successive approximation converter. The ADC0809 is an 8-bit, microprocessor-compatible converter with an 8-channel analog multiplexer on the input. The block diagram is shown in Fig. 7-5. The DAC that converts the digital approximation to an analog value uses an $R-2R$ resistive ladder network. The output latch is a three-state register that can safely be connected to a microprocessor bus. An address must be presented to the three input address lines prior to the application of an address latch enable (ALE) pulse. The rising edge of the ALE signal latches the address of the selected input channel into a register. This address remains unchanged until a different address and an ALE signal is presented.

The clock frequency can range from 10 kHz to 1.28 MHz, however, due to propagation delays in various components, this frequency is divided by 8 to allow 8 clock

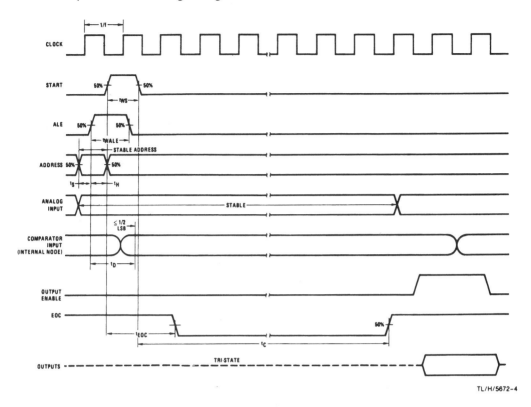

Figure 7-6 Timing chart for the ADC0809. Reprinted with permission of National Semiconductor.

cycles per iteration. A timing chart is shown in Fig. 7-6. The conversion time for a 1.28-MHz clock is typically 50 μs.

In a microprocessor application, the ADC might be used in two different ways. One mode requires the SC signal to be driven high for a minimum time of 200 ns and then returned to a low level. This initiates the conversion, which is completed in a time period given by conversion time. The microprocessor might be programmed to read the output register after a delay from the SC falling edge that exceeds the conversion time. Rather than wait for a specified time, the microprocessor can check the status of the EOC line to determine if the conversion is complete.

The second mode of operation is called the free-running mode. In this method, the SC line is asserted by the microprocessor initially, then the EOC line drives the SC input from this point on. The circuit of Fig. 7-7 can be used for this method. When conversions are to be initiated, the microprocessor sets the SR latch. Since the EOC line will initially be high, the AND gate asserts the SC line. After a short delay, specified as 8 clock cycles plus 2 μs, the EOC line goes low. This also drives the SC line low, which initiates the conversion. After the conversion is completed, the EOC line goes high, which again asserts the SC line. This action is repeated continually until the SR latch is reset. Each conversion is available in the output register until the next conversion is set into the output register. This occurs just prior to the EOC line going high. Often the EOC line

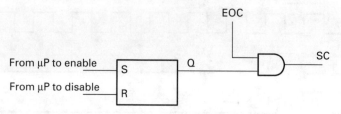

Figure 7-7 Free-running mode initiated by microprocessor.

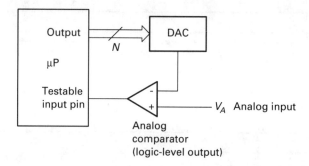

Figure 7-8 A successive approximation ADC using a microprocessor.

signals the microprocessor via the interrupt input, resulting in immediate read-in of the ADC data each time a conversion is completed.

A microprocessor successive approximation ADC. For lower-speed ADC applications involving a microprocessor, the successive approximation ADC of Fig. 7-4 can be implemented with fewer components. All elements in this block diagram are digital components with the exception of the analog comparator and the DAC. Consequently, a microprocessor can be used to implement the remaining digital circuits such as the holding register, output register, and control section. A block diagram of this implementation is shown in Fig. 7-8.

The voltage to be converted is applied to the analog comparator. When a conversion is to be made, the microprocessor inserts a 1 in the MSB of its internal holding register, then outputs the result to the DAC. The output of the DAC is compared to V_A by means of the comparator, which generates either a logic 1 or 0, depending on which input is larger. The microprocessor reads in the value of the comparator output and responds accordingly. If the guess is too large, resulting in a 0 value of comparator output, the microprocessor resets the bit in the holding register to 0 and places a 1 in the next bit position. If the guess is too small, the holding register number is increased by placing a 1 in the next bit position without resetting the original bit. This process is repeated until the least significant bit of the holding register is determined. A flowchart for an 8-bit conversion appears in Fig. 7-9.

7.2.3 Iterative Converter

The iterative converter combines the flash converter with the successive approximation technique. In this case, however, a low-resolution conversion is made followed by suc-

Chap. 7 The Analog-to-Digital Converter

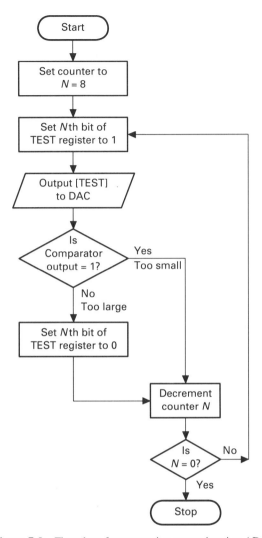

Figure 7-9 Flowchart for successive approximation ADC.

cessively higher resolution conversions. A two-stage iterative converter based on two 6-bit parallel converters is shown in Fig. 7-10.

The quantum interval of ADC1 is $q_1 = 2^6 \times q = 64q$, where q is the quantum interval of ADC2. All quantum levels of ADC1, including the first level, are separated from the preceding one by q_1. Normally the first level of an ADC is set at half the quantum interval, but for ADC1 this is not the case. The quantum interval of ADC2 determines the quantum interval of the overall ADC. When the START CONVERT signal is asserted, ADC1 converts V_A to a 6-bit code with a resolution of $64q$, as shown in Fig. 7-11(a).

In the situation depicted by Fig. 7-11, the first ADC will cause an output of 100001. This output drives a DAC with a conversion factor of $q_D = q_1 = 64q$. This component produces an output of $33q_1$, which is subtracted from the input signal V_A to

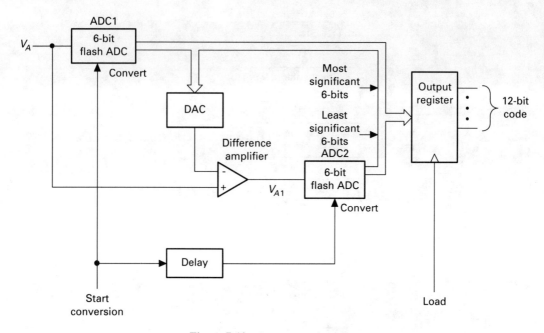

Figure 7-10 A two-stage iterative converter.

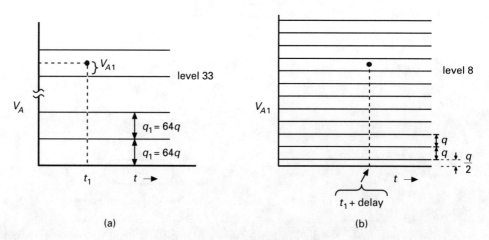

Figure 7-11 Two-stage conversion.

generate V_{A1}. After a time delay that must exceed the conversion time of ADC1 plus the settling time of the DAC plus the rise time of the difference amplifier, the signal that initiates the conversion of ADC2 is asserted. By this time, the voltage V_{A1} appears at the input of ADC2. This input is immediately converted to 001000 for this case, and the output is combined with the output of ADC1. The first ADC generates the most significant 6 bits of the result, and the second ADC generates the least significant 6 bits of the result. The 12-bit result is loaded into the output register after both ADCs have completed their conversion. In the example of Fig. 7-11, the resultant output code would be 100001001000.

Chap. 7   The Analog-to-Digital Converter

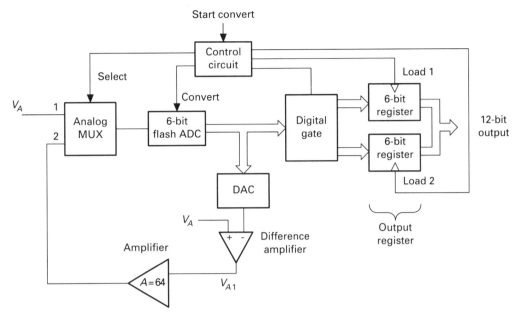

Figure 7-12   An iterative converter using one parallel ADC.

It is possible to extend this architecture to three or four stages if necessary. It is not required to use ADCs with an equal number of bits. It is always required that

$$q_m = 2^n q_{m+1}$$

where n is the number of bits of the $(m + 1)$st ADC. While the conversion time of the iterative converter is longer than the parallel converter, the overall complexity of the iterative converter is much less than that of the parallel system. A 12-bit converter can easily be realized in the iterative architecture but is too complex for implementation in the parallel converter.

An alternate version of a 12-bit iterative ADC uses multiplexing to eliminate the second parallel ADC, as shown in Fig. 7-12. When the start convert line is asserted, the control circuit selects channel 1 of the analog MUX and then asserts the convert line of the flash ADC. The digital gate is driven to pass the ADC output data to the upper 6 bits of the output register. After the conversion is completed, the control circuit provides a load 1 pulse to the upper half of the register. Channel 2 of the analog MUX is then selected resulting in the application of $64V_{A1}$ to the flash ADC. Multiplication of V_{A1} by a gain of 64 results in a flash ADC conversion that is 64 times less significant than the first conversion. When this value is converted, it is loaded into the lower half of the output register to reflect its smaller significance. The control circuit connects the proper lines of the digital gate and provides the load 2 input to complete the 12-bit conversion.

The accuracy of the first ADC must be compatible with the total number of bits converted by the system. For a 12-bit converter, the threshold of each reference level applied to the comparators of the first ADC must have an accuracy that is compatible with a 12-bit conversion. The DAC that then converts this signal and drives the differ-

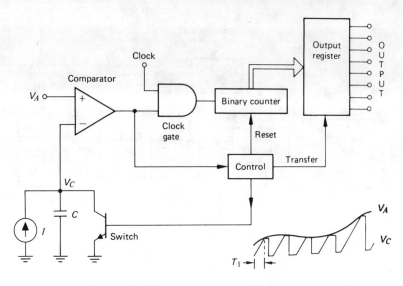

Figure 7-13 Ramp converter.

ence amplifier must also be accurate to a level indicated by the total number of bits converted by the iterative converter.

7.2.4 The Ramp, Staircase, and Tracking Converters

Ramp, staircase, and tracking converters are all slower than the three types considered previously. They are becoming less popular because the improvements in integrated circuit fabrication techniques make parallel and successive approximation converters very attractive. The circuits to be discussed in the following paragraphs do find application in a limited number of less critical systems.

One common characteristic of these converters is a variable conversion time. The time between the start and end of conversion depends on the magnitude of the signal to be converted for the ramp and staircase circuits. For the tracking converter, this time depends on the magnitude of signal change since the last conversion.

Ramp converter. Figure 7-13 shows the block diagram of a ramp converter. When the analog input is applied, the comparator output goes high to open the clock gate to the counter. The transistor switch is turned off at this time to allow the capacitor to charge. At time T_I the capacitor voltage V_C reaches the same value as the analog input signal. The clock gate now closes, and the control circuit strobes the transfer line to move the contents of the counter to the output register. The control circuit then resets the counter and turns the transistor switch on to discharge the capacitor. Although the clock gate again opens as the capacitor is discharged, the counter does not begin to count until the reset signal is removed by the control circuit. This takes place at the same time the transistor switch is turned off to again initiate the ramp.

The time taken for the capacitor to charge is proportional to the analog input signal, because the charging is linear. The counter then develops a binary number that is a measure of the charge time and, hence, is proportional to V_A. The scale factor or propor-

Chap. 7 The Analog-to-Digital Converter

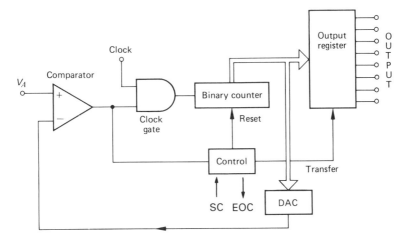

Figure 7-14 Staircase converter.

tionality constant can be controlled by the clock frequency and the charging rate of the capacitor.

The ramp comparator ADC is relatively simple, but it has the disadvantage of being slow. Since the capacitor discharges to 0 after each conversion, when V_A is large the ramp must make a large excursion and conversion time is great.

A staircase converter is similar to a ramp converter, but the ramp generator is replaced by a DAC and a counter. Figure 7-14 shows the block diagram of this ADC. When V_A is applied, the comparator output opens the clock gate to the counter input. The counter continues to count upward, increasing the DAC output until this voltage equals V_A. The comparator then shuts the clock gate, and the control circuit transfers the count to the output register and then resets the counter. The process begins again when the reset signal becomes inactive. The basic idea of the staircase converter is that when the digital output is such that it causes the DAC to generate an analog output equal to V_A, the digital output must be proportional to V_A.

No discharge time is required for the staircase converter, but the DAC itself may be more complex than the ramp generator. The major disadvantage is, again, the slow conversion time for large values of V_A, since the counter runs consecutively from zero up to some high value for each conversion.

An improved version of the staircase ADC is obtained by replacing the simple binary counter with an up-down counter and modifying the control circuitry, as shown in Fig. 7-15. In the tracking converter, the counter is not reset after each conversion. When a conversion is made, the code on the counter generates an analog signal through the DAC that equals the input. The control circuit transfers this count to the output register and disables the clock gate. After a predetermined delay, the clock gate is again opened. By this time V_A will have changed. If V_A has increased over the previously converted value, the counter begins counting up from the value reached at the last conversion. If V_A has decreased, the down-count input will be activated by the comparator, causing the counter to be decremented until the comparator output changes polarity. The control circuit must sense transitions of either polarity, rather than a level.

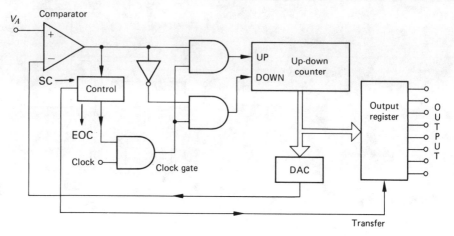

Figure 7-15 Block diagram of tracking converter.

The small signal response of this ADC is much better than that of the staircase converter, since the counter is not reset to zero after each conversion. Again, the conversion time is a function of the analog input signal, and if a large, rapid change in V_A occurs, conversion time can be high. Normally, this situation does not occur, since this converter is used in relatively low-frequency applications.

In the ramp, staircase, and tracking converters the initiation of each conversion can be controlled internally or externally. If a fixed rate of conversion is required, a clock can govern the control circuitry to set the conversion rate. For example, a conversion may be initiated every 1 ms by a 1-kHz clock. This assumes that conversion time plus reset time never reaches 1 ms. In some applications the conversion is governed by an external command. This signal to initiate a conversion is directed via the SC line to the control circuit, which responds by causing the conversion to take place.

7.2.5 The Dual-Slope Digital Voltmeter

The dual-slope digital voltmeter (DVM) uses a form of analog-to-digital conversion that could have been considered in the preceding section. However, because of the importance of the dual-slope conversion process in DVMs, an entire section is devoted to this subject.

The basic function of the DVM is to measure a constant or slowly varying analog voltage and convert the result to a decimal output. For AC measurement, a rectifier is first used to convert to a DC signal. The basic dual-slope voltmeter of Fig. 7-16 is based on the assumption that 7-segment LEDs are used for the display. The counter is designed such that the maximum number of pulses that can be counted before overflow is 1999. When the next pulse is applied to the counter input, the device resets to 0 and generates an overflow or carry pulse.

Prior to $t = t_0$ the counter has been reset and the ramp generator output voltage is near zero or slightly positive. The ramp generator input is connected to ground through the switch. At $t = t_0$, the control circuit switches the ramp generator input to V_x, the voltage to be measured. The output of the generator is a negative-going ramp which immediately causes the comparator to switch to a positive output. When this occurs, the

Chap. 7   The Analog-to-Digital Converter

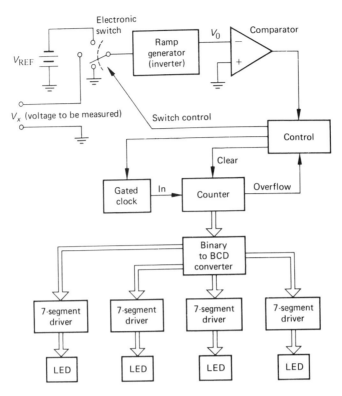

Figure 7-16 Dual-slope voltmeter block diagram.

control circuit directs the gated clock to generate a clock signal. The counter begins counting while the ramp continues to grow in the negative direction, as shown in Fig. 7-17. When the counter reaches 1999, the next pulse resets the counter and signals the control circuit to switch the ramp generator input to a voltage of $-V_{REF}$. The peak voltage reached by the ramp is

$$V_P = -kV_x(t_1 - t_0)$$

where k is a constant relating the output slope to the input signal.

Since the input to the ramp generator is now negative, the ramp changes slope and moves toward 0 V. The counter was reset to 0 as the ramp changed slope at t_1, but it continues counting. The time taken for the ramp to reach 0 is determined by V_{REF}, since this value determines the slope of the ramp. When the ramp reaches 0 V and the comparator output switches polarity, the control circuit stops the clock and switches the ramp generator input to ground.

If we let N be the number of clock periods occurring between t_0 and t_1 and n be the number occurring between t_1 and t_2, then we can write

$$NT = t_1 - t_0$$

and

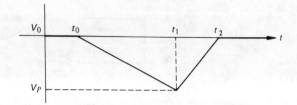

Figure 7-17 Ramp generator output.

$$nT = t_2 - t_1$$

where T is the clock period. From the ramp generator relations we know that

$$V_P = -kV_x(t_1 - t_0) = -kV_{REF}(t_2 - t_1)$$

Substituting NT for $t_1 - t_0$ and nT for $t_2 - t_1$ gives

$$kV_x NT = kV_{REF} nT$$

or

$$V_x = \frac{n}{N} V_{REF}$$

If N is chosen as 2000 and $V_{REF} = 2$ V, then

$$V_x = \frac{n}{1000}$$

The count on the counter is n; thus this number displays the number of millivolts of V_x. It is a simple matter to place the decimal correctly to read in volts. The binary number on the counter is converted to BCD to drive the BCD to 7-segment code converters.

Although there are four LED displays associated with the DVM discussed, only three of these displays take on the full range of digits. The most significant digit can take on only the values 0 or 1, since the maximum measurable voltage is 1.999 V. This is referred to as a 3 1/2-digit meter.

The concepts discussed in the preceding paragraphs lead to a rather simple DVM; however, its capability can be easily expanded. A precision attenuator or amplifier can be used to extend the voltage ranges. In fact, it is possible to add an automatic range-selection feature generally referred to as autoranging. Very complex DVMs with autoranging, self-calibration, output signals for printing or control, and other sophisticated features can be controlled by a microprocessor.

To demonstrate the design considerations that are necessary for the DVM, we will consider the following example.

Chap. 7 The Analog-to-Digital Converter

Example 7-1

Assume a DVM with a maximum display value of 1.99 is to be constructed (2 1/2-digit meter). The output display should appear at 100 ms or less after the read switch is closed.

Solution: We will choose $V_P = -5$ V when $V_x = 1.99$ and $V_{REF} = 2$ V. To display 1.99 V the counter must count to 199 giving $N = 200$ and $n_{max} = 199$. The longest read time will occur when the input signal is maximum or $V_x = 1.99$ V. This will require that 399 clock periods take place. If this must occur in 100 ms, then the clock period is found from

$$399T = 100 \text{ ms}$$

giving

$$T = 250.6 \text{ }\mu\text{s}$$

and

$$f = 3.99 \text{ kHz}.$$

If V_P is to reach -5 V during the time period $t_1 - t_0$, or 200 clock periods, then the slope of the ramp must be

$$\frac{dV}{dt} = \frac{-5}{200 \times 250.6 \times 10^{-6}} = 99.76 \text{ V/s}$$

The ramp generator constant k is then

$$k = \left| \frac{dV}{dt} \right| \times \frac{1}{V_x} = \frac{99.76}{1.99} = 50.13 \text{ s}^{-1}$$

This constant would be achieved by proper integrator design.

Exercises

E7.2a How many comparators are needed for an 8-bit flash converter?

E7.2b Of the various types of ADC considered in this section, which ones have a constant conversion time regardless of the magnitude of input signal?

E7.2c Rework Example 7-1 if the output display should appear in 80 ms or less after read switch closure.

GENERAL EXERCISES

7.1 An 8-bit ADC outputs a digital code corresponding to 255 for a minimum voltage of 8.364 V. Calculate the quantum interval and the output code for an input voltage of 3.764 V.

7.2 A 10-bit ADC has a quantum interval of 0.012 V. What is the output code for an input voltage of 4.126 V? At what minimum voltage will the output code be 1111111111?

7.3 Calculate the number of comparators needed for a 7-bit flash ADC. If $q = 0.025$, what reference voltage connects to the highest-level comparator?

7.4 An 8-bit successive approximation ADC has a quantum interval of 0.020 V. If an input voltage of 3.124 V is converted, indicate all codes that the successive approximation register will sequence through as the conversion is made.

7.5 Repeat problem 7.4 for a 10-bit conversion.

7.6 Design the digital gating circuitry required by the converter of Fig. 7-12.

7.7 Design a 10-bit iterative converter based on the configuration of Fig. 7-10. Use the 6-bit parallel ADC to generate a 6-bit first iteration, adding 4 more bits with the second iteration.

7.8 Design a 12-bit, three-stage iterative converter. If the maximum voltage to be converted is 12.000 V, specify the three values of quantum interval.

7.9 Select the values of the current source, I_1, and the capacitor, C, in the ramp ADC of Fig. 7.13. The maximum conversion time for a 10-V input signal should be 2 ms.

7.10 Design the control circuit for a ramp ADC. State all assumptions made.

7.11 Design the control circuit for a tracking ADC. State all assumptions made.

7.12 Design a 3 1/2-digit DVM using dual-slope techniques. The maximum display value should be 19.99, and the output display should appear within 180 ms from the time of read switch closure.

REFERENCES

1. TRW Electronics Component Group, *TRW LSI Products Division*, TRW, La Jolla, CA, 1985.
2. Engineering Staff, *Data Acquisition Databook Update and Selection Guide*, Analog Devices, Norwood, MA, 1986.
3. Loriferne, B., *Analog-Digital and Digital-Analog Conversion*, Heyden and Sons, London, 1982.
4. Comer, D. J., *Electronic Design with Integrated Circuits*, Addison-Wesley, Reading, MA, 1981.

8

Sample-and-Hold Circuits

The sample-and-hold (S/H) circuit is described very accurately by its name; that is, it samples an analog input voltage or a digital input code at a precise time and holds the corresponding analog value at its output. A sample pulse acts as the control input to determine when the sample is taken. This input is driven by a digital signal, while the input sampled is typically driven by the analog signal or by a digital code being converted by a DAC.

8.1 USES OF THE SAMPLE-AND-HOLD CIRCUIT

The S/H circuit is often used in connection with an ADC. In this instance the S/H module is inserted between V_A, the analog voltage to be converted, and the ADC. For slow converters or those with variable conversion time, the S/H module can lead to ADC output values corresponding to precise sample times. The sample pulses are applied to the S/H module at the desired times of conversion. This value is then held while the ADC makes the conversion. As long as the ADC completes the conversion prior to the next sample point, the converted value is taken to correspond to the last sampled value of V_A. This is shown in Fig. 8-1. The sample taken at t_1 is converted by the ADC after some finite time. The value appearing at the ADC output is associated with the sample time t_1. When the ADC output appears, the voltage V_A may have changed considerably from the value at t_1.

The same result can be achieved in a different way, as pointed out in connection with the parallel converter of Fig. 7-3. The comparator outputs are latched on the leading edge of the CONVERT signal. Even if V_A changes during conversion time, the latched

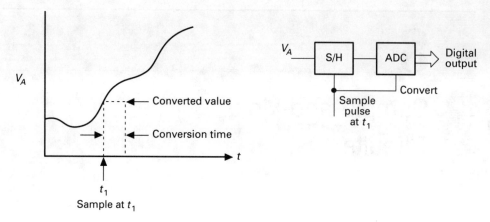

Figure 8-1 Sample/hold followed by conversion.

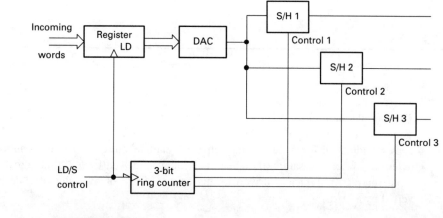

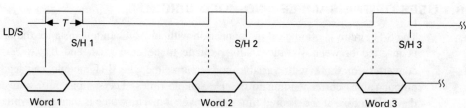

Figure 8-2 A demultiplexed output.

comparator values do not change. In effect, this amounts to a sample-and-hold operation, but instead of a single value being sampled, several values (comparator outputs) are sampled and held.

A second use of the S/H module is that of removing glitches in DAC outputs when necessary. This use has been discussed in Chapter 6 and demonstrated in Fig. 6-13. In this case, the DAC input is a digital code, while the analog output is sampled and held.

A third use of this module occurs when demultiplexing a DAC output to multiple channels, as shown in Fig. 8-2. The DAC converts each sequential code or word to an

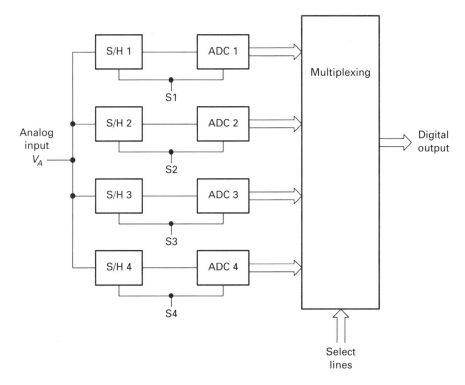

Figure 8-3 A high-speed conversion system.

analog value and then stores each value in a S/H module. The sample pulses applied to the S/H modules must be synchronized with the conversion of each code. In Fig. 8-2, the incoming words are loaded into the register on the leading edge of the LD/S signal. The DAC converts this signal, and the ring counter is updated on the trailing edge of the LD/S signal. The ring counter drives the sample input of an S/H module to store the DAC output. Each of three successive pulses of the LD/S control line converts a word and stores the analog value in a different S/H module. The modules in this example are assumed to sample and hold on the leading edges of their control inputs.

Several S/H modules can be used to speed up the conversion of an analog signal. If each ADC has a conversion time of 400 ns, the system shown in Fig. 8-3 can achieve a conversion time that approaches 100 ns. The sample pulses occur at intervals that slightly exceed the conversion time of the DACs, proceeding in the repetitive sequence S1, S2, S3, S4, S1, S2, S3, S4,..... Each ADC converts a value that is then multiplexed to the output. The select lines are synched with the sample pulses, but they select the following channel. That is, when sample pulse S2 is applied to start a conversion on channel 2, the MUX should output channel 3, which has now completed an earlier conversion. This configuration speeds the overall conversion rate by a factor of 4.

The S/H module can also be used in applications having an ADC that converts much faster than the several individual analog inputs require. In this situation the total system cost may be minimized by using a single ADC with several S/H modules and an analog multiplexer, as shown in Fig. 8-4. The three channels are sampled simultaneously,

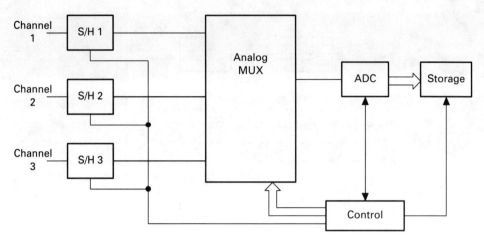

Figure 8-4 A multiplexed system with simultaneous sampling.

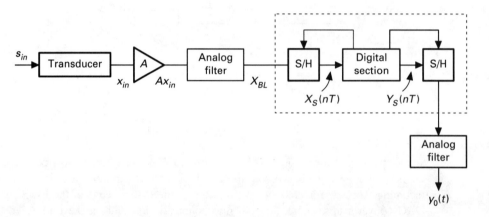

Figure 8-5 Expansion of signal processor to include S/H modules.

with each output being connected sequentially to the ADC through the analog MUX. The ADC converts each channel with the result being digitally stored before the next channel is converted.

With the preceding ideas in mind, we can then expand our data acquisition system to that shown in Fig. 8-5. In this figure $X_s(nT)$ and $Y_s(nT)$ are signals sampled at time $t = nT$. The time T is the time between samples, corresponding to a sample frequency of $f_s = 1/T$.

We have previously mentioned the fact that there are two main types of S/H modules that depend upon the storage or hold method: analog and digital. In the analog hold a capacitor is used to store the amplitude of the analog signal. With the digital hold the sample-and-hold module contains an A/D converter, and the analog amplitude is converted and held in digital form. This type may be slower than analog, but it is more stable over long sampling periods.

8.2 WHEN TO USE THE S/H MODULE

In a data acquisition system, an S/H module may not be needed. In general, this unit is added to decrease the error in the system, but in some applications it may introduce more error than it eliminates. This section considers when it is appropriate to use the S/H module.

Perhaps the most common application of the S/H module is to hold the analog signal while the ADC performs a conversion. Most designers consider an S/H module necessary if the analog input changes significantly during the conversion time of the ADC. Some designers consider a change of one quantum interval, or q, significant. This criterion is actually only a rule-of-thumb that needs closer scrutiny before application.

First of all, a change of input signal by an amount q or less guarantees that only one level is crossed during conversion time. This will ensure that the output code at the end of conversion will be within one quantum level of the analog input that was present at the beginning of conversion. However, some converters produce an output that is within q of the analog input at the end of conversion even if the signal varies by several quantum levels during conversion. Staircase, tracking, or ramp converters have relatively long conversion times, but the output value at the end of conversion represents the analog input very accurately at that time. These converters finish the conversion when a comparison voltage is approximately equal to the analog input value. Thus if the converted digital value must represent the analog value accurately at the end of conversion, these types of ADC may be used without an S/H module even if the input changes by several levels during conversion.

In applications that require an ADC output at aperiodic intervals, these types of converters can be used with no S/H module. For example, a microprocessor may send a start conversion command to an ADC which then interrupts the microprocessor when the end-of-conversion signal is asserted. The microprocessor reads the converted value, which is representative of the analog input existing at end-of-conversion time.

The majority of applications require conversions at fixed, periodic intervals. For such cases, the converted value is often associated with the analog input existing at the times of the periodic sample signal. For high-speed converters that can generate an output before the input changes by one quantum interval, an S/H module is again unnecessary. If the input changes more than this amount during conversion time, the S/H module is placed in front of the ADC. The periodic control signal then drives the sample input of the module, and the start conversion input is delayed slightly to allow the S/H output to stabilize. As long as the input value can be sampled, converted, and used or stored by the signal processing system before the next sample pulse occurs, the acquired data can be assumed to apply to the preceding sample pulse.

The equation that describes the ADC conversion time necessary to avoid the use of an S/H module can be derived in terms of the maximum input slope. Since the input must change less than q during conversion, we can write

$$t_c \cdot \left.\frac{dV_A}{dt}\right|_{max} \leq q \approx \frac{V^{span}}{2^N} \tag{8-1}$$

where V^{span} is the input voltage span of the ADC.

The evaluation of dV_A/dt is straightforward for well-defined signals such as pure sinusoids. For more complex waveforms, this value may be difficult to evaluate. In certain applications, a transducer or amplifier might impose an upper limit on dV_A/dt. An amplifier slew rate sets an upper limit on this slope that would result in a very conservative design if used in Eq. (8-1).

Example 8-1

A 10-bit successive approximation ADC has a conversion time of 1 μs. The maximum slope of the analog signal presented to the ADC is 0.04 V/μs, and the span is 10 V. Determine whether an S/H module is required. Assume that a fixed-period pulse is to control the conversions and that the converted values will be associated with the leading edge of this pulse.

Solution: The left side of Eq. (8-1) is calculated to be

$$t_c \cdot \left.\frac{dV_A}{dt}\right|_{max} = 10^{-6}\text{s} \times 4 \times 10^4 \frac{\text{V}}{\text{s}} = 4 \times 10^{-2} \text{ V}$$

The right side of the equation is found to be

$$\frac{V^{span}}{2^N} = \frac{10 \text{ V}}{2^{10}} = 9.77 \times 10^{-3} \text{ V}$$

The inequality is not satisfied, thus an S/H module must precede the ADC.

If the converter had a faster conversion time, the S/H module could be eliminated. This could occur if the conversion time satisfies

$$t_c \leq \frac{q}{\left.\frac{dV_A}{dt}\right|_{max}} = \frac{9.77 \times 10^{-3}}{4 \times 10^4 \text{ V/s}} \text{ V} = 0.244 \text{ μs}$$

It is possible to estimate the maximum slope of V_A by taking the highest-frequency component of interest, ω_h, and assuming this component has the maximum possible amplitude $V^{span}/2$. The maximum slope of a sinusoid appears at $t = 0$ and is then

$$\left.\frac{dV_A}{dt}\right|_{max} = \left.\omega_h \frac{V^{span}}{2} \cos \omega_h t \right|_{t=0} = \omega_h \frac{V^{span}}{2} \text{ V/s} \qquad (8\text{-}2)$$

Equation (8-2) represents a very conservative estimate in most cases, because the amplitudes of higher-frequency components in most engineering waveforms are considerably smaller than those of the lower-frequency components. Thus the component ω_h will typically have an amplitude that is much less than $V^{span}/2$.

8.3 OPERATING CHARACTERISTICS OF SAMPLE-HOLD DEVICES [3]

The ideal performance of an S/H module is shown in Fig. 8-6. Usually, logic 1 on the control input is sample (track), and logic 0 is hold. Ideally the S/H module has a gain of

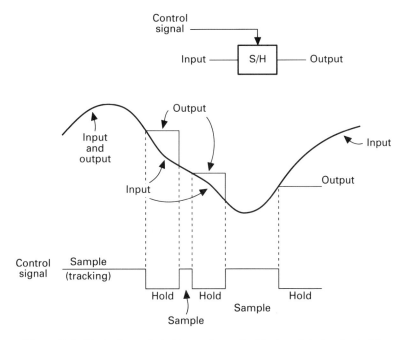

Figure 8-6 Typical sample-hold waveforms for ideal module (analog hold).

+1 in the sample mode so that the output is an exact replica of the input. When the hold command is given, the device would maintain a constant value output corresponding to the input signal value at the exact moment of application of the hold command. Some S/H modules may have gain if the input signals are small, so that the sampled output can span the entire A/D voltage range.

There are four operating regions of interest for practical S/H devices that must be characterized. These are, as shown in Fig. 8-7,

1. sample time
2. the transition from sample to hold
3. hold time
4. the transition from hold to sample

We next discuss important parameters associated with each region.

Region a. This sample region is shown in Fig. 8-8.

1. Offset is the extent by which the output deviates from zero for zero input. This is a function of time and temperature, so there is a drift problem as with offset in an amplifier. At a given temperature any steady state offset can be nulled out by adjustment, and it must be adjusted so that the residual output offset V_{os}^r is minimized. A maximum upper limit is

$$V_{os}^r = \frac{1}{2} \frac{V^{span}}{2^N} \qquad (8\text{-}3)$$

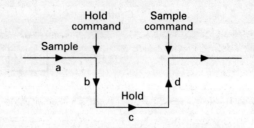

Figure 8-7 Regions of interest in S/H operation.

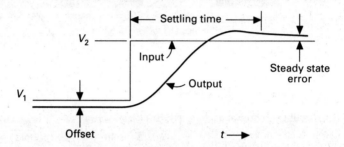

Figure 8-8 Tracking errors during sample.

As a function of time and temperature changes, the drifts of the output offset voltage must satisfy

$$\frac{dV_{os}}{dt}\Delta t + \frac{dV_{os}}{dt}\Delta T \leq 1/2 \frac{V^{span}}{2^N} \quad (8\text{-}4)$$

where t = time and T = temperature. For input voltages other than zero (as V_1 or V_2 in Fig. 8-8) deviations from the true value may be due to offset, gain (scale factor) error, and/or nonlinearity.

2. Nonlinearity (or linearity) is the same parameter as discussed for transducers in Chapter 1. All definitions and results there apply to this characteristic. If the input signal to the S/H were doubled, one hopes that the sampled value would also double. If not, the module is nonlinear.

3. Scale factor error occurs during the sample interval when the ideal S/H unit would have unity gain. The scale factor error is the amount by which the gain deviates from the ideal or specified value.

Suppose the ideal gain value were unity and that the actual value were G. Then we would have

$$scale\ factor\ error = = SFE = |G - 1|$$

$$output\ error = |V_{in}(G-1)| = |V_{in} \cdot SFE|$$

Again, this error should be minimized and should always be less than

$$\frac{V^{span}}{2^{N+1}}$$

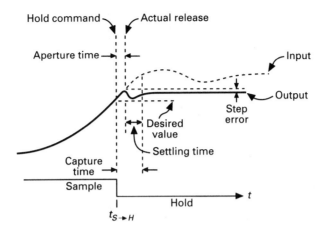

Figure 8-9 Sample-to-hold transition errors.

4. Settling time is a measure of how the output of the S/H module follows the input signal changes that occur while the module is in the sample mode. This specification is based on the worst case. It is defined as the time required by the output to move within a specified fraction (or percentage) of the final full-scale value when a step input is applied.

Since in the sample mode we have essentially an amplifier (often of gain 1), all of the amplifier parameters discussed in Chapter 2 apply and should be examined in the design process or during selection of an S/H module.

Region b. Transition from sample to hold (Fig. 8-9).

1. Aperture time is the time that elapses between the logic hold signal application and the time that the hold "switch" actually activates. This delay is caused by finite transistor switching time. There are two important components of aperture time:

   a. Constant time delay. This could be removed by simply offsetting the time scale by a fixed amount in any computer algorithms, so that one knows that the sample value is really at a time $t_{S \to H} + t_a$, not at time $t_{S \to H}$.
   b. Aperture jitter. The aperture time is not actually a constant. This is shown in Fig. 8-10. Note that if jitter were excessive the signal may not be sampled where one thinks it is, even if a correction is made for nominal constant delay. Jitter is usually specified as a ± deviation from t_a as $t_a \pm t_j$. The jitter is also sometimes called the aperture uncertainty.

2. Sample to hold offset is a step-type offset that occurs when one charged capacitor (such as a transistor switch element capacitor) is connected to the storage capacitor. This can be reduced by a signal from an externally connected compensating capacitor if necessary, although it is a very sensitive compensation.

3. Settling time is similar to the settling time discussed under the "during sample" interval, in that it represents the time to reach within a specified fraction of final value (usually to the maximum value of the system, since this would be the worst case). This effect is measured from the application of the hold command, as opposed to being mea-

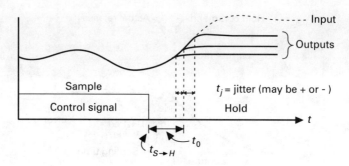

Figure 8-10 Jitter in aperture time.

sured from an input signal change as it is during sample. Notice that for this time error, allowance can be made, since the effect always starts at the hold command. Thus time can always be allowed for adequate settling before a conversion. A quantity called capture time is often defined for the sample-to-hold transition as follows:

Capture time = aperture time + settling time

Region c. During hold (Fig. 8-11).

1. Droop: This is one of the more critical characteristics. It is caused by the hold capacitor charge leaking off through other circuit elements, either a resistance or else as a bias current in the following amplifier stage. The droop can be positive or negative slope, depending upon the required bias. Units that have digital hold, of course, have no droop, since the digital code is fixed and will not change with time if the input analog signal is removed. This assumes there is an A/D converter as an integral part of the system. Droop is expressed in V/S. A figure of merit can be defined in terms of droop and settling time. Suppose the droop is 0.05 V/S and the settling time is 10 μs for a 10-bit data word system operating 10 V full scale. One-half quantum interval corresponds to

$$1/2 \text{ LSB} = 1/2q = \frac{10}{2^{11}} = 4.88 \text{ mV}$$

The droop time interval that would lead to an error of this magnitude is

$$\frac{4.88 \text{ mV}}{0.05 \text{ V/S}} = 98 \text{ ms}$$

If the hold time is less than this value, the corresponding error will be proportionately less than 1/2 LSB.

2. Feedthrough: The schematic diagram of Fig. 8-14 shows a switch that opens to hold the voltage on the capacitor C. In practice these switches are semiconductor devices such as JFETs. Such a hold switch is not perfect and there is a small capacitor across the terminals of the device. This produces a series capacitive voltage divider which allows a small portion of the voltage v_a to appear across the hold capacitor C. This varying signal component will be superimposed upon the voltage being held by C. This is the variational component show in Fig. 8-11.

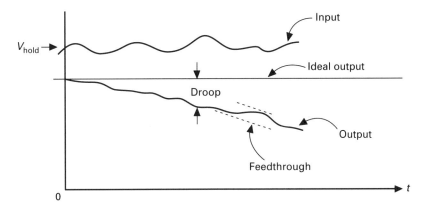

Figure 8-11 Hold errors: droop and feedthrough.

Feedthrough is usually expressed as a percent of the input signal and is denoted by *FT*. To compute the feedthrough error one first estimates the largest expected input signal. If no a priori knowledge of the signal size is known, one can estimate the worst case by using the largest permissible signal that can be applied to the A/D converter that follows the sample-and-hold circuit. Call this $V_{s_{max}}$. Then

$$V_{FT} = \frac{FT}{100} \times V_{s_{max}}$$

The largest deviation from the ideal output would then be

$$\Delta V = V_{error} = \left(V_{FT} + \text{droop} \times t \right) - V_{Ideal}$$

The worst-case estimate of time is then obtained by setting the error equal to or less than 1/2 quantum interval:

$$\left| \frac{FT}{100} \times V_{s_{max}} + \text{droop} \times t - V_{Ideal} \right| \leq \frac{1}{2} \frac{V_o^{span}}{2^N}$$

3. Dielectric Absorption: When the voltage on a capacitor is changing due to current flow the charge may not distribute evenly over the plates. Thus the charge on the capacitor does not have the same effective plate area for a rapid charge input as it does for a slower input charge rate. The effective capacitance is then not the same for all current changes. This is caused by the dielectric which must be polarized, and the polarization cannot change instantaneously. However, the polarization continues to change even when the external current is removed, and the charge redistributes itself over the entire plate area. Thus for the same charge, a different effective plate area is present at two different times. Since capacitance is of the basic form $C = \varepsilon A/d$, where A is the plate area, d the plate separation, and ε the dielectric constant, the initial area will be smaller than the final area after charge has become distributed over the plates. Let C_1 and V_1 be

TABLE 8-1 DIELECTRIC ABSORPTION OF COMMON DIELECTRICS

Material	Temperature range (°C)	Dielectric absoprtion (% change)
Polyester		0.22
Polycarbonate	−55 to +125	0.05 to 0.09
Metalized polycarbonate	−55 to +125	0.05
Polypropylene	−55 to +105	0.03
Metalized polypropylene	−55 to +105	0.03
Polystyrene	−55 to +85	0.02
Metalized teflon	−55 to +200	0.02
Teflon	−55 to +200	0.01
Ceramic	poor range	0.01 to 0.02

the initial values of capacitance and voltage and C_2 and V_2 be the final capacitance and voltage values. If Q_T is the capacitor charge at the time the current is removed then

$$V_2 = Q_T/C_2 = (C_1 V_1)/C_2.$$

Since the initial effective plate area is less than the final area, C_1 is less than C_2 so the voltage appears to decrease.

A similar analysis shows that if charge was being removed from a capacitor the final voltage would tend to be larger than the initial value.

This can also be explained from the point of view that the molecules of the dielectric cannot change polarization direction instantaneously with the result that the charge on the plates does not have an immediate effect on the state of the dielectric. This will not be discussed in detail here.

Generalizing on this analysis we say that the effect of dielectric absorption is to return the capacitor voltage toward the value it was before current (charge) changes occurred. This effect can be demonstrated by taking a large power supply capacitor which has a "slow" dielectric and charging it up and then momentarily shorting the terminals (with the resulting arc). If the capacitor is allowed to rest a moment a voltage of the same polarity (but small value) can be measured across the terminals. Thus as one tries to reduce the capacitor voltage the dielectric absorption tends to return the voltage toward the initial value.

Table 8-1 lists the dielectric absorption effects of several popular dielectrics.

Region d. Hold-to-sample transition (Fig. 8-12).

1. Acquisition time: During the hold interval the charge on the hold capacitor remains fairly constant but the input signal is changing. On the receipt of the sample command, the capacitor must reacquire the new input value and then track it. The left portion of Fig. 8-12 illustrates this. The acquisition time is the length of time the input must be applied in order for the capacitor to be within a specified percentage of the final value. This is not the same as settling time, since settling time is based on the output V_O getting to a specified percentage. The difference is that the amplifier following the capacitor also has a slew rate that may be the limiting factor. Of course, in most cases the

Chap. 8   Sample-and-Hold Circuits

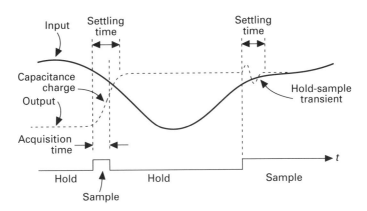

Figure 8-12 Hold-to-sample transition errors.

capacitor terminal may not be available, so that acquisition time is often used as settling time. In fact, if there is any feedback between the amplifier and capacitor, the two times are essentially identical. Note also that any capacitance at the input of the A/D converter will also affect this parameter as well as any interconnection stray capacitance.

As was discussed in the topic of sample-to-hold transition, this acquisition time can be allowed for by first ensuring that the sample interval or period is longer than the acquisition time and second by ensuring that during the transition the input signal does not change significantly.

2. Hold-to-sample transients, as illustrated in the right portion of Fig. 8-12. Such transients are caused by stray inductances and capacitances. If these are very large, it is possible that a glitch may be produced. The best approach is to ensure that the time to the next hold command is longer than the transient period. Since these transients are strongly influenced by the elements connected external to the S/H module, these transients should be checked with the module in its operating environment. One should also observe that even though the input signal is band-limited, the S/H module may still be subject to step changes, since the capacitor held at one value suddenly must attempt to change to a new input value.

As a final note to this section it is important to remember that the time intervals discussed have only accounted for times in the sample-and-hold module. There is also a finite time for the A/D converter that follows the sample and hold. Thus when considering the system performance one will need to add the times of the sample and hold to the conversion time of the A/D converter. This total time will determine the fastest sampling rate, which, of course, must be consistent with the Nyquist sampling theorem.

8.4 FUNDAMENTAL CONFIGURATIONS AND CIRCUIT OPERATION

Analog signal values can be sampled and held by both analog and digital techniques, as previously mentioned. The lower-cost conventional S/H types generally employ a capacitor storage technique that holds the value of the input voltage directly as the analog value. These analog types are usually faster and simpler in construction, but they have higher droop rates than the digital type.

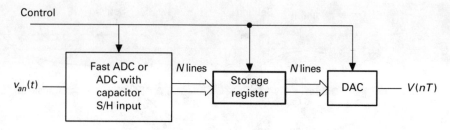

Figure 8-13 Basic digital sample and hold.

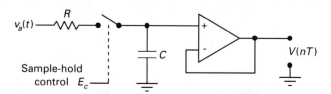

Figure 8-14 Basic capacitor analog hold.

Whenever samples must be held for longer periods of time or when several analog signals are being multiplexed through a single A/D converter, implying a long hold time, a digital S/H should be considered. The basic configuration is shown in Fig. 8-13. The prices one pays for the long hold time are slower conversion times and increased complexity. Once the storage register content is held, the D/A output will remain at the same value, subject, of course, to D/A converter time and temperature drifts.

The basic open-loop analog S/H module is shown in Fig. 8-14. The switch is commonly an FET switch operated by a control voltage E_C. The switch is closed during sample, and the acquisition and settling times are functions of the equivalent Thevenin source impedance R and the maximum current that can be delivered by the preceding circuit, which is frequently a filter. This filter may or may not have an output amplifier. Additional delays are introduced by the output amplifier as well. Once the capacitor has reached the desired precision (within 1/2 LSB of true value), the hold control signal can be applied to open the switch (FET). Droop is caused by the capacitor losing or gaining charge through the input positive terminal of the operational amplifier (bias or leakage currents) or the switch if it is imperfect. Note that the switch can be opened as soon as the capacitor has acquired the analog value, even if the output amplifier has not settled to its final value. Here is an example of a situation wherein acquisition time is not the same as settling time. Even though the switch is opened, the value held on the capacitor will continue to force the output $V(nT)$ to its proper value (neglecting capacitor droop).

An input driving circuit can be added to the basic hold circuit. One possibility is the voltage-isolated driver shown in Fig. 8-15. Since many of the S/H modules are driven by a filter, the large transient currents in the capacitor may cause oscillations in the filter, which can be transferred to the output $V(nT)$ before the switch opens. The advantages of this scheme include a low-impedance drive to the capacitor and greater current capacity than most passive filter structures allow. Both advantages tend to reduce the acquisition and settling time. One disadvantage of this configuration is that when the system goes

Chap. 8 Sample-and-Hold Circuits

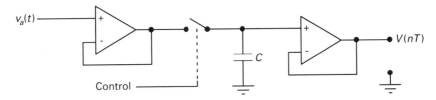

Figure 8-15 S/H module with voltage isolation, voltage drive.

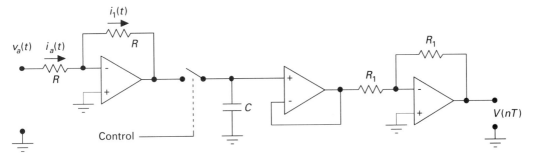

Figure 8-16 S/H module with voltage isolation, current drive.

back to sample (switch closed) and $v_a(t) \neq v_c$ (the capacitor voltage), a momentary saturation of the output of the driving op amp may occur. This can slow the system response.

A second scheme involves driving the input as a current source rather than a voltage source. The scheme shown in Fig. 8-16 involves a voltage-to-current conversion at the input amplifier. If the input op amp had an infinite voltage gain, the voltage at the negative terminal would be zero and no current would flow into that terminal. This would lead to a current

$$i_1(t) = i_a(t) = \frac{v_a(t)}{R}$$

which is independent of $v_c(t)$. The disadvantages of this scheme are first that the current drive is dependent upon the signal amplitude so that the drive is small for small voltages, and, second, the voltage inversion requires an output voltage inverter. The unity gain buffer is still required, since the capacitor needs to see high resistance during hold rather than seeing R_1.

In terms of the relationship of the S/H module to the number of bits used for the data words, one would have to determine components so that:

1. The cumulative effect of common mode output voltages will not produce more than 1/2 LSB at $V(nT)$.
2. The resistor tolerances and amplifier finite open loop gains will not deviate from overall unity (or selected gain) by more than 1/2 LSB.

These ideas have been covered by examples in the earlier chapter on amplifiers, so they will not be repeated here.

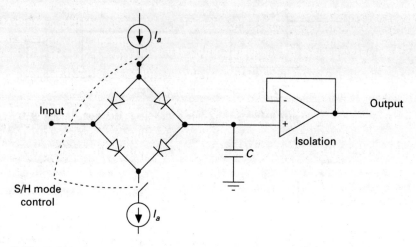

Figure 8-17 Switched current sources for shorter acquisition time (switches shown in hold mode).

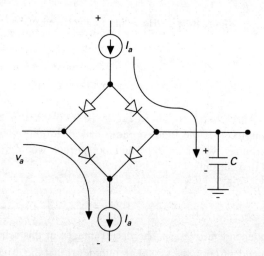

Figure 8-18 Sample mode for $v_c < v_a$.

Since the capacitor charge can be more precisely controlled by current driving directly, it is desirable to develop a scheme that drives the capacitor with a constant current until the proper voltage is reached. The acquisition time can then be more accurately predicted. Such a scheme is given in Fig. 8-17.

The two switches are closed for the sample (tracking) mode. Suppose the capacitor is initially at zero volts and a positive signal $v_a(t)$ is applied. The current I_a is a constant value. The situation is shown in Fig. 8-18. Since v_c is initially at 0 and v_a is positive, diode 1 is cut off and diode 2 is conducting. Thus

$$\Delta v_c(t) = \frac{\Delta q(t)}{C} = \frac{1}{C}\int_0^{\Delta t} I_a dt = \frac{I_a}{C}\Delta t$$

Chap. 8 Sample-and-Hold Circuits

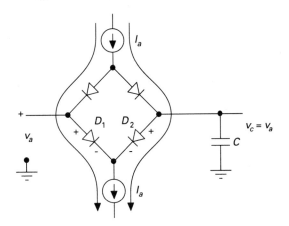

Figure 8-19 Sample mode for $v_c = v_a$.

As soon as $v_c(t)$ equals $v_a(t)$, the current is diverted down through diode 4 or through the left diodes, and v_c has acquired v_a as shown in Fig. 8-19. Note that as long as the diodes are matched (a relatively simple task in IC configurations), $v_c = v_a$ since the diode drops cancel. This can be shown by writing

$$v_a = v_{D1} - v_{D2} + v_c = v_c$$

A disadvantage of this arrangement is that the source v_a must be capable of delivering the full current value I_a regardless of voltage v_a.

Example 8-2

Suppose we have 5-mA current sources that supply the diode bridge. For a ±5-V system, what capacitor value is required for a 10-bit system to have an acquisition time of 50 ns?

Solution: The longest time will be required for the capacitor to charge from −5 V to +5 V or,

$$\Delta v_c = 10 - \frac{10}{2^{11}} \approx 10 \text{ V}.$$

Using

$$\Delta v_c = \frac{1}{C} I_a \Delta t,$$

we require

$$\Delta t > \frac{\Delta v_c C}{I_a},$$

or

$$C \leq \frac{I_a \Delta t}{\Delta v_c} = \frac{5 \times 10^{-3} \times 50 \times 10^{-9}}{10} = 25 \text{ pF}$$

which is a small value but good for an integrated circuit construction. Note that any other voltage changes will occur much faster than this. The only other problem will be the settling time of the amplifier due to the capacitance which shunts the amplifier input resistance.

The advantage of small capacitor size is obvious. The not-so-obvious advantage is the droop. For an FET amplifier following this arrangement, a typical input resistance is 10^{11} Ω. Thus in the hold mode we have a time constant of

$$RC = 25 \times 10^{-12} \times 10^{11} = 2.5 \text{ s}$$

which is a respectable value.

We can use a simple RC network to analyze the preceding design for droop. For this simple circuit,

$$v(t) = V_o e^{-t/RC}$$

The slope is

$$\frac{dv}{dt} = -\frac{1}{RC} V_o e^{-t/RC} = -\frac{1}{RC} v(t) = -\frac{1}{C} \frac{v(t)}{R} = -\frac{i(t)}{C}$$

or,

$$\left| \frac{dv}{dt} \right| = \frac{i(t)}{C} \text{ V/s}$$

The steepest slope will be at $t = 0$ and is shown as follows:

$$\left| \frac{dv}{dt} \right|_{t=0} = \frac{V_o}{RC} \text{ V/s}.$$

Note that this slope depends upon the initial capacitor voltage, and the worst case here will exist for maximum V_o. If we let V_o^{max} equal the voltage span, V^{span}, we obtain

$$\left| \frac{dv}{dt} \right|_{max} = \frac{V^{span}}{RC} \text{ V/s}$$

Thus any calculations using the specification for droop should be based on this worst case.

Chap. 8   Sample-and-Hold Circuits

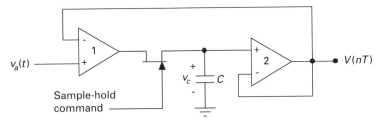

Figure 8-20   Fundamental configuration of feedback S/H module.

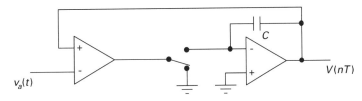

Figure 8-21   Integrating sample and hold.

For the situation of Example 8-2, we find that

$$Droop = \frac{10}{2.5} = 4\,\frac{V}{s} = 4\,\frac{\mu V}{\mu s}$$

which is a good performance figure.

The second basic structure for the sample-and-hold module is the feedback type. The open loop types discussed in the previous paragraphs generally have fast response; that is, they have fast acquisition and settling times. If greater accuracy is desired, as may be required for digital systems using about 12 or more bits for data representation, overall feedback may be employed. The price frequently paid for this greater accuracy is speed, although solid-state technology has reduced this difference.

The fundamental configuration is shown in Fig. 8-20 to analyze the performance. We assume that the amplifier voltage gains are both very high. Suppose that we have a positive value of $v_a(t)$ which is larger than a present value of $v_c(t)$ when the system is in the sample mode (FET shorted). For high gain op amps the output $V(nT)$ will equal v_c. This voltage is then fed back to the negative terminal of amplifier 1. Since v_c is less than V_a, the output of amplifier 1 will be positive, which drives current into C to raise v_c. This continues until $V(nT) = v_a(t)$. The higher the gain, the closer $V(nT)$ will be to the value of $v_c(t)$. The slow response of this type of S/H module is due to the fact that the slew rates of two amplifiers limit the rate of change of output voltage. An advantage of this module is that common mode errors and offset errors are automatically compensated for by the amplifiers.

The switches used in the S/H circuits thus far presented always had one terminal at the higher potential on the capacitor. The modified sample-and-hold circuit shown in Fig. 8-21, called an integrating sample-and-hold, allows the switching and driving circuits to

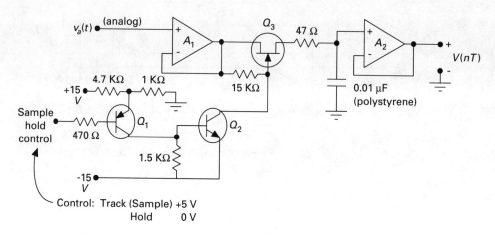

Figure 8-22 Control circuitry for sample/hold module [3].

operate into near 0 volts. Since the high gain of the output amplifier will tend to keep its own negative terminal at ground, the switch always works to 0 volts. The first amplifier is thus a current driver to charge C, the current capability being the limiting factor for acquisition time.

8.5 CONTROL CIRCUITRY FOR S/H MODULE: AN EXAMPLE

There are many ways to obtain the sample-and-hold operations for a module. We present here one way control circuitry could be implemented with discrete components if desired. The low cost of IC modules makes a discrete version of doubtful effort, but it is also a good example of the application of basic circuit theory and electronics. Figure 8-22 is an older version that was available commercially. The circuit was designed to track ±10-V input signals with FET input op amps.

The analysis is simplified if one first redraws the emitter circuitry of Q_1 using Thevenin's theorem, as shown in Fig. 8-23. This gives a clearer picture that shows how the control signal amplitudes given provide control. The Thevenin values are

$$R_{Thev} = \frac{1 \times 4.7}{1 + 4.7} \text{ k}\Omega = 824 \text{ }\Omega$$

$$V_{Thev} = \frac{1}{1 + 4.7} \times 15 = 2.63 \text{ V}$$

When the hold signal is present (0 V) Q_1 is conducting. Current flows out of the collector of Q_1 through the 1.5-kΩ emitter-base resistor, which forward biases Q_2. Q_2 conducts current out of its emitter to the −15-V supply into the 15-V op amp (A_1) supply out of the op amp through the 15-kΩ resistor then back to Q_2. The voltage produced across the 15-kΩ resistor plus that to the left reverse biases Q_3, turning it off. The output amplifier has high input impedance which reduces droop on the capacitor voltage. Any leakage in Q_3 may produce a drift. For example, at 100-pA leakage current, the drift rate of the circuit is 10 mV/s and the rate doubles for every °C increase in temperature.

Chap. 8   Sample-and-Hold Circuits

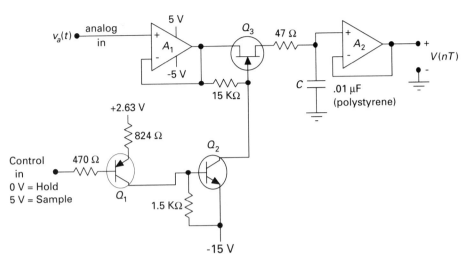

Figure 8-23 Equivalent circuit for S/H module control circuitry [3].

The switching FET, Q_3, has a low pinch-off voltage and allows the circuit to handle ±10-V analog input signals with standard ±15-V supplies. In the sample mode, the +5 V on the control line can be traced in a manner similar to the preceding analysis to show that Q_3 will be conducting with Q_1 and Q_2 both cut off. Since Q_2 is not conducting, there will be no appreciable voltage across the 15 kΩ (only leakage currents that may be present in Q_2 and the gate of Q_3) so that Q_3 will be zero biased for any analog input voltage. The FET resistance is on the order of 100 Ω.

Feedthrough in the hold mode is caused by the small capacitance between source and drain and gate and drain. Thus even though Q_3 may be off, there is coupling from A_1 to the capacitor. At 10-V input the feedthrough is on the order of 10 mV.

The acquisition time is computed by using the 100-Ω FET internal resistance and the 47-Ω series resistor. The time constant would be

$$\tau = RC = 147 \times 1 \times 10^{-8} = 1.5 \text{ μs}$$

The acquisition time would then be the time required to rise to a value depending upon the number of bits. The equation would be

$$\frac{1}{2}\frac{20}{2^n} \leq 20\, e^{-t_a/RC}$$

Since this is an open loop sample-and-hold, there will be additional time required to settle to the final value, due to amplifier slew rate limiting. This example shows the difference between acquisition time and settling time, as shown in Fig. 8-12.

For the dielectric indicated on the circuit, there could be as much as 3 mV of dielectric absorption effect for rapid signal changes occurring when the hold control signal is given.

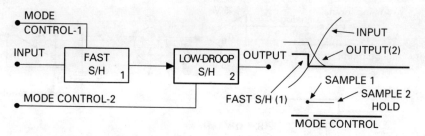

Figure 8-24 Cascaded S/H modules to obtain fast conversion and low droop [3].

8.6 APPLICATIONS OF SAMPLE-HOLD MODULES AND OTHER SYSTEMS

As a first example of an S/H application, suppose the system requirement is that the conversion time be extremely fast and yet it is necessary to hold the value a long time. These are contradictory requirements, since fast acquisition time implies a small capacitor, but a small capacitor with the associated small time constant means large droop for the hold interval. Such requirements often arise in multiplexed data systems. One solution to the problem is shown in Fig. 8-24. The overall sample rate is slow, but for multiplexed systems one channel may need to be held to wait its turn for A/D conversion.

Expanded systems showing typical multiplexing possibilities are given in Fig. 8-25 (a) and (b). In data distribution systems, accurate S/H chips may be less costly than providing several equally fast and accurate converters. In Fig. 8-25 (a) the ADC can be converting while the multiplexer is preparing the next channel. This, of course, requires synchronization for proper operation. The system in (b) can be highly asynchronous if the droop is small enough, even though the acquisition needs to be fast. The S/H modules in (b) can also be used advantageously to remove glitches from the outputs of the DAC.

In some data acquisition systems it is desired to track only the peak value of the analog input. A sample-and-hold may be used to perform this task, as shown in Fig. 8-26. Usually the system is initially biased with some minimum value, e_{min}, which could be a few millivolts. The system is initialized by the reset switch, which moves both switches clockwise. Note that a 1 is interpreted as a sample. In the reset position, e_{min} is applied to the S/H input, which produces e_{min} at the output e_s. The switches are then returned to the positions shown. Whenever e_{in} exceeds e_s the comparator output goes high (sample) and the S/H module follows the input until e_{in} decreases again, at which time the output of the comparator causes a hold signal to be felt at the S/H control input.

A very long hold time with essentially zero droop can be implemented using an ADC and a DAC. The sample-hold structure is shown in Fig. 8-27. The chief disadvantage is that the conversion time is much slower than a regular S/H module. The clock period, τ_s, is initially set by a digital input. The period is dependent upon the time it takes the DAC to get to within 1/2 LSB of the final value, and the number of counts required obviously depends upon the resolution. The full-scale step input is the worst case for time and would be about $2^N \tau_s$. This means that on the average most changes will be tracked much faster, but application design must be based on worst-case time. The mod-

Chap. 8   Sample-and-Hold Circuits

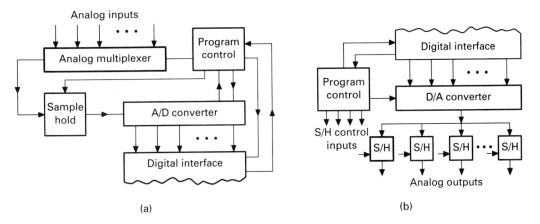

Figure 8-25 Sample-and-hold circuits in multiplexed analog systems [3].

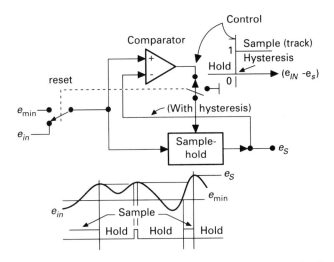

Figure 8-26 A sample-and-hold used in a peak tracker circuit [3].

ule operates by comparing the past DAC output (e_o) with the present input (e_{in}). If $e_{in} > e_o$ the comparator generates a 1 (count up) which, if a sample command 1 is at the control input, enables the count up AND gate, and the clock output is sent to the up terminal on the counter, which increases the digital output count. The DAC output increases with the digital count. This continues until e_o equals or exceeds e_{in} when the comparator switches to a zero, which then enables the down counter. Note that there will be a 1-bit amplitude jitter in general. The module can be used as a peak follower also by disabling the down count. Note that e_o remains without droop when a hold command is given at the control input, since the counter output holds its value indefinitely. However, if the DAC has a convert line, it may need to be refreshed for long hold times, but for this application this would not be a wise choice for a DAC.

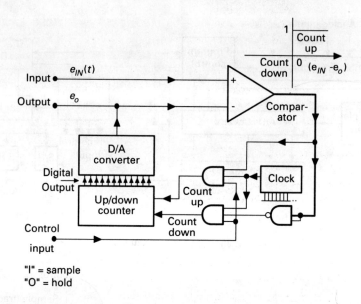

Figure 8-27 Sample-and-hold function using A/D and D/A converters [3].

GENERAL EXERCISES

8.1 For an 8-bit data word system used to sample sinusoidal signals with $V_{max} = \pm 5V$ and an upper frequency of 8 kHz, determine the minimum conversion time of an ADC if an S/H module is not used before the converter. Consider two cases:

 a. Signal is to be sampled at the Nyquist rate.
 b. Signal is to be tracked point-by-point to 8-bit accuracy.

Repeat for 4-bit data words.

8.2 State the criteria for determining whether or not a system requires an S/H module.

8.3 An S/H module having a droop of $\pm 10\ \mu V/ms$ is to be used to hold unipolar signals up to a 5-V maximum. The acquisition time, including 0.005% settling time, is 75 µs. Recommend and justify an appropriate number of bits to use for the digital data words. What is the maximum sample frequency that might be used?

8.4 For the system described in problem 8.1, suppose an S/H module is used just before the ADC. For the 8-bit data words (cases *a* and *b*) determine the time interval between hold commands and necessary specifications for the four regions discussed. Assume that the ADC makes conversions in zero time.

8.5 A digital audio recorder with an 18-kHz low-pass filter operates with a maximum amplitude of 10 V. An S/H module is used to precede a 12-bit ADC. If all error is attributed to aperture jitter, what is the maximum allowable value?

8.6 Assume that a manufacturer specifies all of the important time parameters discussed in the text for an S/H module. Develop an expression for the minimum time interval between convert commands to an ADC. How does the number of bits of the data word affect this answer?

8.7 Design a constant current source of 5 mA for the bridge circuit of Fig. 8-17. Show how you would switch the current away from the bridge.

8.8 Analyze the circuit of Fig. 8-20 to show how the common mode and offset errors are automatically nulled out.

8.9 Determine the maximum allowable droop in an S/H module that drives a 10-bit ADC. The voltage range of the converter is ± 5 V. The maximum hold time is 1 second.

8.10 Show how an instrumentation amplifier could be used to provide current drive to a holding capacitor that is independent of capacitor voltage.

8.11 What is the minimum capture time for the S/H configuration of Fig. 8-21? Consider the amplifier current limit to be I_{sat}.

REFERENCES

1. VanDoren, A. H., *Data Acquisition Systems*, Reston Publishing Co. (Prentice Hall), Reston, VA, 1982.
2. Graeme, J. G., and Tobey, G. E., *Operational Amplifiers: Design and Application*, McGraw-Hill, New York, 1971.
3. Sheingold, D. H., *Analog-Digital Conversion Handbook*, Analog Devices, Norwood, MA, 1972.

Appendix
Laboratory Exercises and Demonstrations

This text was designed for a one semester, three credit hour theory course. There was no formal laboratory period conducted as part of the course. In order to give students some practical experience, we would occasionally assign a laboratory exercise in place of one class period and that period's homework. Additionally, we would assign two students to present a short class demonstration, and the class members could ask the presenters questions about measurement techniques, the results, and so on. For the laboratory exercises we required each student to submit a one-page informational (not descriptive) abstract and a photocopy of the laboratory notebook in which the data, diagrams, calculations, and the like were recorded. For the class demonstrations we rated the manner in which the presentations were made and how well questions were answered. We have found that approximately three laboratory exercises and eight or ten demonstrations usually give students enough experience to solidify the theory discussed in class.

Students who enroll in this course have typically had at least four circuits and electronics courses which have associated laboratory work. Because of that, the laboratory exercises assigned were not step-by-step type exercises, but rather stated in terms of design or measurement objectives. This approach required students to design systems, select appropriate instrumentation, and set up their own laboratory notebooks for schematic diagrams, data tables, and so on. Students responded positively to this approach, since they were at liberty to select power supplies and components.

In this appendix, we simply list some of the laboratory exercises and demonstrations that illustrate the main concepts of the course. We have not identified them specifically as a demonstration or laboratory exercise, since most have been used for either.

Appendix   411

These have been most frequently given as stated, but on occasions "tightened" a bit to give some guidance to the student.

1. Design and evaluate the performance of a code converter for any pair of codes listed in Table 0-8. Speed should be one of your evaluation parameters.
2. For a transducer of your choice, measure any five parameters discussed in Chapter 1. How many bits in the data words can you recommend for each parameter (individually) and for the overall transducer performance?
3. Obtain a coaxial cable or other transmission line and evaluate its performance with pulse excitation using loads having values 1/10, 1, 2, 5, 10, 100 times Z_o. Is the pulse width critical? What is the effect of pulse rate?
4. For an operational amplifier measure any five of the parameters discussed in Chapter 2. See the appendices in reference 4 of Chapter 2 for some ideas.
5. Evaluate the performances of the two integrator configurations shown in Fig. 2-14 and Figure E2.4.1e. Use the same input resistance and feedback capacitor for both. Record the output voltage as a function of time for both when the input is grounded.
6. Evaluate the common mode performance of the unity gain buffer. How many bits could be used for your amplifier?
7. Design and evaluate the operation of the basic instrumentation amplifier of Fig. 2-50. (Set $E_R = 0$ and $E_S = V_0$). Increase the resistors by a factor of 10 and reevaluate. Repeat for a decrease of 10. Use a gain of 100.
8. Design and evaluate a current boost amplifier having a gain of 10, as illustrated in Fig. 2-60. Design the system for a maximum output voltage of 10 V (20 volt span) and a current of 3 A. Measure drift.
9. Design a chopper modulator which will operate at a 10-MHz chopper rate and determine the accuracy of the magnitude of the chopper modulator output. What are the input signal frequency range and amplitude range?
10. Design a passive fourth-order Chebyschev filter terminated in 1000 Ω (resistance) with a 3-dB frequency of 20 kHZ and a ripple of 1 dB. Do not use tables that give component values of the filter. Plot the amplitude and phase responses. Over what frequency range is the phase delay essentially (suggest a reasonable meaning of essential) constant? Compare theoretical and measured responses. Convert your design to the high-pass case and evaluate.
11. Complete assignment 10, but use an active realization. Test your design at two temperatures at least 20 °C different.
12. Design and evaluate a 4-bit binary-weighted DAC (Fig. 6-4) to operate from + 5 V logic voltage. Determine linearity and glitches. Analog range is to be 0 to 10 V. Use JFETs as switches.
13. Design and evaluate an R-2R ladder 4-bit offset binary DAC which is current driven and which has an analog range of -5 to + 5 V. Use JFETs as switches.
14. Design and evaluate the performance of a 3-bit flash ADC.

15. Design a current source that can charge a capacitor linearly from 0 to +5 V. Evaluate the temperature characteristics of your system also.
16. Evaluate the performance of any available DAC.
17. Evaluate the performance of any available ADC.
18. Design a JFET-controlled S/H unit using current switching, and evaluate five parameters from any of the four operating regions. Check temperature performance also.

Index

Accelerometer, 47
   piezoelectric, 47
   strain gauge, 47
Accuracy, 91
Active filters, 245
Additive noise, 110
Additivity, 48
Aliased components, 250
Aliases, 250
Aliasing, 245–52
Amplifier, 110, 119
   gain, 120
   impedance, 121
   instrumentation, 135
   isolation, 135
   noise, 110
   operational, 135
   practical, 122
Amplitude modulation, 246
Analog signal, 320
Analog-to-digital converter (ADC), 331, 365–83
   accuracy, 366
   conversion time, 366
   dual-slope, 367, 380
   free-running mode of, 373
   integrating, 367, 380
   iterative, 367, 374
   LSB value, 366
   microprocessor-compatible, 365, 372
   number of bits, 366
   parallel, 367
   parameters, 365
   quantum interval, 366, 375
   ramp, 367, 378
   staircase, 367, 378
   successive approximation, 367, 370, 374
   tracking converter, 367, 378
Aperture, 393
   jitter, 393
   time, 393
Attenuation constant, 126
   transmission line, 126

Bach filters, 307
Backlash, 58
Band-limited signal, 248, 334
Band-pass filter, 312
   Sallen and Key resonator, 313
Bandwidth, 62
Bessel (linear phase) filters, 269–76
   time delay, 274
Bits, 7
Butterworth filter, 255, 261–64
   functions, 263
   presampling, 278
   response, 261, 264
   three-dB point, 263, 335
Bytes, 7

Capacitive sensor, 34, 47
Capture time, 394
Cauer filter, 285
   first form, 285
   second form, 288

413

Characteristic impedance, 125
Chebychev filter, 255, 264–69
    polynomials, 265
    response, 266
    ripple width, 266
    transition region, 264
Coding, 9–23
    binary, 9
    binary-coded decimal (8-4-2-1 BCD), 10
    binary-coded decimal (2-4-2-1 BCD), 12
    bipolar, 14–23
    complementary, 16
    conversion between types, 21
    digital, 11, 320
    Gray, 14
    natural binary, 9
    offset binary, 17
    one's complement, 20
    reflected, 15
    sign-magnitude, 16
    summary of schemes, 23
    two's complement, 20
    unipolar signals, 9–23
Comparator, 369
Compensation for nonlinearity, 53
Compounding, 330
Conversion factor, 347, 371
    of DAC, 347
Conversion error, 332
Converter, 331, 333, 345–62, 365–83
    analog-to-digital (ADC), 331, 365–83
    digital-to-analog (DAC), 333, 345–62
    size, 331

Data acquisition, 1, 333
    problem, 6
    system, 1, 333
Delay time, 59
Denormalization, 255
    frequency, 256
    impedance, 256
Dielectric absorption, 395
Digital code, 11, 320
Digital-to-analog converter (DAC), 333, 345–62
    accuracy, 347
    BCD, 358
    binary-weighted, 350
    bipolar, 359
    conversion factor, 347
    design, 348
    double-buffered, 348, 356
    flow-through mode, 348, 356
    linearity, 347
    microprocessor compatible, 348, 356
    number of bits, 346
    offset binary, 359
    parameters, 346
    resolution, 347
    settling time, 346
    two's complement, 360
Discrete signal, 246
Dispersion, 132
Distortion, 123
    transmission line losses, 124
    transmission line mismatch, 123–32
D-latches, 349
Drift, 104
    temperature, 104
    time, 104
Driving point function, 284
Droop, 394

Electromagnetic interference (EMI), 109
Encoder, 368
Endpoint-based linearity, 50
Equiripple filter, 255
Error, 6, 32, 279, 319, 332
    analysis for amplifiers, 228
    budgets for amplifiers, 228
    conditioning, 323
    conversion, 319, 332
    effect on word size, 323
    filter, 279
    in data acquisition systems, 319
    limit, 33
    noise, 329
    quantization, 319
    relation to 1/2 LSB, 33, 335
    relationship to signal processing, 32
    root sum square, 33
    scale factor, 392
    sources of, 6
    tracking, 392
    transducer, 323

Feedthrough, 394
FET switch, 398
Filters, 245, 252–318
    active, 245, 305–16
    analog, 245
    antialiasing, 253
    Bach, 307
    band-pass, 312
    Bessel, 269–76
    Butterworth, 255, 261–64
    Cauer, 285
    Chebychev, 255, 264–69
    error, 279
    high-pass, 304
    low-pass, 254, 304
    magnitude response, 254
    need for, 252

Filters (cont'd)
  passband, 254
  passive, 245, 283–304
  phase shift, 314
  presampling, 253, 276
  specifications, 254
  stopband, 254
  switched-capacitor, 316
  synthesis, 283–316
First-order hold, 343
  linear, 343
  predictive, 343
Fourier analysis, 338
Fourier integral, 246
Fourier series, 338
Fourier transform, 246, 248
Frequency response, 66, 246, 250, 257
  maximally flat, 261, 310
  monotonic, 255
  one-half LSB, 279
Frequency spectrum, 246
Friction, 58
Fundamental frequency, 339, 342

Gain, 120
  amplifier, 120
  current, 121
  voltage, 120
Glitch, 361
Group delay, 69

Harmonics, 339
High-pass filter, 304
  transfer function, 311
Homogeneity, 48
Humidity, 40
Hysteresis, 58

Ideal impulse sampling, 249
Impedance, 125, 206
  characteristic, 125
  common mode, 206
  denormalization, 255
  differential mode, 206
Inductive sensor, 37
Instrumentation amplifier, 135, 209–23
  applications of, 217
  properties of, 210
  selection of, 228
Inverting amplifier CMR, 191
Isolation amplifier, 135, 224–27
  applications of, 224
  characteristics of, 224
  offset drift, 226
  temperature drift, 226
  time drift, 226

Jitter, 393
Johnson noise, 106

Ladder network, 295, 354
  filters, 295
  DAC, 354
LC networks, 284
Linearity, 48
  endpoint-based, 50
  zero-based, 50
Linear phase (Bessel) filters, 269–76
  time delay, 274
Linear variable differential transformer (LVDT), 38, 46
Low-pass filter, 254
  transfer function, 308
  transformation to high-pass, 304

Magnitude response, 254
  maximally flat, 255, 310
  monotonic, 255
Measurand, 29
Mu-law (μ-law), 330

Nibbles, 7
Noise, 105, 245, 326
  additive, 110
  amplifier, 110
  bandwidth, 107
  considerations, 326
  factor, 110
  figure, 105–12
  Johnson, 106
  peak, 108
  resistance, 329
  RMS, 108
  voltage, 106
  white, 106, 327
Noninverting amplifier CMR, 192
Nonlinearity, 48
Normalization, 255
  frequency, 256
  impedance, 256
Nyquist sampling theorem, 250, 334

Offset, 391
  in S/H circuit, 391
  in op amps, 195, 199
Operational amplifier or op amp, 135–208
  applications, 137
  common mode impedance, 206
  common mode input, 177
  common mode rejection (CMR), 184–93
  common mode rejection ratio (CMRR), 184
  current offset, 195
  differential impedance, 206

Operational amplifier or op amp (*cont'd*)
　differential input, 177
　frequency response, 157–61
　ideal, 137
　input impedance, 205
　integrator, 142
　inverting configuration, 139–43
　noninverting configuration, 144–46
　offset adjustment, 199
　open loop gain, 151
　parameters, 149, 207
　slew rate, 167, 175
　slope limiting, 172
　specifications, 147, 149
　temperature characteristics, 202
　temperature drift, 203
　unity gain buffer, 152
　voltage offset, 195

Passive filters, 245
Phase shift, 269, 314
　filters, 314
Phase velocity, 126
Physically realizable circuit, 258
Piezoelectric sensor, 40
Poles of a transfer function, 257
　LHP, 260
　removal, 286
　RHP, 260
Precision, 87
Presampling filter, 276–80
Pressure transducer, 35–46
Probability density function, 327
Probability distribution function, 328

Quantization, 319
　error, 319, 322
Quantum interval, 319, 322, 330
　variable, 321, 330

Radio frequency interference (RFI), 109
Ramp generator, 380
Reconstruction of signal, 333
Recovery sample frequency, 336
Reflection coefficient, 127
Reliability, 46
Reluctance sensor, 38
Repeatability, 87
Resistance, 40, 329
　noise, 329
　temperature dependent, 40
Resistive sensor, 39
　optoelectric, 40
　photoelectric, 40
　photoresistor, 40
Resolution, 84
　relation to number of bits, 86

Ring counter, 387
Ripple factor, 71
Ripple width, 71, 266
Rise time, 62
Root sum square error, 33

Sallen and Key resonator, 313
Sample-and-hold (S/H) circuit, 333, 369, 385–407
　aperture time, 393
　capture time, 394
　control circuitry, 404
　droop, 394
　feedthrough, 394
　hold time, 391
　jitter, 393
　module, 386
　operating characteristics, 390
　tracking error, 392
　when to use, 389
Sample mode, 391
Sample rate, 250, 332
Sample time, 385
Sampling, 245–52
　frequency, 247, 251
　ideal impulse, 249
　rate, 334
　recovery frequency, 336
Selectivity, 99
Sensitivity, 94
Sensor, 29
　accelerometer, 47
　capacitive, 34, 47
　inductive, 37
　LVDT, 38, 46
　optoelectric, 40
　piezoelectric, 42
　PN junction, 40
　resistive, 39
　thermoelectric, 40
Settling time, 393
Shifting theorem, 248
Signal processing system, 3, 4
Signal reconstruction, 333
Signal-to-noise ratio, 111, 320
Span, 322, 325
　of an analog signal, 322
　transducer, 325
Specifications, 45
　dynamic, 45
　linearity, 48
　static, 45
　transducer, 45
Spectrum, 248, 327
　amplitude, 327
Strain gauge, 39, 47
Superposition theorem, 49
Switched-capacitor filter, 316

Index

Synthesis of filters, 255
  active, 305–16
  passive, 283–304
System, 3
  signal processing, 3

Temperature drift, 104
Thermocouple, 40
Three-dB frequency, 75
Time delay of a filter, 269
Time drift, 104
Time response, 59
Tracking error, 392
Transadmittance, 121
Transducer, 28–113, 329
  accelerometer, 47
  accuracy, 91
  capacitive, 34, 47
  DC transfer characteristics, 70
  drift, 104
  elements of, 28
  frequency response, 66
  inductive, 37
  LVDT, 38, 46
  optoelectric, 40
  parameters, 28
  piezoelectric, 40
  precision, 87
  pressure, 35
  repeatability, 87
  resistive, 39
  resolution, 84
  ripple factor, 71
  selectivity, 99
  sensitivity, 94
  sensor, 29–45
  specifications, 45, 48–113
  strain gauge, 39
  thermoelectric, 40
  time response, 59
  unipolar, 329
  unit step response, 60
Transfer function, 261
  of LC network, 284
  synthesis, 293

Transient state, 361
Transimpedance, 121
Transition region, 255
Transmission line, 124–35
  attenuation constant, 126
  characteristic impedance, 125
  dispersion distortion, 132
  distortion, 123–32
  high frequency, 128
  impedance, 127
  losses, 124
  low frequency, 128
  matched lossless, 129
  mismatch distortion, 123–32
  models, 125
  phase velocity, 126
  pulses on, 121, 130
  reflection coefficient, 127
Transmission zeros, 294
  finite, 300
Two-port parameters, 294
  h parameters, 291
  y parameters, 290
  z parameters, 290

Unit step response, 60

Voltmeter, 380
  dual-slope, 380

Waveform reconstruction, 333
Wheatstone bridge, 54
White noise, 106
Word, 7, 8
  data, 8
  digital, 7
  instruction, 8

Zero-based linearity, 50
Zero-order hold, 333, 336
  reconstruction, 341
  with filter, 341
Zero shifting, 300
Zeros of transmission, 294